Glassy Materials Based Microdevices

Glassy Materials Based Microdevices

Special Issue Editors

Giancarlo C. Righini
Nicoletta Righini

MDPI • Basel • Beijing • Wuhan • Barcelona • Belgrade

MDPI

Special Issue Editors
Giancarlo C. Righini
"Enrico Fermi" Historical Museum of Physics and Study & Research Centre
Italy

Nicoletta Righini
National Autonomous University of Mexico (UNAM)
Mexico

Editorial Office
MDPI
St. Alban-Anlage 66
4052 Basel, Switzerland

This is a reprint of articles from the Special Issue published online in the open access journal *Micromachines* (ISSN 2072-666X) from 2018 to 2019 (available at: https://www.mdpi.com/journal/micromachines/special_issues/Glassy_Materials_based_Microdevices)

For citation purposes, cite each article independently as indicated on the article page online and as indicated below:

LastName, A.A.; LastName, B.B.; LastName, C.C. Article Title. *Journal Name* **Year**, *Article Number*, Page Range.

ISBN 978-3-03897-618-9 (Pbk)
ISBN 978-3-03897-619-6 (PDF)

Contents

About the Special Issue Editors

Giancarlo C. Righini, Physicist, studied at the University of Florence, Italy, and made his scientific career at National Research Council of Italy (CNR) and Enrico Fermi Center (Rome, Italy), becoming director of research in both the institutions. After retiring, he keeps holding a position of associate scientist in both of them. He always did experimental research, mostly on fiber and integrated optics, with focus on glass materials. His recent interests deal with photoluminescent materials and optical microresonators. He published more than 500 papers and is co-editor of a few books. He was co-founder and then President of the Italian Society of Optics and Photonics (SIOF) and Vice-President of ICO (International Commission for Optics). Currently he is chair of Technical Committee TC-20 (Glasses for Optoelectronics) of the International Commission on Glass. G. C. Righini is Fellow of EOS, OSA, SIOF and SPIE, and Meritorious Member of SIF.

Nicoletta Righini, ecologist, is a postdoctoral fellow at the National Autonomous University of Mexico. She obtained her degrees at the University of Florence (Natural Sciences, BSc), Instituto de Ecología A.C. (Ecology, MSc), and University of Illinois (Anthropology, PhD). Her research is interdisciplinary and spans the areas of integrative biology, environmental science, animal eco-physiology, nutrition and health. She is member of the Mexican National System of Researchers (SNI).

Preface to "Glassy Materials Based Microdevices"

Galileo Galilei, in his book titled *Discorsi e Dimostrazioni Matematiche Intorno a Due Nuove Scienze* (in English: Discourses and Mathematical Demonstrations Concerning Two New Sciences) [1], published in Holland in 1638, when writing "Whereas, if the size of a body be diminished, the strength of that body is not diminished in the same proportion; indeed the smaller the body the greater its relative strength" seemed to someway anticipate by more than three centuries the visionary lecture delivered by Richard Feynman in 1959 titled "There's plenty of room at the bottom" [2], which has definitely provided inspiration to many scientists for the field of micro and nanotechnology. The attempt to understand physical and chemical processes at a smaller and smaller scale, as well as to produce instruments and devices more and more compact, dates back to the beginning of the past century, but microtechnology started to become relevant in 1960s, with the first semiconductor microchips, and exploded in the 1970s with microelectronics.

Glass has been part of the development of microtechnologies since the very beginning; even in microelectronics, glass films are useful to protect the underlying silicon substrate and conductive paths in semiconductors and help with device planarization. Glass ceramics have proven to offer a valid alternative to alumina for microwave integrated circuits. Thin glass substrates may be used in high density interconnection integrated circuit (IC) packages. Optics and photonics, however, remain the main field of application of glasses and glass devices, even at a micrometer level: optical fibers, integrated optical circuits, optical microcomponents, and microresonators are excellent examples of the capabilities of this material.

Polymers, on the other hand, have been fundamental for the development of a critical step in the fabrication process of integrated circuits, namely photolithography and electron-beam-lithography. Moreover, they, too, have found wide application in microphotonics, sometimes competing with glasses, due to their excellent material properties and usually lower cost both in purchase price and cost of processing.

Glasses and most of the polymers are substances frozen in a liquid-like structure and are characterized by a glass transition temperature; from this comes the common definition of glassy materials. The processing of glassy materials (among others) by ultrashort laser pulses has evolved significantly over the last decades and has opened new ways to microfabrication; it now reveals all its scientific, technological and industrial potential. Another revolutionizing technology, first introduced during the 1980s, is represented by the additive manufacturing or 3D printing (the former term being more used by engineers, the latter being more common in the media). Almost all the 3D printing techniques utilize thermoplastic or thermoset polymers as build materials. The development of 3D printing has introduced a break-through in the computerized fabrication of complex objects and multifunctional materials systems; the spatial resolution is usually in the range of tens of micrometers but may go down to a fraction of micron for some materials, like acrylate polymers. Thus, the prospects for glassy materials are excellent. Both glasses and polymers are widely used in sensing and biomedical applications; for example, microfluidics and optofluidics, which are fundamental components of many lab-on-chip architectures, are usually based on etching (and/or laser processing) of these materials. Glass-polymer hybrid materials represent a novel frontier allowing significant improvements in properties that are impossible to achieve from classical materials. Flexible electronics, flexible photonics, metamaterials with extraordinary properties are being developed also thanks to the combination of glasses and polymers.

This book, which is the printed copy of the Special Issue of *Micromachines* on Glassy Material Based Microdevices [3], seeks to collect a set of works on the application of glassy materials to microdevice fabrication. Characterization and processing of materials are treated, together with the design and application (especially in microfluidics and in photonics) of glassy microdevices. We hope that this collection will be useful to both newcomers and researchers in this field.

We would like to thank all the authors and the reviewers who contributed to ensure the quality of the published papers. The efficient assistance of the editorial office of *Micromachines*, and in particular of Ms. Mandy Zhang, is also gratefully acknowledged.

References

1. Galileo Galilei, Dialogues Concerning Two New Sciences, New York: Macmillan, 1914; p. 109.
2. The transcript of the talk that Richard Feynman gave on December 29, 1959 at the annual meeting of the American Physical Society at the Caltech is available on the web at https://www.zyvex.com/nanotech/feynman.html (accessed on 21 January 2019).
3. https://www.mdpi.com/journal/micromachines/special_issues/Glassy_Materials_based_Microdevices.

Giancarlo C. Righini, Nicoletta Righini
Special Issue Editors

micromachines

MDPI

Editorial

Editorial for the Special Issue on Glassy Materials Based Microdevices

Giancarlo C. Righini [1,2,*] **and Nicoletta Righini** [3,*]

1   "Enrico Fermi" Historical Museum of Physics and Study & Research Centre, 00184 Roma, Italy
2   "Nello Carrara" Institute of Applied Physics (IFAC), CNR. 50019 Sesto Fiorentino, Italy
3   Research Institute on Ecosystems and Sustainability (IIES), National Autonomous University of Mexico (UNAM), 58190 Morelia, Mexico
*   Correspondence: giancarlo.righini@centrofermi.it (G.C.R.); nrighini@cieco.unam.mx (N.R.)

Received: 7 January 2019; Accepted: 7 January 2019; Published: 8 January 2019

Glassy materials, i.e., glasses and most polymers, play a very important role in microtechnologies and photonics. Both are substances frozen in a liquid-like structure and are characterized by a glass transition temperature. The excellent properties of glasses have made them fundamental for the development of optical and photonic components, from very large sizes (e.g., telescope lenses) down to the micrometric scale (e.g., microlenses and microresonators). Polymers, too, generally do not crystalize but form amorphous solids: their glassy properties are important for many applications such as nanolithography. This special issue of *Micromachines,* entitled "Glassy Materials Based Microdevices", contains 19 papers (five reviews and 14 research articles) which highlight recent advances in microdevices and microtechnologies exploiting the properties of glassy materials. Contributions were solicited from both leading researchers and emerging investigators.

Several of these papers deal with the fabrication, physics and applications of glass and polymer microspheres, which constitute a very simple but very intriguing type of microdevice. A broad overview of the smart uses of solid and hollow glass microspheres in the field of energy, from solar cells to hydrogen storage and nuclear fusion, is presented in the review paper by Righini [1]. Polystyrene microspheres are used in the article by Piccolo et al. to construct a two-dimensional grating for the development of a low-cost chromatic sensor able to simultaneously determine the vectorial strain–stress information in the x and y directions [2]. As it is well known, microspheres may also operate as resonating cavities, where the light is trapped at the surface in whispering gallery modes (WGM). Yu et al. review the fabrication methods of microspherical resonators using various compound glasses, including heavy metal oxide glasses and chalcogenide glasses, and present some applications, e.g., lasing and sensing [3]. The critical issue of the robust coupling of light into a glass microspherical resonator is discussed in the article by Chiavaioli et al., who present a comprehensive model for designing an all-in-fiber sensing set-up and validate it by comparing the simulated results with the experimental ones [4]. Another aspect of light coupling in and out a WGM resonator is examined in the article by Konstantinou et al., where the implementation of a three-port, light guiding and routing T-shaped configuration based on the combination of a WGM microresonator and micro-structured optical fibers is demonstrated [5]. To complete this group of papers, the work by Li et al. illustrates the potential of an optofluidic hollow microsphere (microbubble) resonator for the highly sensitive label-free detection of small molecules and drug screening [6].

Microfluidics is fundamental for the development of biomedical sensing and analysis microsystems. Glass has proved to be a very convenient substrate for microfluidic chips thanks to its insulating properties, mechanical resistance and high solvent compatibility. The prototyping of microfluidic devices in low quantities may be time-consuming and expensive; the article by Wlodarczyk et al. describes a laser-based process that enables the fabrication of a fully-functional microfluidic device in less than two hours by using two thin glass plates [7]. The femtosecond laser irradiation followed by chemical etching (FLICE) technique was used by Italia et al. to fabricate a buried microfluidic device in a silica substrate; the design was optimized to minimize the diffusive mass transfer between two laminar flows [8]. The fabrication of glass microfluidic chips by a molding process that requires only tens of minutes and therefore appears to be a promising method for fast prototyping and mass production of microfluidic chips is described in the article by Wang et al. [9]. A microfluidic flow cytometer fabricated in polydimethylsiloxane (PDMS) by using a SU-8 photoresist mold was employed by Fan et al. for single cell analysis; data about the expression of β-actin proteins in ~10,000 single cells were obtained [10].

Precision glass molding and micropatterning technologies are of paramount importance in other fields, too. Zhou et al. provide a review of the fabrication technique of infrared aspherical lenses and microstructures in chalcogenide glass through precision glass molding [11]. Micropatterning of metal substrates, in particular the forming of microdomes on an aluminum substrate, is described in the article by Kim et al. [12]. The silicon–glass platform is at the base of the MEMS technology, which has been exploited by Knapkiewicz for the construction of high-vacuum self-pumping cells that are fully suitable for atomic spectroscopy [13]. The scribing of glass for subsequent dicing may be critical in the manufacturing of some microcircuits and microdevices; Zhang et al. discuss and experimentally test a method involving micro-crack-induced severing in order to realize the rapid and precision cleaving of the hard quartz glass in chip materials [14].

Many microdevices find application in the field of photonics. The paper by Amiri et al. provides a basic introduction to optical waveguides and to their applications, with special attention to fiber Bragg gratings for sensing applications [15]. Materials are very important both in microelectronics and in photonics: Falcony et al. provide an overview of the spray pyrolysis technique, with the focus on the research work performed in relation to the synthesis of high-K dielectric and luminescent materials in the form of coatings and powders as well as multiple layered structures [16]. The sol-gel technique is also extensively used to synthesize optical materials, especially glassy materials doped with rare earths; in the article by Trejo-García et al., the synthesis and spectroscopic characterization of Eu^{3+}-doped hybrid silica–poly(methyl methacrylate) (PMMA) material is presented [17]. The photoluminescent properties of rare earth-doped glasses are discussed in the article by Enrichi et al., with reference to the broadband sensitization effect of Yb^{3+} ions due to the energy transfer from silver dimers/multimers [18]. Frequency conversion processes, based on efficient light excitation and re-emission in rare earth ions, may be exploited to increase the performance of silicon solar cells, and the article by Quandt et al. discusses the modelling of up-conversion processes, in particular, in the context of solar cell device simulations, showing their potential for the proper design of new types of highly efficient solar cells [19].

We would like to thank all the authors for their submissions to this special issue. We also thank all the reviewers for dedicating their time and helping ensure the quality of the submitted papers, and, last but not least, the staff at the editorial office of *Micromachines* for their efficient assistance.

Conflicts of Interest: The authors declare no conflicts of interest

References

1. Righini, G.C. Glassy Microspheres for Energy Applications. *Micromachines* **2018**, *9*, 379. [CrossRef] [PubMed]
2. Piccolo, V.; Chiappini, A.; Armellini, C.; Barozzi, M.; Lukowiak, A.; Sazio, P.-J.A.; Vaccari, A.; Ferrari, M.; Zonta, D. 2D Optical Gratings Based on Hexagonal Voids on Transparent Elastomeric Substrate. *Micromachines* **2018**, *9*, 345. [CrossRef] [PubMed]
3. Yu, J.; Lewis, E.; Farrell, G.; Wang, P. Compound Glass Microsphere Resonator Devices. *Micromachines* **2018**, *9*, 356. [CrossRef] [PubMed]
4. Chiavaioli, F.; Laneve, D.; Farnesi, D.; Falconi, M.C.; Nunzi Conti, G.; Baldini, F.; Prudenzano, F. Long Period Grating-Based Fiber Coupling to WGM Microresonators. *Micromachines* **2018**, *9*, 366. [CrossRef] [PubMed]

5. Konstantinou, G.; Milenko, K.; Kosma, K.; Pissadakis, S. Multiple Light Coupling and Routing via a Microspherical Resonator Integrated in a T-Shaped Optical Fiber Configuration System. *Micromachines* **2018**, *9*, 521. [CrossRef] [PubMed]
6. Li, Z.; Zhu, C.; Guo, Z.; Wang, B.; Wu, X.; Fei, Y. Highly Sensitive Label-Free Detection of Small Molecules with an Optofluidic Microbubble Resonator. *Micromachines* **2018**, *9*, 274. [CrossRef] [PubMed]
7. Wlodarczyk, K.L.; Carter, R.M.; Jahanbakhsh, A.; Lopes, A.A.; Mackenzie, M.D.; Maier, R.R.J.; Hand, D.P.; Maroto-Valer, M.M. Rapid Laser Manufacturing of Microfluidic Devices from Glass Substrates. *Micromachines* **2018**, *9*, 409. [CrossRef] [PubMed]
8. Italia, V.; Giakoumaki, A.N.; Bonfadini, S.; Bharadwaj, V.; Le Phu, T.; Eaton, S.M.; Ramponi, R.; Bergamini, G.; Lanzani, G.; Criante, L. Laser-Inscribed Glass Microfluidic Device for Non-Mixing Flow of Miscible Solvents. *Micromachines* **2019**, *10*, 23. [CrossRef] [PubMed]
9. Wang, T.; Chen, J.; Zhou, T.; Song, L. Fabricating Microstructures on Glass for Microfluidic Chips by Glass Molding Process. *Micromachines* **2018**, *9*, 269. [CrossRef] [PubMed]
10. Fan, B.; Li, X.; Liu, L.; Chen, D.; Cao, S.; Men, D.; Wang, J.; Chen, J. Absolute Copy Numbers of β-Actin Proteins Collected from 10,000 Single Cells. *Micromachines* **2018**, *9*, 254. [CrossRef] [PubMed]
11. Zhou, T.; Zhu, Z.; Liu, X.; Liang, Z.; Wang, X. A Review of the Precision Glass Molding of Chalcogenide Glass (ChG) for Infrared Optics. *Micromachines* **2018**, *9*, 337. [CrossRef] [PubMed]
12. Kim, J.; Hong, D.; Badshah, M.A.; Lu, X.; Kim, Y.K.; Kim, S.-M. Direct Metal Forming of a Microdome Structure with a Glassy Carbon Mold for Enhanced Boiling Heat Transfer. *Micromachines* **2018**, *9*, 376. [CrossRef] [PubMed]
13. Knapkiewicz, P. Alkali Vapor MEMS Cells Technology toward High-Vacuum Self-Pumping MEMS Cell for Atomic Spectroscopy. *Micromachines* **2018**, *9*, 405. [CrossRef] [PubMed]
14. Zhang, L.; Xie, J.; Guo, A. Study on Micro-Crack Induced Precision Severing of Quartz Glass Chips. *Micromachines* **2018**, *9*, 224. [CrossRef] [PubMed]
15. Amiri, I.S.; Azzuhri, S.R.B.; Jalil, M.A.; Hairi, H.M.; Ali, J.; Bunruangses, M.; Yupapin, P. Introduction to Photonics: Principles and the Most Recent Applications of Microstructures. *Micromachines* **2018**, *9*, 452. [CrossRef] [PubMed]
16. Falcony, C.; Aguilar-Frutis, M.A.; García-Hipólito, M. Spray Pyrolysis Technique; High-*K* Dielectric Films and Luminescent Materials: A Review. *Micromachines* **2018**, *9*, 414. [CrossRef] [PubMed]
17. Trejo-García, P.M.; Palomino-Merino, R.; De la Cruz, J.; Espinosa, J.E.; Aceves, R.; Moreno-Barbosa, E.; Moreno, O.P. Luminescent Properties of Eu^{3+}-Doped Hybrid SiO_2-PMMA Material for Photonic Applications. *Micromachines* **2018**, *9*, 441. [CrossRef] [PubMed]
18. Enrichi, F.; Cattaruzza, E.; Ferrari, M.; Gonella, F.; Ottini, R.; Riello, P.; Righini, G.C.; Enrico, T.; Vomiero, A.; Zur, L. Ag-Sensitized Yb^{3+} Emission in Glass-Ceramics. *Micromachines* **2018**, *9*, 380. [CrossRef] [PubMed]
19. Quandt, A.; Aslan, T.; Mokgosi, I.; Warmbier, R.; Ferrari, M.; Righini, G. About the Implementation of Frequency Conversion Processes in Solar Cell Device Simulations. *Micromachines* **2018**, *9*, 435. [CrossRef] [PubMed]

micromachines

MDPI

Article

Study on Micro-Crack Induced Precision Severing of Quartz Glass Chips

Long Zhang, Jin Xie * and Aodian Guo

School of Mechanical and Automotive Engineering, South China University of Technology, Guangzhou 510640, China; zhanglong1226@126.com (L.Z.); aodianguo@163.com (A.G.)
* Correspondence: jinxie@scut.edu.cn; Tel.: +86-20-2223-6407

Received: 11 April 2018; Accepted: 5 May 2018; Published: 8 May 2018

Abstract: It is difficult to cut hard and brittle quartz glass chips. Hence, a method involving micro-crack-induced severing along a non-crack microgroove-apex by controlling the loading rate is proposed. The objective is to realize the rapid and precision severing of the hardest quartz glass in chip materials. Firstly, micro-grinding was employed to machine smooth microgrooves of 398–565 μm in depth; then the severing force was modelled by the microgroove shape and size; finally, the severing performance of a 4-mm thick substrate was investigated experimentally. It is shown that the crack propagation occurred at the same time from the microgroove-apex and the loading point during 0.5 ms in micro-crack-induced severing. The severing efficiency is dominated by the severing time rather than the crack propagation time. When the loading rate is less than 20–60 mm/min, the dynamic severing is transferred to static severing. With increasing microgroove-apex radius, the severing force decreases to the critical severing force of about 160–180 N in the static severing, but it increases to the critical severing force in the dynamic severing. The static severing force and time are about two times and about nine times larger than the dynamic ones, respectively, but the static severing form error of 16.3 μm/mm and surface roughness of 19.7 nm are less. It is confirmed that the ideal static severing forces are identical to the experimental results. As a result, the static severing is controllable for the accurate and smooth separation of quartz glass chips in 4 s and less.

Keywords: micro-crack propagation; severing force; quartz glass; micro-grinding

---

1. Introduction

Rolling scribing with a tungsten carbide or polycrystalline diamond (PCD) wheel is widely used to separate silicate glass substrates without any coolant and material removal. In order to improve the life of tool micro-tips, a chemical vapor deposition (CVD) diamond roller was developed for scribing instead of PCD and carbide alloy rollers [1]. Lateral and radial cracks were, however, produced on severing surfaces due to the mechanical force of mechanical wheel rolling [2,3]. Generally, a follow-on smooth profile grinding and polishing were needed along with inefficiency and pollution. Recently, researchers have focused on the crack generation on severing surfaces. From the in-process estimation of fracture surface morphology, severing surface cracks and breakages were produced during wheel scribing of a glass sheet [4]. It has been known that the median crack depth decreased with decreasing loading force, but it still reached about 90 μm with the load of 22 N in scribing alumina ceramics [1]. In scribing LCD glass, it reached about 45 μm with the load of 5.5 N [2] and about 90 μm with the cutting pressure of 0.16 MPa [3], respectively. Until now, the scribing of harder quartz glass chips has not yet been reported.

In order to improve scribing performance, vibration-assisted scribing was used to increase the median cracks for severing [5], but uneven cracks existed on microgroove-apex edges and the severing surface. It has been reported that laser beams could be used to irradiate the scribing [6–9]. Although laser beam irradiation enhanced scribing speed, edge cracks, severing form deviation and the bevel surfaces were produced [6]. Moreover, the laser irradiation made the mechanical breaking easier in scribing [7], but thermal damage accumulated at the severing edges. Because the laser scribing produced micro-cracks and burrs on the machined microgrooves, it led to the cracks and burrs on the severing surface edge [8]. Although a hybrid of laser beam, water jet coolant and pre-bending were employed to eliminate the micro-cracks [9], form deviation happened along the beam moving direction. A picosecond Ultraviolet (UV) laser was also used to induce the scribing of polyethylene terephthalate films to control the local bending flexibility [10], but it has not yet been applied to severing.

To predict severing force, Filippi first proposed the existence of linear elastic stress fields in the neighborhood of rounded-tip V-shaped notches [11]. The linear elastic stress was also used to derive two brittle fracture criteria such as mean stress (MS) and point stress (PS) criteria [12]. Moreover, these criteria were used to predict compressive notch fracture toughness [13,14]. The minimum fracture loading of a U-notches plate was introduced by means of MS and PS criteria [15]. In the case of low and high loading rates, it was found that the loading rate produced little influence on the maximum load for the V-notch on fracture [16]. However, these workpieces only concerned easy-to-cut polymeric and metallic materials. Until now, these criteria have not been applied to difficult-to-cut quartz glass due to the fabrication difficulty of microgroove. Although the fracture of ceramic-metal joint surface has been divided into static and dynamic states [17], the critical loading rate and force have not yet been studied in detail.

As for the fabrication of microgroove, laser and etching approaches have been used to fabricate the microgroove with 7.5 μm and less in depth on Si surface and ceramic cylinder [18,19], but it was irregular and rough. It would lead to cracks on the severing surface when it was used for the induced severing. Moreover, the micro-grinding with a sharpened diamond wheel micro-tip may be employed to fabricate accurate and smooth microgrooves on difficult-to-cut silicon, carbide alloy and glass surfaces [20], but it has not yet been applied to the crack propagation for precision and smooth severing of difficult-to-cut materials.

In this paper, a new micro-crack-induced severing with static loading and dynamic loading is proposed for the crack propagation along an accurate and smooth microgroove-apex. The objective is to realize rapid and precision severing of difficult-to-cut quartz glass. Firstly, the trued diamond wheel micro-tip was employed to grind the accurate and non-crack microgroove on workpiece surface; then the severing force was modelled in micro-crack induced severing by microgroove parameters and loading rate; finally, severing force, severing time, cracking propagation time, severing form errors and severing surface roughness were experimentally investigated.

2. Micro-Crack Induced Severing of Brittle Workpiece

Figure 1 shows the stress field model in micro-crack induced severing along a microgroove-apex. The microgroove is parameterized by height h_v, angle β_v and microgroove-apex radius r_v. Under mode I loading condition, $\sigma_{\theta\theta}$, $\sigma_{r\theta}$, and σ_{rr} are tangential stress, shear stress and radial stress, respectively, and r_c^* is critical distance (see Figure 1a) [11]. When the loading force F increases to the critical value called severing force F_c, the tangential stress $\sigma_{\theta\theta}$ $(r_c^*, 0)$ reaches the ultimate tensile strength σ_u and the micro-cracks are produced from the microgroove-apex (see Figure 1b). It leads to the crack propagation along the microgroove-apex.

Figure 2 shows the scheme of micro-crack-induced severing. The working sizes of the workpiece substrate are given by the thickness W and the width B. The workpiece is supported by two supporting rods with an interval L. The arc-shaped loading rod is loaded on the upper surface of substrate. The microgroove-apex is positioned on the opposite side of substrate. The vertical loading direction aims to the microgroove-apex. The loading rod moved vertically with the loading rate v. When the loading force F reaches the severing force F_c, the micro-crack occurs at the microgroove-apex. It leads to the severing for the separation of workpiece.

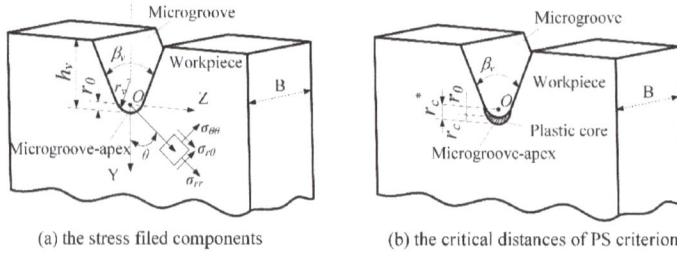

(a) the stress filed components   (b) the critical distances of PS criterion

Figure 1. The stress field model in micro-crack induced severing along a microgroove-apex: (a) the stress filed components and (b) the critical distances of point stress (PS) criterion.

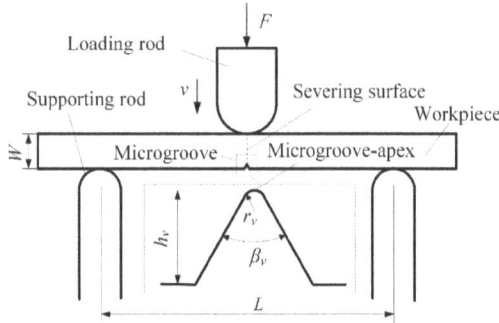

Figure 2. The scheme of micro-crack induced severing.

3. Modelling of Severing Force

The elastic stresses were described at the neighborhood of microgroove-apex in polar coordinate system (see Figure 1a). r_0 is the distance between coordinate origin O and microgroove-apex. The critical distances r_c^* and r_c are defined from the coordinate origin O and the microgroove-apex, respectively (see Figure 1b). The ideal severing force F_c^* was described as follows [15]:

$$\sigma_{max} = \sigma_{nom} K_t = \frac{3 K_t F_c^* L}{2B(W - h_v)^2} \tag{1}$$

where σ_{max} and σ_{nom} are the maximum stress and the nominal stress at microgroove-apex, respectively. K_t is the stress concentration factor.

At the neighborhood of microgroove-apex under pure mode I loading, the elastic stresses are described in the polar coordinate system as follows [11]:

$$\begin{Bmatrix} \sigma_{\theta\theta} \\ \sigma_{rr} \\ \sigma_{r\theta} \end{Bmatrix} = \frac{K_I^{V,r_v}}{\sqrt{2\pi} r^{1-\lambda_1}} \left[\begin{Bmatrix} m_{\theta\theta}(\theta) \\ m_{rr}(\theta) \\ m_{r\theta}(\theta) \end{Bmatrix} + \left(\frac{r}{r_0} \right)^{\mu_1 - \lambda_1} \begin{Bmatrix} n_{\theta\theta}(\theta) \\ n_{rr}(\theta) \\ n_{r\theta}(\theta) \end{Bmatrix} \right] \tag{2}$$

where the $m_{\theta\theta}(\theta)$ is expressed as follows:

$$\begin{Bmatrix} m_{\theta\theta}(\theta) \\ m_{rr}(\theta) \\ m_{r\theta}(\theta) \end{Bmatrix} = \frac{1}{[1 + \lambda_1 + \chi_{b_1}(1 - \lambda_1)]} \left[\begin{Bmatrix} (1+\lambda_1)\cos(1-\lambda_1)\theta \\ (3-\lambda_1)\cos(1-\lambda_1)\theta \\ (1-\lambda_1)\sin(1-\lambda_1)\theta \end{Bmatrix} + \chi_{b_1}(1-\lambda_1) \begin{Bmatrix} \cos(1+\lambda_1)\theta \\ -\cos(1+\lambda_1)\theta \\ \sin(1+\lambda_1)\theta \end{Bmatrix} \right] \tag{3}$$

The $n_{\theta\theta}(\theta)$ is expressed as:

$$\left\{\begin{matrix} n_{\theta\theta}(\theta) \\ n_{rr}(\theta) \\ n_{r\theta}(\theta) \end{matrix}\right\} = \frac{1}{4(q-1)\left[1+\lambda_1+\chi_{b_1}(1-\lambda_1)\right]}\left[\chi_{d_1}\left\{\begin{matrix} (1+\mu_1)\cos(1-\mu_1)\theta \\ (3-\mu_1)\cos(1-\mu_1)\theta \\ (1-\mu_1)\sin(1-\mu_1)\theta \end{matrix}\right\} + \chi_{c_1}\left\{\begin{matrix} \cos(1+\mu_1)\theta \\ -\cos(1+\mu_1)\theta \\ \sin(1+\mu_1)\theta \end{matrix}\right\}\right] \tag{4}$$

The K_I^{V,r_v} is the mode I notch stress intensity factor (NSIF). It is described as follows:

$$K_I^{V,r_v} = \frac{\sqrt{2\pi}r_0^{1-\lambda_1}\sigma_{max}}{1+\omega_1} \tag{5}$$

where ω_1 is an auxiliary parameter. They are expressed as follows:

$$r_0 = \frac{q-1}{q}r_v \tag{6}$$

$$\omega_1 = \frac{q}{4(q-1)}\left[\frac{\chi_{d_1}(1+\mu_1)+\chi_{c_1}}{1+\lambda_1+\chi_{b_1}(1-\lambda_1)}\right] \tag{7}$$

where q is a real positive coefficient ranging as:

$$q = \frac{2\pi-\beta_v}{\pi} \tag{8}$$

λ_1, μ_1, χ_{b1}, χ_{c1} and χ_{d1} are the values of auxiliary parameters for different microgroove angle [11]. Under pure mode I loading, the tangential stress $\sigma_{\theta\theta}(r, 0)$ from Equation (2) can be written as follows:

$$\sigma_{\theta\theta}(r,0) = \frac{K_I^{V,r_v}}{\sqrt{2\pi}r^{1-\lambda_1}}\left[1+\left(\frac{r}{r_0}\right)^{\mu_1-\lambda_1}n_{\theta\theta}(0)\right] \tag{9}$$

Substituting Equation (5) into Equation (9), the tangential stress can be expression as follows:

$$\sigma_{\theta\theta}(r,0) = \frac{r_0^{1-\lambda_1}\sigma_{max}}{r^{1-\lambda_1}(1+\omega_1)}\left[1+\left(\frac{r}{r_0}\right)^{\mu_1-\lambda_1}n_{\theta\theta}(0)\right] \tag{10}$$

According to PS criterion, the brittle fracture takes place when the tangential stress $\sigma_{\theta\theta}(r, 0)$ reaches critical value σ_u at specified critical distance [21]. Hence, r_c^* (see in Figure 1b) can be expressed as follows:

$$r_c^* = r_c + r_0 \tag{11}$$

For brittle materials, the critical distance r_c can be written as follows [22]:

$$r_c = \frac{1}{8\pi}\left(\frac{K_{IC}}{\sigma_u}\right)^2 \tag{12}$$

where K_{IC} is the material attribute called fracture toughness.

According to Equations (1) and (9)–(12), the ideal severing force F_c^* is deduced as follows:

$$F_c^* = \frac{2\sigma_u B(W-h_v)^2(1+\omega_1)(r_0+r_c)^{1-\lambda_1}}{3K_t Lr_0^{1-\lambda_1}\left(1+\left(1+\frac{r_c}{r_0}\right)^{\mu_1-\lambda_1}n_{\theta\theta}(0)\right)} \tag{13}$$

In Equation (13), the K_t is achieved in the case of the U-notch with $\beta_v = 0$ [15], but it was calculated by the fitting of experimental data in this study. This is because a microgroove with $\beta_v > 0$ was employed in micro-crack-induced severing.

4. Micro-Grinding of Microgroove on Workpiece Surface

Figure 3 shows the micro-grinding of microgrooves on a workpiece surface with a diamond wheel micro-tip. It was difficult to dress the wheel micro-tip due to its high hardness. Before micro-grinding, the truing of the #600 metal-bond diamond wheel micro-tip (Changxing technology Co. LTD., Shenzhen, China) was performed by the Numerical Control (NC) mutual wearing between the diamond wheel and the #800 Green Silicon Carbide (GC) dresser using Computer Numerical Control (CNC) grinder (Fuyu Machine Tool co. LTD., Zhang Hua, Taiwan) (see Figure 2a) [20]. Finally, the diamond wheel micro-tip angle α_v was identical to the angle of the NC tool paths. The truing conditions of the wheel micro-tip are shown in Table 1.

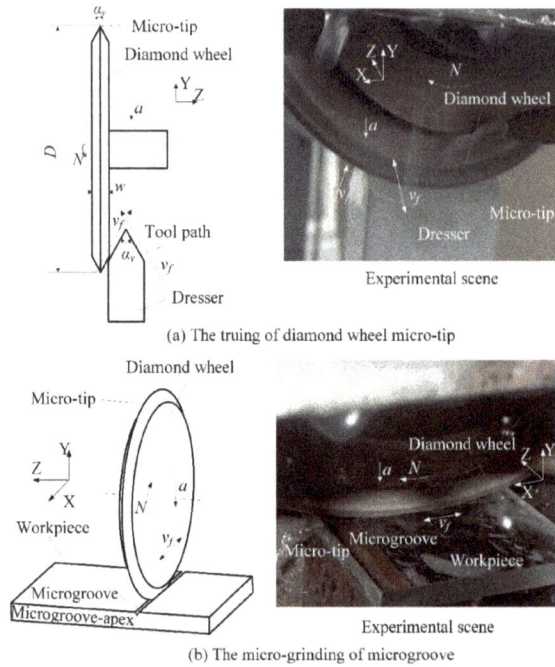

(a) The truing of diamond wheel micro-tip

(b) The micro-grinding of microgroove

Figure 3. Micro-grinding of microgroove on quartz glass: (**a**) the truing of diamond wheel micro-tip and (**b**) the micro-grinding of microgroove.

Table 1. The truing conditions of diamond wheel micro-tip.

CNC grinder	SMART B818
Diamond wheel	SD600 (Metal-bond, Grain size: 24 μm), D = 150 mm, w = 4 mm
Dresser	#800 GC, Ceramic bond
Tool paths	V-shaped symmetrical, α = 60°, 90°, 120°
Truing parameters	v_f = 500 mm/min, N = 2400 rpm, a = 20 μm, Σa = 5 mm
Coolant	Water

Then, the diamond wheel micro-tip was used to grind a microgroove on the workpiece surface (see Figure 2b). A quartz glass substrate was chosen as workpiece. The diamond wheel micro-tip angle α_v was equal to the microgroove angle β_v [19]. The micro-grinding conditions are shown in Table 2.

Table 2. Micro-grinding conditions of microgroove on workpiece substrate.

CNC Grinder	SMART B818
Diamond wheel	SD600 (Metal-bond, grain size: 24 µm), D = 150 mm, w = 4 mm
Workpiece	Quartz glass
Grinding parameters	v_f = 500 mm/min, N = 2400 rpm, a = 20 µm, $\sum a$ = 500 µm,
Coolant	Water

5. Experiment and Measurement

Figure 4 shows the experimental setup of micro-crack induced severing. It is based on its working principle (see Figures 1 and 2). A WDW-05 electronic universal testing machine (Jinan Kason Testing Equipment Co., Ltd., Qingdao, Shandong, China) was employed to perform the loading and measure the on-line loading force F. A high-speed camera was used to record the propagation of micro-cracks during the loading process (see Figure 4). In the experiments, the microgroove angle β_v was set as 60°, 90° and 120°, respectively. The loading rate v was set as 5 mm/min, 10 mm/min, 20 mm/min, 60 mm/min, 100 mm/min, 200 mm/min and 300 mm/min, respectively. In order to calculate ideal severing force F_c^*, the values of the auxiliary parameters such as λ_1, μ_1, χ_{b1}, χ_{c1} and χ_{d1} in Equation (13) were given in Table 3. The mechanical properties of quartz glass were given in Table 4.

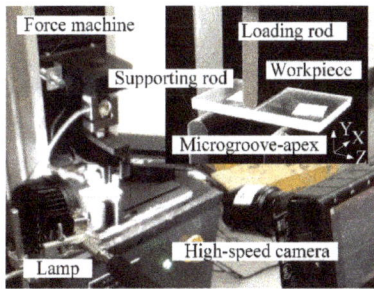

Figure 4. The experimental setup of micro-crack induced severing.

Table 3. The auxiliary parameters.

β_v	λ_1	μ_1	χ_{b1}	χ_{c1}	χ_{d1}
60°	0.5122	−0.4057	1.3123	3.2832	0.0960
90°	0.5448	−0.3449	1.8414	2.5057	0.1046
120°	0.6157	−0.2678	3.0027	1.5150	0.0871

Table 4. Mechanical properties of quartz glass.

Property	Value
Young's modulus [GPa]	77
Poisson's ratio	0.17
Ultimate tensile strength σ_u [MPa]	50
Fracture toughness K_{IC} [MPa·m$^{0.5}$]	0.81

In order to compare the traditional scribing with the micro-crack induced severing, an SFT-QG500A glass-cutter machine (Shufeng technology co. Ltd., Shenzhen, China) was employed to perform the scribing-and-breaking severing experiments of quartz glass. The rolling scribing conditions were given by 120° in tungsten carbide wheel V-tip angle, 0.19 MPa in scribing pressure, 1 mm in setting depth and 30 mm/s in scribing speed.

6. Results and Discussions

6.1. Profile of Micro-Ground Microgroove on Workpiece Surface

Figure 5 shows the microgroove profile of quartz glass after micro-grinding. It is shown that the microgroove was regular and its edges were smooth. No micro-cracks existed on the microgroove. This is because the grain cutting depths may be controlled to be less than the critical cutting depth transferred from brittle cutting to ductile cutting in micro-grinding, leading to a no-crack microgroove. According to the measured results of VHX-1000 microscope (Keyence, Osaka, Japan), the microgroove angle β_v averagely reached 61.7°, 91.8° and 119.4° in contrast to the designed ones of 60°, 90° and 120°, respectively. The microgroove angle error was ±1.8°. Correspondingly, the microgroove heights h_v were 545 μm, 495 μm and 479 μm, respectively. The microgroove-apex radius r_v were 39.3 μm, 41.3 μm and 39.7 μm, respectively. The microgroove surface roughness R_a was 80–100 nm. As a result, the micro-grinding was able to fabricate an accurate and smooth microgroove without any cracks around its apex. In contrast, the scribing produced the cracks along the scratch [2,3].

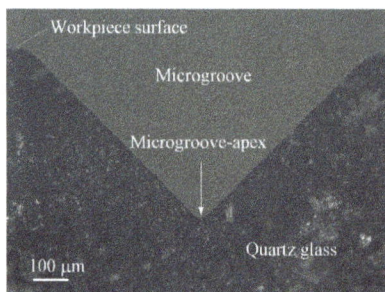

Figure 5. The microgroove profile of quartz glass after micro-grinding.

6.2. Severing Surface Topography

Figure 6 shows the severing surface topographies of quartz glass in mechanical rolling scribing and micro-crack-induced severing. It is shown that breakages happened on the scratch microgroove edges in mechanical rolling scribing (see Figure 6a). This is because the pure mechanical compression produced the radial cracks, leading to edge cracks. It also produced median cracks, radial cracks and lateral cracks, leading to an uneven severing surface in break-severing. It was identical to the results in the scribing-and-breaking severing of ceramics and silica glass substrates [1,3]. In contrast, no breakages happened on the severing workpiece edges along the micro-ground microgroove-apex in micro-crack-induced severing (see Figure 6b).

The severing surface edges were undamaged. The severing surface was flat and smooth. This is because the crack propagation was precisely induced along the micro-ground microgroove-apex in micro-crack-induced severing (see Figure 5). This also means that the accurate and smooth microgroove-apex without any micro-cracks was able to induce the accurate and smooth crack propagation in severing.

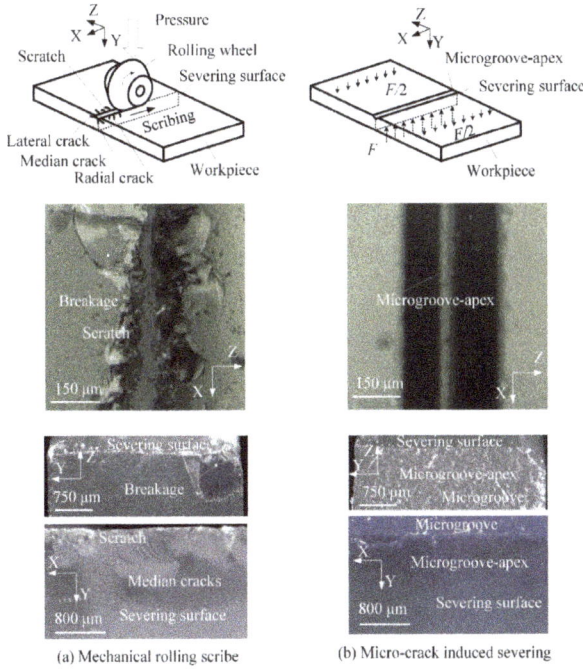

(a) Mechanical rolling scribe   (b) Micro-crack induced severing

Figure 6. The severing surface topographies of quartz glass substrate: (**a**) mechanical rolling scribe (**b**) micro-crack induced severing.

6.3. Loading Force and Loading Time versus Loading Rate

Figure 7 shows the loading force F and loading time T versus loading rate v. Experimental results showed that the loading force F increased with increasing loading time T at beginning, but it rapidly decreased after the micro-crack propagation happened on the microgroove-apex.

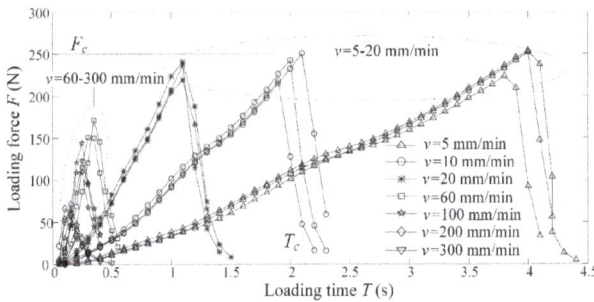

Figure 7. Loading force F and loading time T versus loading rate v.

It was identical to the relationship between loading force and displacement in tensile fracture of notched polycrystalline graphite [23]. The critical loading force and time were regarded as severing force F_c and severing time T_c, respectively. In the case of loading rate v = 60–300 mm/min, the severing force rapidly increased with increasing loading time T, but it slowly increased in the case of v = 5–20 mm/min. This mean that there existed two different mechanisms in micro-crack induced severing. Their critical loading rate v_c was distributed between 20 mm/min and 60 mm/min. As a result, the micro-crack induced severing may be distinguished by static severing ($v < v_c$) and dynamic severing ($v > v_c$), respectively.

6.4. Severing Force versus Microgroove-Apex Radius and Loading Rate

Figure 8 shows the severing force F_c versus microgroove-apex radius r_v for different loading rate v and microgroove angle β_v. Three experiments were accomplished for the same loading rate. The singularity was removed. It is shown that the severing force F_c rapidly increased and slowly approach the critical severing force F_{cc} with increasing microgroove-apex radius r_v in dynamic severing, but it slowly decreased and gradually approach the critical severing force F_{cc} in static severing. This mean that the static severing and dynamic severing produced different influence on severing force with reference to microgroove-apex radius. Moreover, the microgroove angle β_v produced little influence on the severing force F_c. The severing force F_c averagely reached 226.5 N in static severing, but it averagely reached 116.2 N in dynamic severing. As a result, the static severing increased the severing force by 95% compared to the dynamic severing.

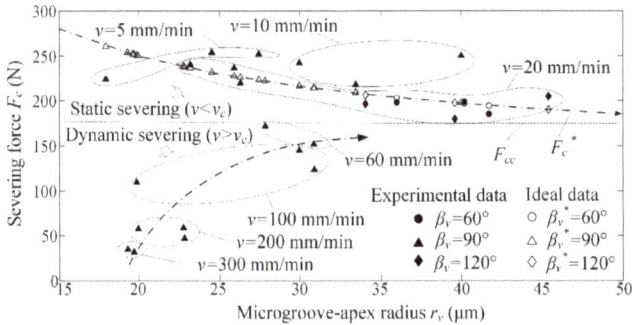

Figure 8. Severing force F_c versus microgroove-apex radius r_v for different loading rate v and microgroove angle β_v.

It is also seen that the severing force F_c rapidly decreased with increasing loading rate v in dynamic severing. In contrast, it slowly decreased in static severing. However, the loading rate produced little influence on maximum loading force in low and high loading rate when the workpiece was thermoset epoxy resin [14]. When the microgroove-apex radius was larger than the critical value of 40–50 μm, the severing force was not dominated by the microgroove-apex radius and the loading rate. Moreover, the critical severing fore F_{cc} ranged 160–180 N when the dynamic severing was transformed into static severing.

6.5. Prediction of Severing Force

In contrast to the ideal F_c^*, the experimental severing force F_c was fitted to achieve K_t as follows:

$$K_t = -0.3741\beta_v + 4.787 \tag{14}$$

According to Equation (13), the ideal severing force F_c^* may be described as follows:

$$F_c^* = \frac{326.06(0.4012r_v + 0.0104)^{0.4878}}{r_v^{0.4878}\left(1 + 0.5791\left(1 + \frac{0.0104}{0.4012r_v}\right)^{-0.9179}\right)}, \beta_v = 60° \tag{15}$$

$$F_c^* = \frac{331.66(0.3333r_v + 0.0104)^{0.4552}}{r_v^{0.4552}\left(1 + 0.54\left(1 + \frac{0.0104}{0.3333r_v}\right)^{-0.8897}\right)}, \beta_v = 90° \tag{16}$$

$$F_c^* = \frac{313.83(0.2481r_v + 0.0104)^{0.3843}}{r_v^{0.3843}\left(1 + 0.4319\left(1 + \frac{0.0104}{0.2481r_v}\right)^{-0.8835}\right)}, \beta_v = 120° \tag{17}$$

Using Equations (15)–(17), the results of ideal severing force F_c^* were plotted in Figure 8. It is seen that the experimental cutting force was identical to the ideal severing force F_c^* (see Figure 8). When the microgroove-apex radius was larger than the critical value of 40–50 μm. The ideal severing force F_c^* was stabilized at 159.5–190.0 N, which was identical to the experimental results. This mean that the Equations (13)–(16) may be used to predict and control the severing force according to the microgroove height h_v, angle β_v and microgroove-apex radius r_v in static severing.

6.6. Ideal Severing Force versus Microgroove Angle and Height

Figure 9 shows the ideal severing force F_c^* versus microgroove angle β_v and height h_v in the static severing. It is shown that the microgroove angle β_v produced little influence on the severing force F_c^* in the case of $\beta_v = 60$–120°, but the severing force F_c^* decreased with increasing microgroove β_v in the case of $\beta_v > 120°$ and $h_v > 30$ μm, respectively (see Figure 9a). It is also seen that the severing force F_c^* decreased with increasing microgroove height h_v and microgroove-apex radius r_v (see in Figure 9b). When the microgroove-apex radius r_v was larger than 40–50 μm, the severing force was not dominated by the microgroove angle and microgroove-apex radius.

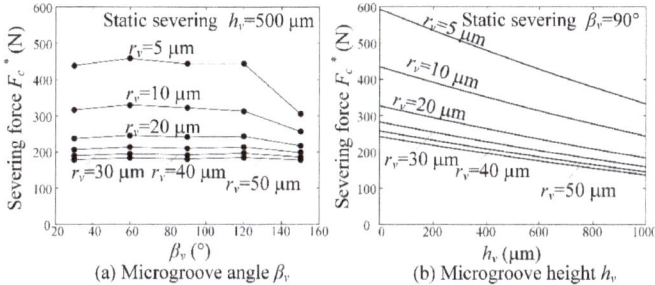

Figure 9. Ideal severing force F_c^* versus microgroove parameters: (**a**) microgroove angle β_v and (**b**) microgroove height h_v.

6.7. Severing Time versus Microgroove-Apex Radius

Figure 10 shows the severing time T_c versus microgroove-apex radius r_v. The severing time T_c was regarded as the mean value of experimental data at the same loading rate v. Experimental results showed that the severing time T_c rapidly decreased with increasing microgroove-apex radius r_v in static severing, but the microgroove-apex radius r_v produced little influence on severing time in dynamic severing. Moreover, the severing time T_c averagely reached 2.43 s in static severing, but it averagely reached 0.27 s in dynamic severing. Hence, the static severing increased the severing time by about 900% compared to the dynamic severing.

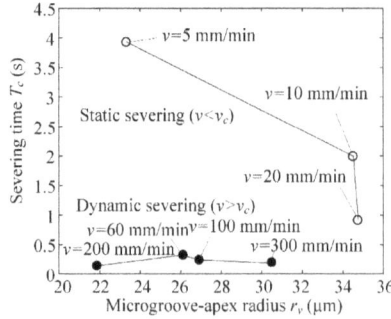

Figure 10. Severing time T_c versus microgroove-apex radius r_v.

6.8. Severing form Errors versus Loading Rate

Figure 11 shows the microgroove-direction e_m and the loading-direction severing form errors e_l versus loading rate v in micro-crack induced severing. Experimental results showed that the severing form errors gradually increased with increasing loading rate v in both static severing and dynamic severing. The microgroove-direction severing form error e_m averagely reached 8.8 μm/mm (see Figure 11a), which was much less than the loading-direction severing form error of 31.7 μm/mm (see Figure 11b). This is because the microgroove direction was dominated by the accurate micro-ground microgroove-apex, but the loading direction depended on precision positioning and loading rate. Moreover, the static severing form error averagely reached 16.2 μm/mm, which was less than the dynamic severing form error of 25.3 μm/mm. The reason is that high loading rate easily produced the position deviation between workpiece and loading rod. Hence, the static severing may decrease the severing form error by 36% compared to the dynamic severing.

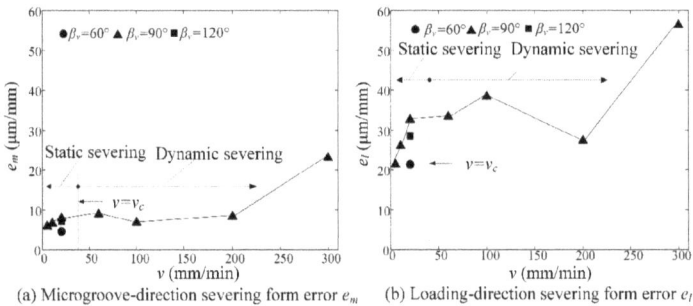

(a) Microgroove-direction severing form error e_m (b) Loading-direction severing form error e_l

Figure 11. Severing form errors versus loading rate v: (**a**) microgroove-direction severing form error e_m and (**b**) loading-direction severing form error e_l.

6.9. Severing Surface Roughness versus Loading Rate

Figure 12 shows the severing surface roughness R_a versus loading rate v. Experimental results showed that the loading rate v had little influence on the severing surface roughness R_a in both static severing and dynamic severing. However, the microgroove-direction severing surface roughness of 13.7 nm was much less than the loading-direction severing surface roughness of 29.6 nm. This mean that severing surface roughness could be dominated by material properties. Moreover, the static severing surface roughness of 19.69 nm was less than the dynamic severing surface roughness of 22.34 nm. Hence, the static severing surface roughness decreased by 12% compared to the dynamic severing. As a result, the micro-crack induced severing may produce smooth severing surface of quartz glass without any polishing.

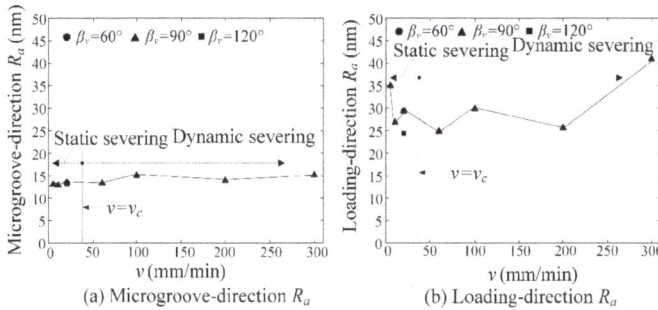

Figure 12. Severing surface roughness R_a versus loading rate v: (**a**) microgroove-direction severing roughness and (**b**) loading-direction severing roughness.

6.10. Cracking Propagation Time and Loading Positions

Figure 13 shows the micro-crack propagation process in static severing. It is shown that crack propagation occurred at the same time from the microgroove-apex and the loading point in 0.5 ms (see Figure 13a,b). This mean that the micro-crack was induced from the microgroove-apex to extend to the loading point. Hence, the loading positions dominated the severing form errors in micro-crack induced severing. This may explain why the severing form errors increased with increasing loading rate v (see Figure 11).

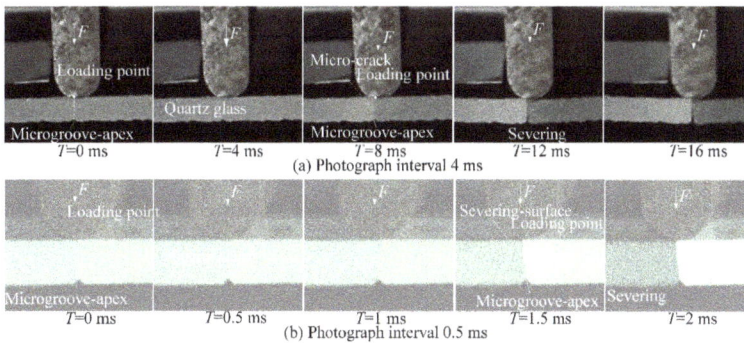

Figure 13. Micro-crack propagation process in static severing (v = 20 mm/min).

It is also found that the cracking propagation time reached 0.5 ms and less to achieve the smooth surface (see Figure 13b). In contrast, the severing time T_c averagely reached 2430 ms in static severing (see Figure 10). This also mean that the severing time was much larger than the crack propagation time. Hence, the efficiency of micro-crack induced severing is dominated by the severing time rather than the crack propagation time.

7. Conclusions

1.   Compared to a mechanical rolling scribe, the micro-crack-induced severing by a non-cracked microgroove-apex produces smooth severing edges without any breakages, median cracks,

radial cracks and lateral cracks. The micro-grinding may machine the accurate and smooth microgroove-apex without any cracks.

2. In micro-crack-induced severing, the crack propagation occurred at the same time from the microgroove-apex and the loading point. The severing efficiency is dominated by the severing time rather than the crack propagation time. The severing energy and quality depend on the loading rate.

3. When the loading rate was less than 20–60 mm/min, the dynamic severing changes to static severing. In static severing, the severing force slowly decreases with increasing microgroove-apex radius, but it rapidly increases in dynamic severing. For the critical severing force of 160–180 N, the static severing force is about two times larger than the dynamic one.

4. The severing time rapidly decreases with increasing microgroove-apex radius in static severing, but it slowly increases in dynamic severing. It reaches on average 2.43 s and 0.27 s, respectively. The static severing increases the severing time by about 9 times compared to the dynamic severing. In contrast, the cracking propagation time reaches 0.5 ms and less.

5. The severing form error and the severing surface roughness reach 4.6–32.5 μm/mm and 12.8–34.9 nm in static severing and 6.9–56.3 μm/mm and 13.3–40.9 nm in dynamic severing, respectively. The static severing may decrease the severing form error by 36% and the severing surface roughness by 12% compared to the dynamic severing, respectively.

6. In static severing, the severing force may be modelled and predicted by microgroove-apex radius, microgroove angle and height. Theoretically, it decreases with increasing the microgroove height. In dynamic severing, it is little influenced by microgroove-apex radius and microgroove angle when the microgroove angle is less than 120°.

Author Contributions: L.Z. and J.X. conceived and designed the research ideas and experimental methods; L.Z. and A.G. performed the experiments; L.Z. and J.X. analyzed the data; J.X. contributed reagents/materials/analysis tools; L.Z. wrote the paper.

Acknowledgments: The project was sponsored by the Chinese Natural Science Foundation (Grant No. 61475046), the Guangdong Science Foundation of China (Grant No. 2015A030311015).

Conflicts of Interest: The authors declare no conflict of interest.

References

1. Tomei, N.; Murakami, K.; Hashimoto, T.; Kitaichi, M.; Hirano, S.; Fukunishi, T. Development of a scribing wheel for cutting ceramic substrates and its wheel scribing and breaking technology. *JSAT* **2015**, *59*, 705–710.

2. Ono, T.; Tanaka, K. Effect of scribe-wheel dimensions on the cutting of AMLCD glass substrate. *J. Soc. Inf. Disp.* **2001**, *9*, 87–94. [CrossRef]

3. Pan, C.T.; Hsieh, C.C.; Su, C.Y.; Liu, Z.S. Study of cutting quality for TFT-LCD glass substrate. *Int. J. Adv. Manuf. Technol.* **2008**, *39*, 1071–1079. [CrossRef]

4. Hasegawa, R.; Matsusaka, S.; Hidai, H.; Chiba, A.; Morita, N.; Onuma, T. In-process estimation of fracture surface morphology during wheel scribing of a glass sheet by high-speed photoelastic observation. *Precis. Eng.* **2016**, *48*, 164–171. [CrossRef]

5. Liao, Y.S.; Yang, G.M.; Hsu, Y.S. Vibration assisted scribing process on LCD glass substrate. *Int. J. Mach. Tools Manuf.* **2010**, *50*, 532–537. [CrossRef]

6. Tsai, C.H.; Lin, B.C. Laser cutting with controlled fracture and pre-bending applied to LCD glass separation. *Int. J. Adv. Manuf. Technol.* **2007**, *32*, 1155–1162. [CrossRef]

7. Tsai, C.H.; Huang, B.W. Diamond scribing and laser breaking for LCD glass substrates. *J. Mater. Process. Technol.* **2008**, *198*, 350–358. [CrossRef]

8. Yamamoto, K.; Hasaka, N.; Morita, H.; Ohmura, E. Crack propagation in glass by laser irradiation along laser scribed line. *J. Manuf. Sci. Eng.* **2009**, *131*, 051002. [CrossRef]

9. Jiao, J.; Wang, X. Cutting glass substrates with dual-laser beams. *Opt. Lasers Eng.* **2009**, *47*, 860–864. [CrossRef]

10. Kang, M.G.; Kim, C.; Lee, Y.J.; Kim, S.Y.; Lee, H. Picosecond UV laser induced scribing of polyethylene terephthalate (PET) films for the enhancement of their flexibility. *Opt. Laser Technol.* **2016**, *82*, 183–190. [CrossRef]
11. Filippi, S.; Lazzarin, P.; Tovo, R. Developments of some explicit formulas useful to describe elastic stress fields ahead of notches in plates. *Int. J. Solids Struct.* **2002**, *39*, 4543–4565. [CrossRef]
12. Ayatollahi, M.R.; Torabi, A.R. Brittle fracture in rounded-tip V-shaped notches. *Mater. Des.* **2010**, *31*, 60–67. [CrossRef]
13. Torabi, A.R.; Firoozabadi, M.; Ayatollahi, M.R. Brittle fracture analysis of blunt V-notches under compression. *Int. J. Solids Struct.* **2015**, *67*, 219–230. [CrossRef]
14. Ayatollahi, M.R.; Torabi, A.R.; Firoozabadi, M. Theoretical and experimental investigation of brittle fracture in V-notched PMMA specimens under compressive loading. *Eng. Fract. Mech.* **2015**, *135*, 187–205. [CrossRef]
15. Barati, E.; Alizadeh, Y. A notch root radius to attain minimum fracture loads in plates weakened by U-notches under Mode I loading. *Sci. Iran.* **2012**, *19*, 491–502. [CrossRef]
16. Kanchanomai, C.; Rattananon, S.; Soni, M. Effects of loading rate on fracture behavior and mechanism of thermoset epoxy resin. *Polym. Test.* **2005**, *24*, 886–892. [CrossRef]
17. Hsieh, C.C.; Yao, S.C. Evaporative heat transfer characteristics of a water spray on micro-structured silicon surfaces. *Int. J. Heat Mass Transf.* **2006**, *49*, 962–974. [CrossRef]
18. Dhupal, D.; Doloi, B.; Bhattacharyya, B. Pulsed Nd: YAG laser turning of micro-groove on aluminum oxide ceramic (Al_2O_3). *Int. J. Mach. Tools Manuf.* **2008**, *48*, 236–248. [CrossRef]
19. Xie, J.; Zhuo, Y.W.; Tan, T.W. Experimental study on fabrication and evaluation of micro pyramid-structured silicon surface using a V-tip of diamond grinding wheel. *Precis. Eng.* **2011**, *35*, 173–182. [CrossRef]
20. Ritchie, R.O.; Knott, J.F.; Rice, J.R. On the relationship between critical tensile stress and fracture toughness in mild steel. *J. Mech. Phys. Solids* **1973**, *21*, 395–410. [CrossRef]
21. Susmel, L.; Taylor, D. The theory of critical distances to predict static strength of notched brittle components subjected to mixed-mode loading. *Eng. Fract. Mech.* **2008**, *75*, 534–550. [CrossRef]
22. Li, L.; Guo, W.G.; Yu, X.; Fu, D.X. Mechanical behavior of ceramic-metal joint under quasi-static and dynamic four point bending: Microstructures damage and mechanisms. *Ceram. Int.* **2017**, *43*, 6684–6692. [CrossRef]
23. Ayatollahi, M.R.; Torabi, A.R. Tensile fracture in notched polycrystalline graphite specimens. *Carbon* **2010**, *48*, 2255–2265. [CrossRef]

micromachines

MDPI

Article

Absolute Copy Numbers of β-Actin Proteins Collected from 10,000 Single Cells

Beiyuan Fan [1,2], Xiufeng Li [1,2], Lixing Liu [1,2], Deyong Chen [1,2], Shanshan Cao [3], Dong Men [3], Junbo Wang [1,2,*] and Jian Chen [1,2,*]

[1]   State Key Laboratory of Transducer Technology, Institute of Electronics, Chinese Academy of Sciences, Beijing 100190, China; fanbeiyuan@ucas.ac.cn (B.F.); lixiufeng13@mails.ucas.ac.cn (X.L.); liulixing16@mails.ucas.ac.cn (L.L.); dychen@mail.ie.ac.cn (D.C.)
[2]   University of Chinese Academy of Sciences, Beijing 100049, China
[3]   State Key Laboratory of Virology, Wuhan Institute of Virology, Chinese Academy of Sciences, Wuhan 430071, China; cao@hotmails.com (S.C.); d.men@wh.iov.cn (D.M.)
*   Correspondence: jbwang@mail.ie.ac.cn (J.W.); chenjian@mail.ie.ac.cn (J.C.)

Received: 24 April 2018; Accepted: 14 May 2018; Published: 22 May 2018

Abstract: Semi-quantitative studies have located varied expressions of β-actin proteins at the population level, questioning their roles as internal controls in western blots, while the absolute copy numbers of β-actins at the single-cell level are missing. In this study, a polymeric microfluidic flow cytometry was used for single-cell analysis, and the absolute copy numbers of single-cell β-actin proteins were quantified as $9.9 \pm 4.6 \times 10^5$, $6.8 \pm 4.0 \times 10^5$ and $11.0 \pm 5.5 \times 10^5$ per cell for A549 ($n_{cell} = 14,754$), Hep G2 ($n_{cell} = 36,949$), and HeLa ($n_{cell} = 24,383$), respectively. High coefficients of variation (~50%) and high quartile coefficients of dispersion (~30%) were located, indicating significant variations of β-actin proteins within the same cell type. Low p values ($\ll 0.01$) and high classification rates based on neural network (~70%) were quantified among A549, Hep G2 and HeLa cells, suggesting expression differences of β-actin proteins among three cell types. In summary, the results reported here indicate significant variations of β-actin proteins within the same cell type from cell to cell, and significant expression differences of β-actin proteins among different cell types, strongly questioning the properties of using β-actin proteins as internal controls in western blots.

Keywords: microfluidics; single-cell analysis; polymeric microfluidic flow cytometry; single-cell protein quantification

---

1. Introduction

As housekeeping proteins, β-actins are obligatory parts of cell cytoskeletons, playing important roles in the maintenance of cellular shapes, migrations, and signal transductions [1]. Due to their constitutive expressions, β-actin proteins are commonly used as internal controls in western blots, based on the assumptions of constant expressions from cell to cell and sample to sample. However, recent studies indicate varied expressions of β-actin proteins; thus, the use of β-actin proteins as internal controls is under question [2,3].

As pioneering studies, in 2005, Banks et al. reported varied expressions of β-actin proteins with (1) coefficients of variation of 28% among 10 renal cancer cell lines; (2) higher levels in tumor versus normal renal tissues; and (3) 4-fold differences between stomach and adrenal tissues [4]. Furthermore, in 2006, Liu et al. reported a 2.5-fold increase in β-actin proteins in injured spinal cords in comparison to normal counterparts [5]. In addition, in 2008, significant differences in β-actin proteins were observed in skeletal muscle tissues of early symptomatic, symptomatic, and terminal stages [6].

More recently, in 2014, Gupta et al. reported higher levels of β-actin proteins in gastric tumor tissues in comparison to normal counterparts [7]. Also in 2014, Deybboe et al. reported decreases in β-actin proteins in human skeletal muscles with aging [8]. Furthermore, in 2015, Nam et al. reported that β-actin proteins were dramatically lower in the proximal duodenum relative to the rest of the small intestines [9]. In addition, in 2016, Chen et al. reported differences of β-actin proteins in the submandibular glands of male and female mice [10].

All of these previous data about the expressions of β-actin proteins were obtained from western blots. As a semi-quantitative approach, it cannot report absolute copy numbers of β-actin proteins, and thus, data reported by different groups cannot be effectively compared with each other. In addition, the previously reported data were derived from population studies, which cannot be used to address questions of whether there exist different expressions of β-actin proteins from cell to cell even within the same cell type.

In order to address this issue, in this study, absolute copy numbers of β-actin proteins were obtained, leveraging a recently reported polymeric microfluidic flow cytometry [11]. More specifically, lung, liver, and cervical tumour cell lines of A549, Hep G2 and HeLa were characterized by the microfluidic platform, yielding absolute copy numbers of β-actin proteins from ~10,000 single cells. Varied expressions of β-actin proteins among individual cells within the same cell type and among different cell types were located. The data reported here may be used as references for future studies of β-actin proteins.

2. Materials and Methods

2.1. Materials

All cell-culture reagents were purchased from Life Technologies Corporation (Grand Island, NY, USA). Materials required for device fabrication included SU-8 photoresist (MicroChem Corporation, Westborough, MA, USA) and polydimethylsiloxane (PDMS, 184 silicone elastomer, Dow Corning Corporation, Midland, MI, USA).

More specifically, materials in cell culture and staining include RPMI-1640 medium (GIBICO, Life Technologies Corporation, Grand Island, NY, USA), DMEM medium (GIBICO, Life Technologies Corporation, Grand Island, NY, USA), fetal bovine serum (GIBICO, Life Technologies Corporation, Grand Island, NY, USA), penicillin and streptomycin (GIBICO, Life Technologies Corporation, Grand Island, NY, USA), trypsin (GIBICO, Life Technologies Corporation, Grand Island, NY, USA), phosphate buffer saline (GIBICO, Life Technologies Corporation, Grand Island, NY, USA), FITC labelled anti-β-actin antibody (ABCAM, ABCAM Corporation, Cambridge, UK), paraformaldehyde (Sigma, Sigma-Aldrich Corporation, St. Louis, MO, USA), triton x-100 (Sigma, Sigma-Aldrich Corporation, St. Louis, MO, USA), and bovine serum albumin (Sigma, Sigma-Aldrich Corporation, St. Louis, MO, USA).

2.2. Working Flowchart

The characterization of the absolute copy numbers of single-cell β-actin proteins mainly includes four steps: device fabrication, cell preparation, device operation & data processing, and data analysis (see Figure 1). In this study, single cells stained with fluorescence labeled antibodies are forced to deform through a polymeric constriction channel (microfabricated channel with a cross-sectional area smaller than a cell) where fluorescent profiles are collected as a function of time, which are further translated to cellular sizes and absolute copy numbers of specific intracellular proteins. Coefficients of variation and quartile coefficients of dispersion were quantified to determine the varied expressions of β-actin proteins among individual cells within the same cell. Statistical analysis and neural network based pattern recognition were conducted to determine the varied expressions of β-actin proteins among different cell types.

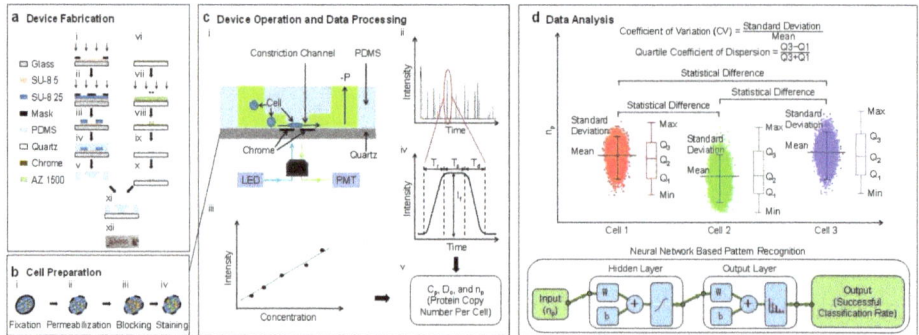

Figure 1. Methodology. Working flowchart for the characterization of the absolute copy number of β-actin proteins at the single-cell level. Key steps include device fabrication (**a**); cell preparation (**b**); device operation & data processing (**c**) and data analysis (**d**). In this study, single cells stained with fluorescence labelled antibodies are forced to deform through a polymeric constriction channel (microfabricated channel with a cross-sectional area smaller than a cell) where the obtained fluorescent profiles are translated to cellular sizes and absolute copy numbers of specific intracellular proteins. Coefficients of variation and quartile coefficients of dispersion were quantified to determine the varied expressions of β-actin proteins among individual cells within the same cell. Statistical +analysis and neural network based pattern recognition were conducted to determine the varied expressions of β-actin proteins among different cell types.

2.3. Device Design and Fabrication

In this study, a constriction channel with a cross-sectional area of 8 μm × 8 μm and a chrome gap of 2.5 μm in width was chosen for single-cell protein characterization [11]. The cross-sectional area of 8 μm × 8 μm ensures that cells with a mean diameter of 15 μm deform through and fully fill the constriction channel. In order to divide fluorescent pulses of traveling cells into rising domains, stable domains and declining domains, the gap of the chrome window should be as small as possible; 2.5 μm was used in this study.

As shown in Figure 1a, the proposed device was fabricated based on conventional microfabrication, including key steps of SU-8 mould fabrication (see Figure 1(a-i)–(a-iii)), PDMS replication (see Figure 1(a-iv),(a-v)), chrome layer patterning (see Figure 1(a-vi)–(a-x)), and bonding (see Figure 1(a-xi),(a-xii)). Detailed fabrication steps can be found from [11].

2.4. Cell Preparation

All cell lines were purchased from China Infrastructure of Cell Line Resources and cultured in a cell incubator (3111, Thermo Scientific, Waltham, MA, USA) at 37 °C in 5% CO_2. More specifically, a lung tumor cell line of A549, a liver tumor cell line of Hep G2, and a cervical tumor cell line of HeLa were cultured with RPMI-1640, DMEM and DMEM media, respectively, which were supplemented with 10% Fetal Bovine Serum (FBS) and 1% penicillin and streptomycin. Prior to experiments, cells were trypsinized, centrifuged, and resuspended in phosphate buffer saline with 0.5% bovine serum albumin at a concentration of ~1 million cells per mL.

Intracellular staining of β-actin proteins was conducted, following well-established protocols used in flow cytometry [12,13], which included key steps of fixation (see Figure 1(b-i)), membrane permeabilization (see Figure 1(b-ii)), blocking (see Figure 1(b-iii)), and antibody staining (see Figure 1(b-iv)). Firstly, the cell suspension was mixed with a 2% formaldehyde solution and incubated for 15 min at 4 °C for fixation. Then, triton x-100 (0.05% for A549 cells, 0.03% for Hep G2 cells, and 0.1% for HeLa cells) was added for an incubation of 15 min at 4 °C, in order to penetrate cellular membranes. Then blocking was conducted based on 5% vs. 1% bovine serum albumin for 30 min at room temperature. Both FITC labelled anti-β-actin antibodies (1:100) and isotype controls (1:34.5 the same final concentration as anti-beta actin antibody) were used to stain cells in suspension for 1, 2, 4, or 8 h at 37 °C for comparison. After the step of staining, cells were divided into two portions which were placed on an inverted fluorescence microscope (IX 83, Olympus, Tokyo, Japan) for imaging, and applied to the microfluidic constriction channel for fluorescent detections, respectively.

2.5. Device Operation and Data Processing

In operations, the microfluidic constriction channel was first filled with phosphate buffer saline with 0.5% bovine serum albumin. Then, suspended cells stained with FITC labelled anti-β-actin antibodies or isotype controls were applied to the entrance of the cell loading channel where a negative pressure of roughly 10 kPa generated from a pressure calibrator (DIP-610 pressure calibrator, Druck, UK) was used to aspirate cells continuously through the constriction channel (see Figure 1(c-i)). Fluorescence of single cells travelling in the constriction channel was captured by a photomultiplier tube (PMT, H10722-01, Hamamatsu, Japan), and sampled by a data acquisition card (PCI-6221, National Instruments, Austin, TX, USA) at a sampling rate of 100 kHz (see Figure 1(c-ii)). In calibrations, solutions with FITC labelled anti-β-actin antibodies were applied into the constriction channel under the same conditions as experiments (see Figure 1(c-iii)).

The fluorescent pulse of a representative cell was divided into three domains: a rising domain with a time duration of T_r, a stable domain with a fluorescent level of I_f and a time duration of T_s, and a declining domain with a time duration of T_d (see Figure 1(c-iv)). These raw parameters were then translated to the diameters of cells (D_c), concentration of β-actins at the single-cell level (C_p), and the absolute copy number of β-actins (n_p) (see Figure 1(c-v)) [11].

2.6. Data Analysis

The measurement results of the absolute copy numbers of single-cell β-actin proteins of the same cell type were represented as means ± standard deviations with three quantified quartiles (e.g., Q_1, Q_2 and Q_3). Dimensionless parameters, including the coefficients of variation (the ratio of the standard deviation to the mean) and the quartile coefficient of dispersion ($(Q_3 - Q_1)/(Q_3 + Q_1)$), were calculated to evaluate the expression differences of β-actin proteins from cell to cell within the same cell type.

In addition, analysis of variance (ANOVA) was used to locate statistical differences of β-actin proteins among A549, Hep G2, and HeLa cells, where values of $p < 0.01$ (*) were considered as statistically significant. Furthermore, neural network based pattern recognitions were conducted based on a 'Neural Network Pattern Recognition App' (MATLAB 2010, MathWorks, Natick, MA, USA) to differentiate the distribution of β-actin proteins among these three cell types. The app employs a two-layer (hidden and output layer) feed forward neural network, with sigmoid hidden and softmax output neurons [14,15].

3. Results

Figure 2a shows representative fluorescent pictures of stained A549, Hep G2, and HeLa cells where the intensities of single cells stained with fluorescence labelled anti-β-actin antibodies or isotype controls were quantified as a function of time. It was observed that the intensities of stained single cells initially increased with the incubation time, and then showed the signs of saturation at 4 h. Further increases in the incubation time (e.g., eight hours) did not lead to further significant increases in the fluorescent intensities, suggesting that after four hours of incubating cells with fluorescence labelled antibodies, all the intracellular β-actin proteins were bound with fluorescence labelled antibodies.

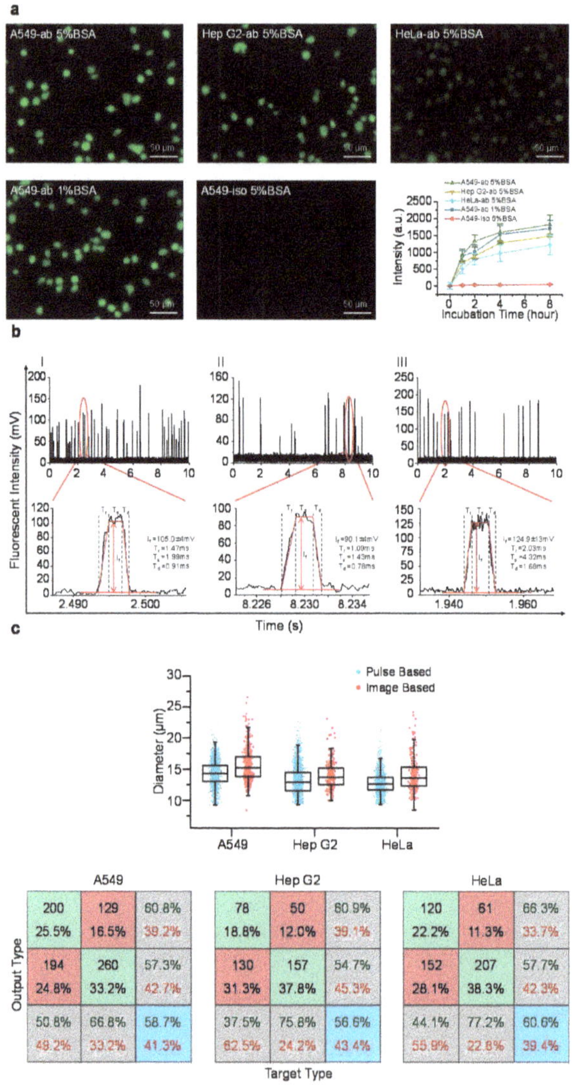

Figure 2. (a) Fluorescent pictures of stained A549, Hep G2, and HeLa cells where the intensities of single cells stained with fluorescence labelled anti-β-actin antibodies or isotype controls were quantified as a function of time under two concentrations of bovine serum albumin (1% vs. 5%) for blocking. These results validated the process of intracellular staining where (1) all the exposed proteins are taken by the fluorescence labelled antibodies and (2) non-specific sites within cells are properly blocked; (b) Fluorescent pulses of travelling A549 (I), Hep G2 (II), and HeLa (III) cells can be effectively divided into rising domains, stable domains and declining domains based on curve fitting; (c) The scatter plots of diameters of cells based on the processing of fluorescent pulses vs. images of microscopy where neural network based pattern recognition produced successful classification rates of 58.7% of A549 cells, 56.6% of Hep G2 cells and 60.6% of HeLa cells. These results indicate that comparable cell diameters were obtained based on curve fitting of fluorescent pulses and processing of microscopic images, validating the processing of fluorescent pulses.

In addition, two blocking parameters of 1% and 5% bovine serum albumin solutions produced comparable fluorescent intensities, indicating that non-specific intracellular sites were properly occupied by bovine serum albumin, and thus, the issue of non-specific binding is not a concern (see Figure 2a). Furthermore, the intensities of isotype controls were two orders lower than the intensities obtained from fluorescence labelled antibodies, further addressing the potential concern of non-specific binding in the step of intracellular staining (see Figure 2a).

Figure 2b shows the preliminary measurement results of travelling A549, Hep G2, and HeLa cells with corresponding pulses effectively divided into rising domains, stable domains and declining domains. By processing these raw parameters, the diameters of cells (D_c) were quantified as 14.3 ± 1.9 µm (A549, $n_{cell} = 14{,}754$), 13.1 ± 2.2 µm (Hep G2, $n_{cell} = 36{,}949$), and 12.7 ± 1.6 µm (HeLa, $n_{cell} = 24{,}383$). These results were consistent with the diameters of cells (D_c) of 15.7 ± 2.6 µm (A549, $n_{cell} = 394$), 13.9 ± 2.5 µm (Hep G2, $n_{cell} = 195$), and 14.1 ± 2.7 µm (HeLa, $n_{cell} = 268$) obtained from image processing of cell pictures, validating the processing of fluorescent pulses (see Figure 2c and Table 1).

Table 1. A summary of quantified key parameters of A549, Hep G2 and HeLa cells including T_r (time duration of the rising domain for a fluorescent pulse representing a traveling cell), T_s (time duration of the stable domain for a fluorescent pulse representing a traveling cell), T_d (time duration of the declining domain for a fluorescent pulse representing a traveling cell), I_f (fluorescent level of the stable domain for a fluorescent pulse representing a traveling cell), D_c (diameter of cells), C_p (concentration of β-actins at the single-cell level) and n_p (absolute copy number of β-actin proteins at the single-cell level).

Cell Type	T_r (ms)	T_s (ms)	T_d (ms)	I_f (mv)	D_c (µm)	C_p (µM)	n_p (/cell)
A549 ($n_{cell} = 14{,}754$)	2.0 ± 1.6	4.5 ± 4.3	1.5 ± 1.2	85.0 ± 24.4	14.3 ± 1.9	1.0 ± 0.3	$9.9 \pm 4.6 \times 10^5$
Hep G2 ($n_{cell} = 36{,}949$)	1.6 ± 2.3	2.9 ± 5.6	1.4 ± 3.0	75.5 ± 26.2	13.1 ± 2.2	0.9 ± 0.3	$6.8 \pm 4.0 \times 10^5$
HeLa ($n_{cell} = 24{,}383$)	2.6 ± 2.9	3.9 ± 5.3	1.9 ± 2.2	132.4 ± 34.5	12.8 ± 1.6	1.7 ± 0.5	$11.4 \pm 5.5 \times 10^5$

Neural network based pattern recognition produced successful classification rates of 58.7% of A549 cells, 56.6% of Hep G2 cells, and 60.6% of HeLa cells, when two groups of cell diameters were compared (see Figure 2c). These values of successful classification rates are within the range of 55–60%, suggesting comparable diameters obtained from fluorescent pulses and microscopic images, which further confirms the processing of fluorescent pulses.

Figure 3 summarizes the quantified single-cell copy numbers of β-actins of A549, Hep G2, and HeLa cells. For A549 ($n_{cell} = 14{,}754$), Hep G2 ($n_{cell} = 36{,}949$) and HeLa ($n_{cell} = 24{,}383$) cells, absolute copy numbers of beta-actins were quantified as $9.9 \pm 4.6 \times 10^5$, $6.8 \pm 4.0 \times 10^5$ and $11.0 \pm 5.5 \times 10^5$ per cell, respectively. The coefficients of variation were quantified as 46.4% for A549, 58.9% for Hep G2, and 47.8% for HeLa cells, which significantly deviated from 0%, and indicated significant variations of β-actin proteins from cell to cell for A549, Hep G2 and HeLa cells, respectively (see Figure 3a and Table 1). Furthermore, three quartiles and the quartile coefficients of dispersion were quantified as 6.53×10^5, 9.27×10^5, 1.25×10^6, and 31.5% for A549 cells, 4.11×10^5, 5.92×10^5, 8.34×10^5, and 33.9% for Hep G2 cells and 7.81×10^5, 1.05×10^6, 1.38×10^6, and 27.7% for HeLa cells, respectively. These values of quartile coefficients of dispersion significantly deviated from 0%, further indicating the significant variations of β-actin proteins within the same cell types (see Figure 3b).

As to the comparisons among three cell types (A549, Hep G2 and HeLa), statistical significances were located based on ANOVA, indicating the existences of expression differences of β-actins among these three cell types (see Figure 3a). Neural network based pattern recognition produced successful classification rates of 73.8% for A549 vs. Hep G2 cells, 63.9% for A549 vs. HeLa cells, and 73.1% for Hep G2 vs. HeLa cells (see Figure 3c). These values of successful classification rates are significantly higher than 50% as an indicator of no distribution difference between two cell types, further confirming expression differences of β-actins among these three cell types.

In this study, the absolute copy numbers of single-cell β-actin proteins of A549 cells were compared to the population approaches based on the conventional enzyme-linked immunosorbent assay (ELISA), producing the results at the same order, which were $1.0 \pm 0.5 \times 10^6$ vs. $3.6 \pm 0.2 \times 10^6$ per cells, respectively [11]. Actually, intracellular staining in flow cytometry has been functioning as a well-established semi-quantitative approach in deep phenotyping [16,17] and signaling state characterization [18–21], which has been demonstrated to be capable of producing trustworthy results.

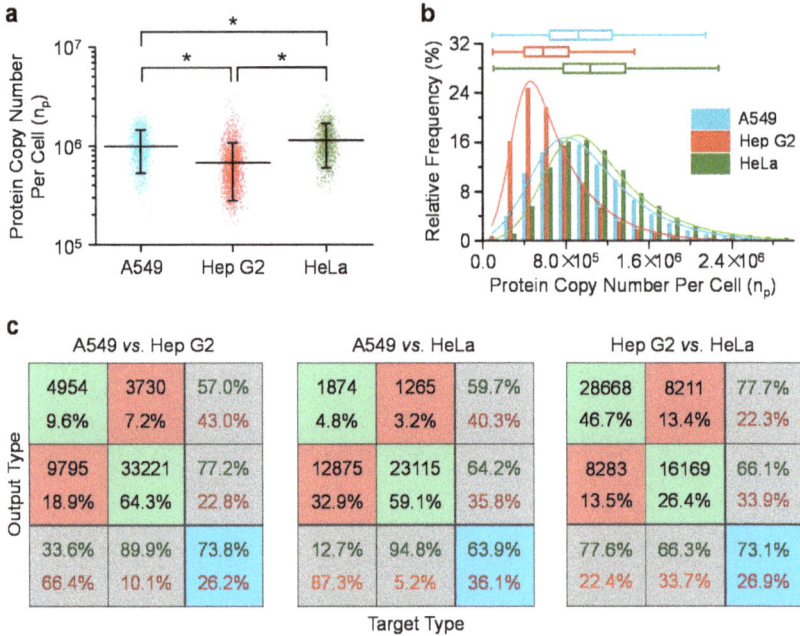

Figure 3. (**a**) Scatter plot of the absolute copy numbers of single-cell β-actin proteins of A549 (n_{cell} = 14,754), Hep G2 (n_{cell} = 36,949) and HeLa (n_{cell} = 24,383) cells with means and standard deviations included (* represents the statistical difference with $p < 0.01$); (**b**) Distributions of absolute copy numbers of β-actin proteins at the single-cell level for A549, Hep G2 and HeLa cells with three quartiles and the quartile coefficients of dispersion included; (**c**) Neural network was used to evaluate the distribution differences of β-actin proteins among A549, Hep G2 and HeLa cells, producing successful classification rates of 73.8% for A549 vs. Hep G2 cells, 63.9% for A549 vs. HeLa cells and 73.1% for Hep G2 vs. HeLa cells.

4. Conclusions

In this study, the copy numbers of β-actin proteins from ~10,000 single cells were reported, based on a previously developed microfluidic platform, and the results were validated by the quality controls in key steps of experimental operations and data analysis. Based on data analysis, significant variations of β-actin proteins within the same cell type from cell to cell and significant expression differences of β-actin proteins among different cell types were located, strongly questioning the use of β-actin proteins as internal controls in western blots, with the assumption of constant expressions of β-actin proteins from cell to cell and sample to sample.

Author Contributions: B.F., J.W. and J.C. conceived and designed the experiments; B.F., X.L. and L.L. performed the experiments; B.F., X.L. and D.C. analyzed the data; S.C. and D.M. contributed to reagents/materials/analysis tools; B.F., J.W. and J.C. wrote the paper.

Acknowledgments: The authors would like to acknowledge financial supports from the National Basic Research Program of China (973 Program, Grant No. 2014CB744600), National Natural Science Foundation of China (Grant No. 61431019, 61671430, 61774157), Chinese Academy of Sciences Key Project Targeting Cutting-Edge Scientific Problems (QYZDB-SSW-JSC011), Instrument Development Program, Youth Innovation Promotion Association and Interdisciplinary Innovation Team of Chinese Academy of Sciences.

Conflicts of Interest: The authors declare no conflicts of interest.

References

1. Gunning, P.W.; Ghoshdastider, U.; Whitaker, S.; Popp, D. The evolution of compositionally and functionally distinct actin filaments. *J. Cell Sci.* **2015**, *128*, 2009–2019. [CrossRef] [PubMed]
2. Li, R.; Shen, Y. An old method facing a new challenge: Re-visiting housekeeping proteins as internal reference control for neuroscience research. *Life Sci.* **2013**, *92*, 747–751. [CrossRef] [PubMed]
3. Ruan, W.; Lai, M. Actin, a reliable marker of internal control? *Clin. Chim. Acta* **2007**, *385*, 1–5. [CrossRef] [PubMed]
4. Ferguson, R.E.; Carroll, H.P.; Harris, A.; Maher, E.R.; Selby, P.J.; Banks, R.E. Housekeeping proteins: A preliminary study illustrating some limitations as useful references in protein expression studies. *Proteomics* **2005**, *5*, 566–571. [CrossRef] [PubMed]
5. Liu, N.K.; Xu, X.M. β-tubulin is a more suitable internal control than β-actin in western blot analysis of spinal cord tissues after traumatic injury. *J. Neurotrauma* **2006**, *23*, 1794–1801. [CrossRef] [PubMed]
6. Calvo, A.C.; Moreno-Igoa, M.; Manzano, R.; Ordovás, L.; Yagüe, G.; Oliván, S.; Muñoz, M.J.; Zaragoza, P.; Osta, R. Determination of protein and RNA expression levels of common housekeeping genes in a mouse model of neurodegeneration. *Proteomics* **2008**, *8*, 4338–4343. [CrossRef] [PubMed]
7. Khan, S.A.; Tyagi, M.; Sharma, A.K.; Barreto, S.G.; Sirohi, B.; Ramadwar, M.; Shrikhande, S.V.; Gupta, S. Cell-type specificity of β-actin expression and its clinicopathological correlation in gastric adenocarcinoma. *World J. Gastroenterol.* **2014**, *20*, 12202–12211. [CrossRef] [PubMed]
8. Vigelsø, A.; Dybboe, R.; Hansen, C.N.; Dela, F.; Helge, J.W.; Guadalupe Grau, A. GAPDH and β-actin protein decreases with aging, making Stain-Free technology a superior loading control in Western blotting of human skeletal muscle. *J. Appl. Physiol.* **2015**, *118*, 386–394. [CrossRef] [PubMed]
9. Yu, S.; Hwang, H.E.; Yun, N.; Goldenring, J.R.; Nam, K.T. The mRNA and protein levels of tubulin and β-actin are greatly reduced in the proximal duodenum of mice relative to the rest of the small intestines. *Digest. Dis. Sci.* **2015**, *60*, 2670–2676. [CrossRef] [PubMed]
10. Chen, G.; Zou, Y.; Zhang, X.; Xu, L.; Hu, Q.; Li, T.; Yao, C.; Yu, S.; Wang, X.; Wang, C. β-Actin protein expression differs in the submandibular glands of male and female mice. *Cell Biol. Int.* **2016**, *40*, 779–786. [CrossRef] [PubMed]
11. Li, X.; Fan, B.; Cao, S.; Chen, D.; Zhao, X.; Men, D.; Yue, W.; Wang, J.; Chen, J. A microfluidic flow cytometer enabling absolute quantification of single-cell intracellular proteins. *Lab Chip* **2017**, *17*, 3129–3137. [CrossRef] [PubMed]
12. Krutzik, P.O.; Nolan, G.P. Intracellular phospho-protein staining techniques for flow cytometry: Monitoring single cell signaling events. *Cytometry A* **2003**, *55*, 61–70. [CrossRef] [PubMed]
13. Lamoreaux, L.; Roederer, M.; Koup, R. Intracellular cytokine optimization and standard operating procedure. *Nat. Protoc.* **2006**, *1*, 1507–1516. [CrossRef] [PubMed]
14. Xu, H.; Lai, J.G.; Liu, J.Y.; Cao, N.; Zhao, J. Neural Network Pattern Recognition and its Application. *Adv. Mater. Res.* **2013**, *756–759*, 2438–2442. [CrossRef]
15. Sanchez, D.V. Neural Network based Pattern Recognition. In *Pattern Recognition from Classical to Modern Approaches*; Pal, S.K., Pal, A., Eds.; World Scientific Publishing: Singapore, 2001; pp. 281–300. [CrossRef]
16. Bendall, S.C.; Simonds, E.F.; Qiu, P.; El-ad, D.A.; Krutzik, P.O.; Finck, R.; Bruggner, R.V.; Melamed, R.; Trejo, A.; Ornatsky, O.I.; Balderas, R.S. Single-cell mass cytometry of differential immune and drug responses across a human hematopoietic continuum. *Science* **2011**, *332*, 687–696. [CrossRef] [PubMed]
17. Levine, J.H.; Simonds, E.F.; Bendall, S.C.; Davis, K.L.; El-ad, D.A.; Tadmor, M.D.; Litvin, O.; Fienberg, H.G.; Jager, A.; Zunder, E.R.; et al. Data-driven phenotypic dissection of AML reveals progenitor-like cells that correlate with prognosis. *Cell* **2015**, *162*, 184–197. [CrossRef] [PubMed]

18. Bodenmiller, B.; Zunder, E.R.; Finck, R.; Chen, T.J.; Savig, E.S.; Bruggner, R.V.; Simonds, E.F.; Bendall, S.C.; Sachs, K.; Krutzik, P.O.; et al. Multiplexed mass cytometry profiling of cellular states perturbed by small-molecule regulators. *Nat. Biotechnol.* **2012**, *30*, 858–867. [CrossRef] [PubMed]

19. Mingueneau, M.; Krishnaswamy, S.; Spitzer, M.H.; Bendall, S.C.; Stone, E.L.; Hedrick, S.M.; Pe'er, D.; Mathis, D.; Nolan, G.P.; Benoist, C. Single-cell mass cytometry of TCR signaling: Amplification of small initial differences results in low ERK activation in NOD mice. *Proc. Natl. Acad. Sci. USA* **2014**, *111*, 16466–16471. [CrossRef] [PubMed]

20. Spitzer, M.H.; Gherardini, P.F.; Fragiadakis, G.K.; Bhattacharya, N.; Yuan, R.T.; Hotson, A.N.; Finck, R.; Carmi, Y.; Zunder, E.R.; Fantl, W.J.; et al. Immunology. An interactive reference framework for modeling a dynamic immune system. *Science* **2015**, *349*, 1259425. [CrossRef] [PubMed]

21. Zunder, E.R.; Lujan, E.; Goltsev, Y.; Wernig, M.; Nolan, G.P. A continuous molecular roadmap to iPSC reprogramming through progression analysis of single-cell mass cytometry. *Cell Stem Cell* **2015**, *16*, 323–337. [CrossRef] [PubMed]

micromachines

MDPI

Article

Fabricating Microstructures on Glass for Microfluidic Chips by Glass Molding Process

Tao Wang [1], Jing Chen [1,*], Tianfeng Zhou [2,*] and Lu Song [1]

[1]   National Key Laboratory of Science and Technology on Micro/Nano Fabrication, Peking University, Beijing 100871, China; t.wang2010@pku.edu.cn (T.W.); l.song@pku.edu.cn (L.S.)
[2]   Key Laboratory of Fundamental Science for Advanced Machining, Beijing Institute of Technology, Beijing 100081, China
*   Correspondence: j.chen@pku.edu.cn (J.C.); zhoutf@bit.edu.cn (T.Z.);
Tel.: +86-10-6726-6595 (J.C.); +86-10-6891-2716 (T.Z.)

Received: 28 April 2018; Accepted: 23 May 2018; Published: 29 May 2018

Abstract: Compared with polymer-based biochips, such as polydimethylsiloxane (PDMS), glass based chips have drawn much attention due to their high transparency, chemical stability, and good biocompatibility. This paper investigated the glass molding process (GMP) for fabricating microstructures of microfluidic chips. The glass material was D-ZK3. Firstly, a mold with protrusion microstructure was prepared and used to fabricate grooves to evaluate the GMP performance in terms of roughness and height. Next, the molds for fabricating three typical microfluidic chips, for example, diffusion mixer chip, flow focusing chip, and cell counting chip, were prepared and used to mold microfluidic chips. The analysis of mold wear was then conducted by the comparison of mold morphology, before and after the GMP, which indicated that the mold was suitable for GMP. Finally, in order to verify the performance of the molded chips by the GMP, a mixed microfluidic chip was chosen to conduct an actual liquid filling experiment. The study indicated that the fabricating microstructure of glass microfluidic chip could be finished in 12 min with good surface quality, thus, providing a promising method for achieving mass production of glass microfluidic chips in the future.

Keywords: glass molding process; groove; roughness; filling ratio

1. Introduction

Microfluidic devices have been drawing a great deal of attention among academic and engineering communities due to their potential to revolutionize analytical measurements in chemical and biomedical areas [1]. They are much smaller, lighter, and cheaper than traditional instruments, and can improve efficiency and reduce reagents' consumption dramatically. Various microfluidic devices are fabricated for various applications, such as capillary electrophoresis [2,3], semen testing [4], electrochromatography [5], and DNA separation [6].

To date, many substrates have been reported to fabricate microfluidic chips, such as silicon [7], polydimethylsiloxane (PDMS) [8], polymethyl methacrylate (PMMA) [9], and glass [10]. Silicon has good chemical and thermal stability, and can obtain complicated 2D and 3D microstructure by photolithography and etching approaches, but the downsides of fragility, high-cost, opacity, poor electrical insulation, and complex surface chemical properties impede its application. Contrastingly, PDMS and PMMA are widely used, in both academic and industrial fields, for biochips due to their high efficiency and low cost of fabrication process. However, they are not appropriate for certain applications, such as operating hydrophobic molecules or when stable surface characteristics are required, due to the issues of dissolution and surface property control [11]. Among them, glass is proved to be a more suitable substrate for microfluidic chips, which can provide advantages over other materials, such as beneficial optical properties, good insulating properties, high resistance to mechanical stress, high surface stability, and high solvent compatibility of the glass [12].

However, fabricating the precise microstructure on glass for microfluidic chips is still challenging and many researchers have been exploring the field of glass microstructure fabrication techniques. Generally, glass microstructure fabrication techniques can be classified into six categories, as shown in Figure 1.

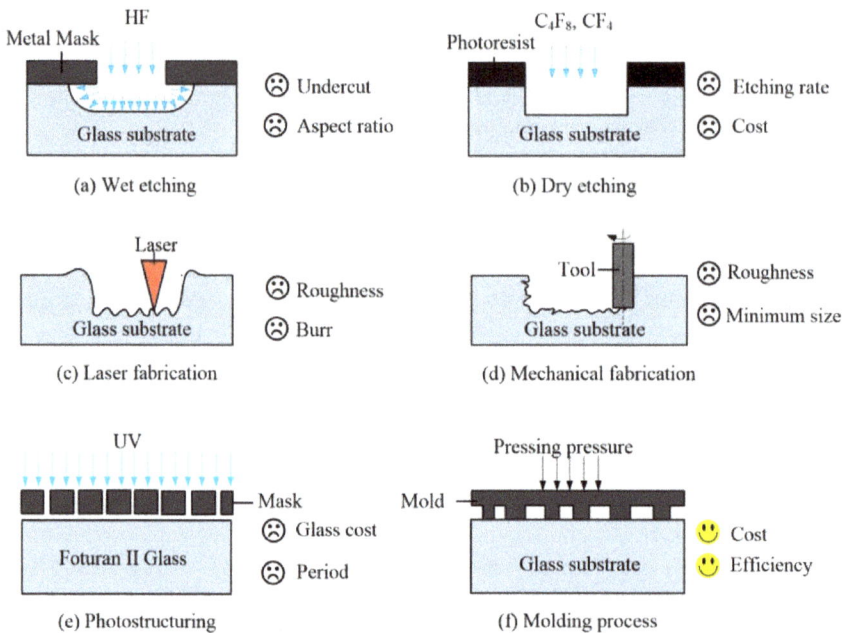

Figure 1. Six typical glass microstructure fabrication techniques. (**a**) Wet etching; (**b**) Dry etching; (**c**) Laser fabrication; (**d**) Mechanical fabrication; (**e**) Photostructuring; (**f**) Molding process.

(1) Wet Etching

Glass is an isotropic material, which can be wet-etched by buffered hydrofluoric acid (HF) in a non-directional manner, as shown in Figure 1a. Wet-etching technique is well developed in fabricating microfluidic channels on glass [13–16] and it can achieve the production of microfluidic glass chips at a commercial scale at a reasonable cost. However, the wet-etching process suffers some inherent limitations, such as undercut, which exerts a challenge to fabricate high aspect ratio microstructure. In addition, the chemical liquid used in the process, especially buffered HF, is harmful to the environment, which needs further disposal before discharging.

(2) Dry Etching

Dry chemical etching of glass is carried out by capacitively coupled plasmas (CCP) [17], CCP/microwave plasma [18] and inductively coupled plasmas (ICP) [19]. It can achieve an anisotropic and precise profile, as shown in Figure 1b. Reactive ion etching results in an anisotropic profile due to the directional nature of the ion bombardment, affecting surface chemical reaction, as well as physical sputtering [20]. The technique for fabricating glass has been reported using different chemistries, such as C_3F_8, CHF_3, CF_4/CHF_3, SF_6/Ar and CF_4/Ar [17–21]. The major weaknesses of glass dry etching are the low etch rate and the low etching selectivity of the glass to the etch mask, which impede its widespread application substantially.

(3) Laser Fabrication

Laser fabrication of microstructure on glass has been reported using CO_2, UV, and ultra-short pulse lasers [22–24], and its mechanism is shown in Figure 1c. Although the efficiency of laser glass micromachining is much higher than those of other conventional methods, its wide application is hindered by the brittleness and poor thermal properties of glass, resulting in a risk of microcracking and poor surface quality on the bottom of groove [25].

(4) Mechanical Fabrication

The major mechanical fabrication techniques include micromilling [26,27], powder blasting [28–30], and micro-ultra-sonic machining [31]. They are superior in fabricating efficiency, despite it usually being a challenge to obtain smooth machined surfaces by mechanical fabrication. In addition, the minimum size of the microstructure mainly depends on the tool.

(5) Photostructuring

Some photosensitive glass is commercially available, with the material itself being sensitive to ultraviolet (UV) light of a wavelength of around 310 nm (e.g., Schott Fortran [32]), which is amenable to anisotropic photostructuring. Therefore, it does not require an intermediate photoresist layer for patterning, as shown in Figure 1e. Although the fabrication process is simpler than the conventional wet-etching process, the whole processing period is still long at over 20 h. This is due to the required heating and cooling processes. In addition, the cost of Schott Fortran glass is greater than for conventional glass.

(6) Molding Process

Glass is a strongly temperature-dependent material. It is hard and brittle at room temperature, while it becomes a viscoelastic body or viscous liquid at high temperate. To date, the GMP has been accepted as a promising technique to efficiently generate microstructure on glass [33–35], as shown in Figure 1f. The main challenges are the microstructure fabrication on mold and the optimization of parameters for GMP.

Although GMP has been investigated in fabricating microstructure on glass for decades, the majority of the published reports focus on optical components [33–38], in which the typical microstructures are arrays of microgrooves, micropyramids, microneedles, microlenses, and microprisms. However, in reference to fabricating microfluidic chips, the typical microstructure is a U or rectangle cross-section groove. Recently, some researchers have started to shift the focus to the fabrication of microfluidic chips. Chen et al. [39,40] used nickel alloy mold to conduct their GMP for fabricating microfluidic chips in a conventional furnace, instead of a large-volume vacuum hot press. Since there is no water cooling system, the entirety of the GMP lasted more than 15 h. Huang et al. [41] utilized silicon molds to conduct the GMP in a GMP-207-HV (Toshiba Machine Co., Ltd., Shizuoka-ken, Japan) optical glass mold press machine for fabricating microfluidic chips, and the time of the GMP could, therefore, be reduced to less than 15 min. Since silicon material is inherently brittle, it is easy for fractures to occur during the GMP.

This paper explored the microstructure fabrication method of GMP on glass for microfluidic chips, which could lay an experimental and theoretical foundation for achieving mass production of microfluidic glass chips in the future. Since the glass–glass bonding process is an important issue for glass microfluidic chips, it will be investigated thoroughly in the future. Therefore, the chip verification in this paper was only conducted on a tape-glass bonding chip for simplification. In the present research, groove molding experiments were conducted and the molded profiles were observed. Moreover, the influence of the molding parameters was investigated, with corresponding experiments investigating the molding of microfluidic chips also being conducted and the molded chips were demonstrated. Finally, issues surrounding the curved side wall and bonding technique were discussed.

2. Experiments

2.1. Experimental Setup and Measurement

All experiments were carried out on an ultraprecision glass molding machine, PFLF7-60A (SYS Corp., Tochigi, Japan), which could operate between 20–750 °C The glass molding process is schematically illustrated in Figure 2. The GMP consists of four steps: Heating, pressing, annealing, and cooling. In the first step, the glass and mold are placed on the bottom platen and heated together until the molding temperature is reached, which is typically 10 °C above the transition temperature of the glass material, as shown in Figure 2a. Next, the upper platen is driven downward to conduct the pressing process, which achieves the replication of the microstructure from the mold onto the preform. Then, the temperature is slowly cooled down to somewhat below the annealing point, while a small pressing pressure is used to maintain high fidelity, as shown in Figure 2c. Finally, the preform experiences the fast cooling step via the water cooling system and the microstructure is obtained by demolding, as shown in Figure 2d. The typical evolution of temperature and pressure during the GMP in experiment is shown in Figure 3.

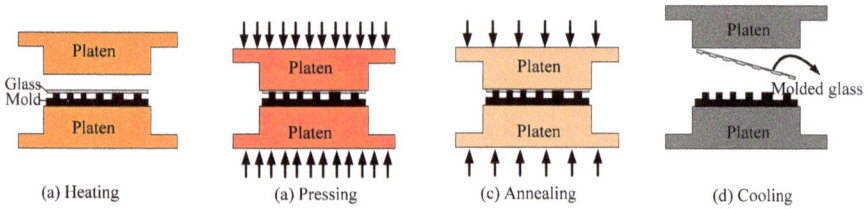

Figure 2. The schematic photo of the GMP. (**a**) Heating; (**b**) Pressing; (**c**) Annealing; (**d**) Cooling.

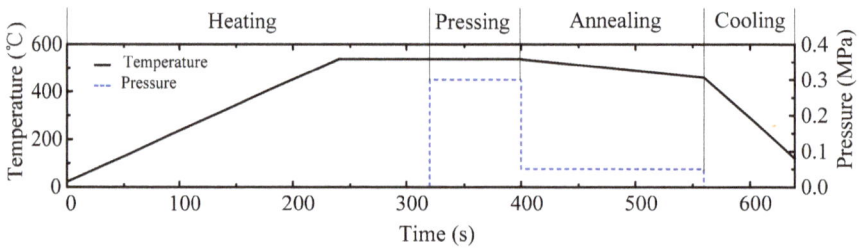

Figure 3. The typical evolution of temperature and pressure during the glass molding process (GMP).

The surface morphology of the mold was observed by a VK-X200 3D Laser Scanning Microscope (Keyence Corporation, Osaka, Japan) and FEI Quanta 250 FEG (Thermo Fisher Scientific, Waltham, MA, USA). The surface morphology of the molded glass was studied by an Olympus Lext OLS4100 (Olympus Corporation, Tokyo, Japan) and all the roughness measurements were conducted by this as well.

2.2. Glass and Mold

The D-ZK3 optical glass wafer, with a 25.4 mm diameter and 0.6 mm thickness, was used in the experiment. The material is dense barium crown optical glass manufactured by CDGM GLASS Co., LTD (Chengdu, China). The major chemical compositions and thermal properties are listed in Tables 1 and 2, respectively. T_g is defined as the transition temperature, above which the volume expanding rate increases abruptly as the temperature increases. T_s is defined as the yielding temperature, at which the glass reaches its maximum expansion point and starts shrinking as temperature increases further.

In order to shorten the paper length, the morphology details of molds are listed in Figures S1–S5 in Supplementary Materials.

Table 1. The major thermal properties of D-ZK3 glass [42].

Content	T_g (°C)	T_s (°C)	$T_{10}^{14.5}$ (°C)	T_{10}^{13} (°C)	$T_{10}^{7.6}$ (°C)	$\alpha_{100/300 \,°C}$ (10^{-7}/K)
Value	511	546	471	499	605	93

Table 2. The chemical compositions of D-ZK3 glass [42].

Content	SiO_2	B_2O_3	CaO	Al_2O_3	BaO	Sb_2O_3
Value	30–40%	20–30%	0–10%	0–10%	10–20%	0–10%

2.3. Design of Experiment

The main factors influencing the molded morphology are the pressing temperature, pressure, and time. The pressing temperature is usually set 10 °C above the transition temperature, T_g, according to previous experiments. In this paper, the pressing temperature range was set between 547 °C and 552 °C, which means that the adhesion between the glass and the mold was avoided and demonstrates the influence of temperature. In regards to pressure, this was set between 0.1 MPa and 0.5 MPa, which is the main molding pressure in the molding machining. Pressing times was set as 80s in all experiments as its influence is insignificant if the time is above 60 s, based on our previous research. The present experiment consisted of two parts. The first part was the fundamental study investigating the influence of molding parameters, such as temperature and pressure, on the morphology of the molded groove, which can help to ascertain the optimal parameters. The second part was the investigation of the microstructure morphology during the fabricating of microfluidic chips by the GMP. The details of the parameters in the experiment design are listed in Table 3.

Table 3. Details of the parameters in the glass molding process (GMP).

Item	No.	Temperature (°C)	Pressure (MPa)
Varying temperature	1	547	0.3
	2	548	0.3
	3	549	0.3
	4	550	0.3
	5	551	0.3
	6	552	0.3
Varying pressure	7	549	0.1
	8	549	0.2
	9	549	0.4
	10	549	0.5
Diffusion mixer chip	11	550	0.5
Flow focusing chip	12	550	0.5
Cell counting chip	13	550	0.5

3. Results and Discussion

3.1. Groove Molding Experiment

The aim of this study was to investigate molding performance during the fabricating of grooves with a 60 μm depth. The molded grooves at different molding temperatures are shown in Figure 4. There are three grooves with a width of 200 μm, 100 μm, and 50 μm, respectively in each photo. All bottom surfaces of the grooves are extremely smooth. In order to get the quantitative value of the surface quality at the groove bottom, the sample molded at the temperature of 547 °C was taken and the surface roughness measurements were taken of all bottom surfaces running parallel to the grooves by the Olympus Lext OLS4100 (Olympus Corporation, Tokyo, Japan), as shown in Figure 4a. The results demonstrate that the surface roughness, R_a, with the width of 200 µm, 100 µm, and 50 µm, are 13 nm, 9 nm, and 7 nm, respectively. Although the results somewhat indicate a correlation between the channel width and the mold surface roughness, its value is mainly determined by the mold surface roughness due to the contact between the mold surface and the groove bottom surface during the molding process. Due to the polishing process, the surface roughness, R_a, of the mold can achieve around 10 nm. However, as an exploring investigation, the molding experiment was not conducted in a clean room and the individual roughness value may be influenced by some dust on the groove bottom surface, leading to small variations in the measurements at different locations.

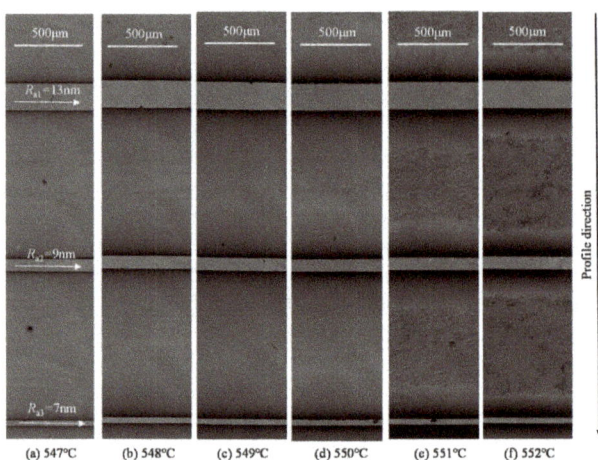

Figure 4. The molded grooves at different molding temperatures ($P = 0.3$ MPa). (**a**) 547 °C; (**b**) 548 °C; (**c**) 549 °C; (**d**) 550 °C; (**e**) 551 °C; (**f**) 552 °C.

It is worth mentioning that the bottom of the groove fabricated by the GMP is much better than those of the conventional fabricating methods, which is beneficial to the reduction of residual bio-samples and the increase of light transmittance performance. In addition, when temperature increases to 551 °C, some rough zones can be observed between grooves, and as temperature increases further, the size of the zones increase correspondingly. Since the surface roughness of the mold during the GMP is relatively rough (R_a around 0.3 µm) due to its fabricating process, it is expected that this zone touched the bottom of the mold during the GMP.

In order to provide more information about the molded grooves, the profiles perpendicular to the grooves were extracted in each sample, as shown in Figure 5. It is evident that the depth of the groove increases with the molding temperature, from 40 µm at 547 °C to 60 µm at 550 °C. The corner radius of the grooves decreases with the molding temperature also. In addition, when the temperature reaches 550 °C, the glass almost touches the mold bottom and when it increases further, the contact zones in the profiles can be observed as the rough and horizontal areas on the top, as shown in Figure 5e,f.

The influence of the pressure is similar to that of the temperature. In order to avoid redundancy, the microstructure images and profiles molded at different pressures are ignored in the paper. The influence of the molding parameters, such as pressure and temperature, on groove depth are shown in Figures 6 and 7, respectively. It indicates that the groove depth increased with both parameters. The depth value reached was maximum at a pressure of 0.4 MPa, with the temperature set at 549 °C, and at 550 °C when the pressure was set at 0.3 MPa. It is worth noting that, although further increases of both parameters would benefit the fill ratio, this increases the risk of cracking in the glass, in addition to adhesion between the mold and the glass. Therefore, the investigation of the optimal parameters for the maximum filling ratio are not included in the scope of the paper. Nevertheless, all the parameters of temperature and pressure used in this experiment were sufficient to avoid adhesion during the demolding process. The phenomenon of typical broken up in glass after demolding is shown in Figure 8. The whole part of the bottom material breaks away from the top material, which should be avoided in the GMP.

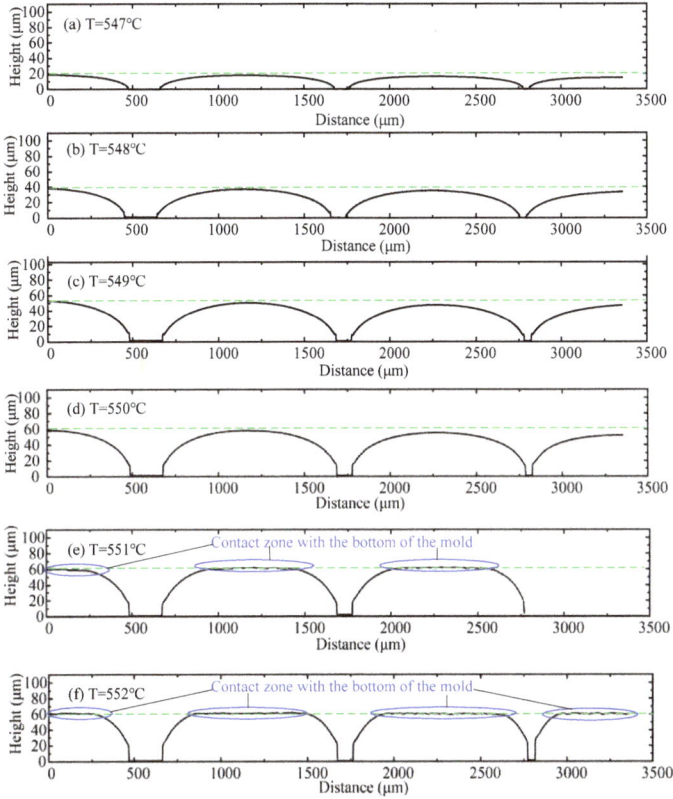

Figure 5. The profiles of the molded grooves at different molding temperatures. (**a**) 547 °C; (**b**) 548 °C; (**c**) 549 °C; (**d**) 550 °C; (**e**) 551 °C; (**f**) 552 °C.

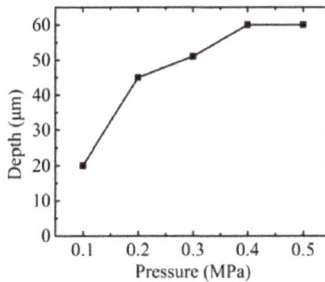

Figure 6. The influence of pressure on depth.

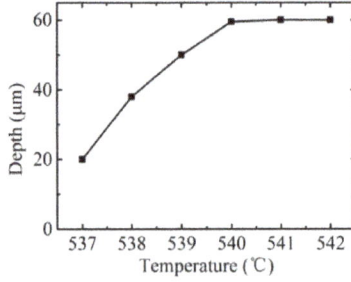

Figure 7. The influence of temperature on depth.

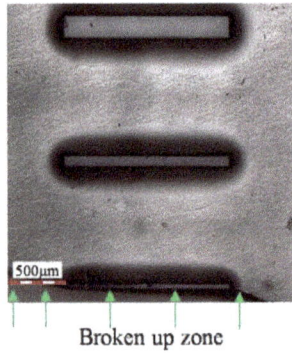

Broken up zone

Figure 8. A broken glass.

In order to reveal the accuracy of the molded glass grooves relative to the actual mold features, the width comparisons between mold protrusions and glass grooves are shown in Figure 9. The sample of glass groove data is extracted from the profile at a temperature of 552 °C, which is demonstrated at the bottom of Figure 5. It indicates that the widths of mold protrusions are all wider than the corresponding widths of glass grooves, and that the molding accuracy was around 4–6 µm. The different widths can be attributed to the fact that, when the force from the protrusions compressing the glass preform is taken off by demolding, the walls of the glass grooves experience springback, therefore, leading to the smaller glass groove width.

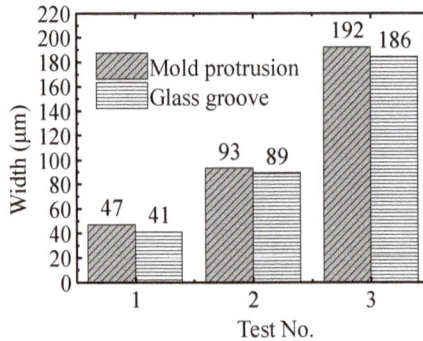

Figure 9. The width comparison between mold protrusions and glass grooves.

3.2. Microfluidic Chips

A typical microstructure of a molded diffusion mixer chip is shown in Figure 10. The macro photo indicates that the basic morphology of the diffusion mixer chip was achieved. In order to provide more details of the molded chip, area A (which is the combination area of the two input channels) and area B (which is in the typical U-shape mixed area) were enlarged, as is shown in Figure 10b,c, respectively. Since it is a small area, the highest depth was only 23.9 μm in Figure 10b. The Y shape was formed, which is suitable for the two liquids combining. In order to provide more details of the morphology, the profile extraction was conducted along the location, as shown in Figure 10b, and the profile is shown in Figure 10d. Although the corner radius is relatively large, it can be reduced by further parameter optimization. In addition, it is also valuable if there are some round corners in the cross-section profile of the channel in microfluidic chips. Regarding the mixing zone, this profile was also extracted, and is shown in Figure 10e. It indicates that the channels running parallel to each other, with a depth of 40μm, can be formed by the GMP, which is deep enough for the mixing function.

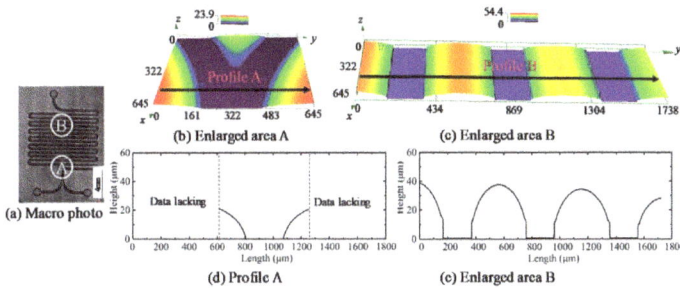

Figure 10. The microstructure of the molded diffusion mixer chip (550 °C, 0.5 MPa). (**a**) Marco photo; (**b**) Enlarged area A; (**c**) Enlarged area B; (**d**) Profile A; (**e**) Enlarged area B.

A typical microstructure of the molded flow focusing chip is shown in Figure 11. The macro photo indicates that the basic morphology of the flow focusing chip was achieved, as shown in Figure 11a. In order to provide more details about the formed microstructure, the area A and area B were enlarged, as shown in Figure 11b,c, which are the combination of the three channels at the bottom and top, respectively. They both demonstrate that the expected channels' morphology was generated, which was deep enough for the flow focusing experiment. The extracted profiles are shown in Figure 11d,e, and the section profiles are both characterized by a large corner radius.

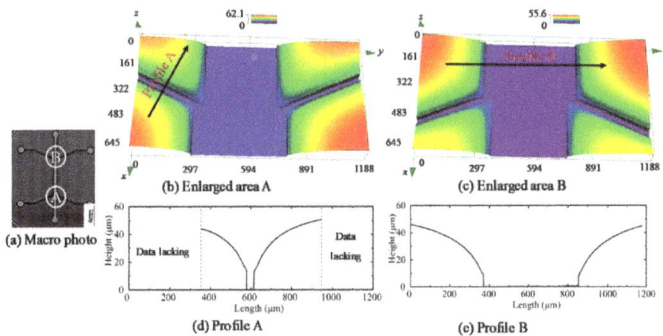

Figure 11. The microstructure of the molded flow focusing chip (550 °C, 0.5 MPa). (**a**) Macro photo; (**b**) Enlarged area A; (**c**) Enlarged area B; (**d**) Profile A; (**e**) Profile B.

A typical microstructure of the molded cell counting chip is shown in Figure 12. The macro photo indicates that the basic morphology of the cell counting chip was achieved, as shown in Figure 12a. The enlarged images of area A and area B are shown in Figure 12b,c, respectively. The profile extracted from Figure 12b is shown in Figure 12d. Although the line width was only 1.5 μm on the mold, the molded groove is clear, with a depth of approximately 2 μm. Due to the contact between the glass and the mold bottom, the top of the chip was relatively rough, which is beneficial for cells to stay inside. According to currently morphology, the molded glass is potentially valuable for cell counting.

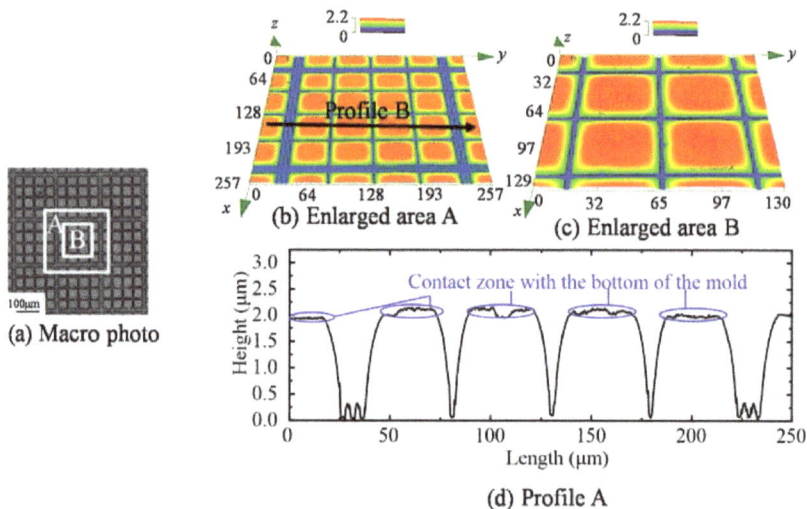

Figure 12. The microstructure of the molded cell counting chip (550 °C, 0.5 MPa). (**a**) Marco photo; (**b**) Enlarged area A; (**c**) Enlarged area B; (**d**) Profile A.

3.3. Chip Application

In order to verify the performance of the molded chips by the GMP, a mixed microfluidic chip was chosen to conduct the filling microfluidic system experiment. In terms of the bonding material, ARseal PSA clear polypropylene film tapes (MH-90697, Adhesives Research, Inc., Glen Rock, PA, USA) were used for their simplicity, and was coated with an inert silicone adhesive. The tapes provide an immediate bond to most materials and can also provide a barrier for preventing evaporation and cross contamination.

The experiment of the filling microfluidic system is shown in Figure 13. Firstly, the corresponding liquid inlet and the outlet hole were fabricated by a laser cutting machining and the bonding process was carried out in a clean room, as shown in Figure 13a. Next, the polyetheretherketone (PEEK) connectors were bonded with the chip by glue, which helped to locate the inlet holes during the filling with liquid, as shown in Figure 13b. Next, two inlet holes were filled with the red and blue ink by a pump at the same filling speed of 2 μL/min. The filling status is shown in Figure 13c and the enlarged image of the mixed channel zone is shown in Figure 13d. It was shown that the two reagents can flow along the channel without cross contamination occurring, and the obvious stratification phenomenon was observed. This proves that the molded chip can be used in a microfluidic system.

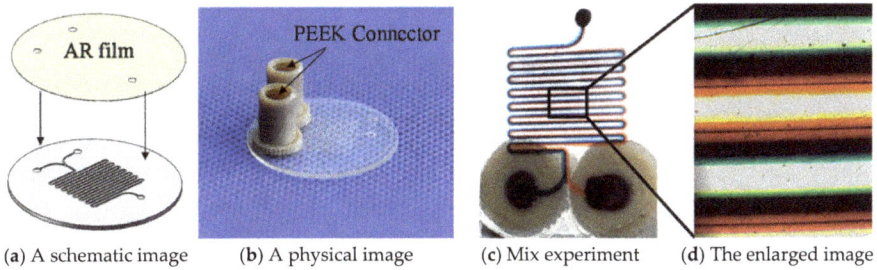

(a) A schematic image   (b) A physical image   (c) Mix experiment   (d) The enlarged image

Figure 13. The experiment of filling the mixing microfluidic system. (a) A schematic image; (b) A physical image; (c) Mix experiment; (d) The enlarged image.

4. Discussions

4.1. Curved Side Wall

Currently, only the curved sidewall of the microchannel could be achieved by the GMP in this experiment. Since the cross-section of the microstructure on the mold is rectangle, it is possible to obtain a side wall of the channels on the molded glass that is close to 90 when appropriate parameters are set. However, this scenario did not occur due to the severe adhesion problem between the mold and the glass in the actual experiment. A promising direction is the utilization of the ultrasonic vibrations during the GMP, which is worth exploring in the future.

Although the curved side wall generated by the GMP is somewhat similar to the morphology obtained by the wet-etching, the GMP has obvious advantages over the wet-etching in several aspects, which are detailed below:

(1) Higher efficiency in the fabricating of a single microfluidic glass chip

The wet-etching process consists of cleaning, lithography, and buffered oxide etch (BOE) etching, which needs several instruments and is longer than 1 h [43]. This process can achieve the production of microfluidic glass chips at a commercial scale on a regular basis and at reasonable costs. However, for the fabricating of a single microfluidic glass chip, the time of the GMP is usually less than 15 min, which is much shorter than that of the hydrofluoric acid (HF) etching.

(2) More eco-friendly

The chemical liquid used in the wet-etching process, especially buffered HF, is harmful to environment, while only nitrogen is utilized in the GMP as a protecting gas, which is significantly more eco-friendly.

(3) Wider fabricating capacity

Some researchers [13–16] have found that it is typically hard for the wet-etching process to generate a microstructure with a depth-to-width ratio that is higher than 0.5, while the GMP has wider fabricating capacity than the wet-etching. For instance, a microstructure with a depth-to-width ratio of one has been achieved by the GMP [44].

In terms of the application of channels with a curved side wall, many researchers have conducted biochemical tests on these. Lin et al. [43] demonstrated a micro flow-through sampling glass chip, which was fabricated by wet-etching, and the side walls of the channels were curved. The chip sampled and separated Cy5-labelled BSA (Nanocs Inc., New York, NY, USA) and anti-BSA successfully. Lee et al. [45] demonstrated a microcapillary electrophoresis (μ-CE) device for DNA separation and detection, and the channels with curved side walls on the plastic chip were fabricated by the hot embossing method. Castaño-Álvarez et al. [46] fabricated glass microfluidic channels with curved side walls by photolithography and wet-etching. The microfluidic chips were successfully used, in combination with a metal-wire end-channel amperometric detector, for capillary electrophoresis (CE). Evander et al. [47] fabricated glass microchannels with curved side walls by wet-etching, and the glass microfluidic chip was used for acoustic force control of cells and particles in a continuous microfluidic process. Nevertheless, although the side walls of the channel fabricated by the GMP were curved, currently, the chips are still valuable for many applications.

4.2. Bonding Techniques

Generally, there are three categories of glass bonding techniques: (1) Direct bonding, involving surface-activated or fusion bonding; (2) Anodic bonding, with silicon as the intermediate layer; and (3) Adhesive bonding, with additional adhesive material. The strengths and weaknesses of each method can be found in [48]. In this paper, since the focus is on investigating the new material performance in the GMP, the bonding process was achieved by a simple method of the glass microchannel being sealed by a polymer adhesive sheet. Compared with the three traditional methods, the bonding process in this paper has the following two advantages:

(1) High efficiency

Compared with direct and anodic bonding, the polymer adhesive sheet bonding can be achieved in room temperature without any additional chemical solution and, subsequently, there is no clogging problem, which usually occurs in adhesive bonding. Thus, it is much faster to achieve the prototyping of the microfluidic chips for research.

(2) Recycling value

The bond of failed or used bonded microchips can easily be broken by soaking the microchips in acetone, with sonication, without damaging the glass microstructure, and they can be reused by the polymer adhesive sheet bonding process.

Although it is a simple method, it is worth pointing out that the bonding process in this paper is assumed to not compromise the main advantages of the glass microfluidic chips over polymer chips, such as biocompatibility, optical transparency, and surface modification. As for biocompatibility, since the anti-reflection (AR) film is designed to be biocompatible, the influence of the AR film is supposed to be small enough. In terms of optical transparency, it is true that the polymer AR film may exert some negative influence on this aspect. However, this issue can be resolved if the test is conducted by placing the chip upside down, as shown in Figure 14. When the surface modification is concerned, since there are still three glass surfaces left in the microchannel, different surface modification procedures can be conducted as well, such as protein immobilization [49]. Therefore, the microfluidic chip assembled by polymer adhesive sheet bonding is still valuable for practical applications.

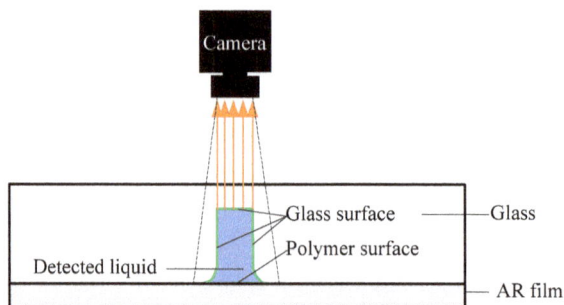

Figure 14. A typical testing diagram of the produced glass and anti-reflection (AR) film chip.

5. Conclusions

This paper investigated the fabricating of glass microfluidic chips by the GMP, which provided a promising method for fast prototyping and mass production of microfluidic chips in the future. The main conclusions were as follows:

(1) A groove with a 60 μm height could be obtained by the GMP in 12 min, and the bottom roughness, R_a, could be as small as 10 nm. The molded groove depth increased dramatically with the molding temperature and pressure before the time when the glass contacted with the bottom of the mold. Precautions should be taken to avoid cracks generated inside the glass.

(2) The microstructure of a diffusion mixer chip and a flow focusing chip were generated with more than 40 μm groove depths by the GMP, and the microstructure of the cell counting chip was fabricated with the groove of a 2 μm depth and 1.5 μm width.

(3) The study of mold wear indicated that the macro shape of the microstructure on the mold changed little over the time of 20 experiments, although the edge became a little blunt due to the weak stiffness in this area. The results from the energy-dispersive X-ray spectroscopy (EDS) analysis found that there was no chemical wear.

(4) The performance verification of the molded chips was conducted on a mixed microfluidic chip and the result indicated that the two reagents could flow along the channel without cross contamination, and that the obvious stratification phenomenon was observed. This proved that the molded chip could be used in a microfluidic system.

Supplementary Materials: The following are available online at http://www.mdpi.com/2072-666X/9/6/269/s1, Figure S1: The morphology of the mold for fabricating channels on glass, Figure S2: The morphology of three different molds for fabricating microfluidic chips, Figure S3: The mold morphology after GMP, Figure S4: The microstructure of molds before GMP, Figure S5: The microstructure of the molds after GMP.

Author Contributions: J.C. conceived and designed the experiments; T.W. performed the experiments and wrote the paper; T.Z. provided theoretical guidance; L.S. contributed to the fabrication of the core mold.

Acknowledgments: This project is supported by the National Natural Science Foundation of China (Grant No. U1537208), the National High Technology Research and Development Program of China (863) Grant No. 2015AA043601 and the National Natural Science Foundation of China (Grant No. 51375050).

Conflicts of Interest: The authors declare no conflict of interest. The founding sponsors had no role in the design of the study; in the collection, analyses, or interpretation of data; in the writing of the manuscript, and in the decision to publish the results.

References

1. Kkr, T.; Vijayalakshmi, M.A. A review on recent developments for biomolecule separation at analytical scale using microfluidic devices. *Anal. Chem. Acta* **2016**, *906*, 7–21.

2. Geigy, C. Capillary electrophoresis and sample injection systems integrated on a planar glass chip. *Anal. Chem.* **1992**, *64*, 1928–1932.

3. Seiler, K.; Harrison, D.J.; Manz, A. Planar glass chips for capillary electrophoresis: Repetitive sample injection, quantitation, and separation efficiency. *Anal. Chem.* **1993**, *65*, 1481–1488. [CrossRef]

4. Kricka, L.J.; Faro, I.; Heyner, S.; Garside, W.T.; Fitzpatrick, G.; McKinnon, G.; Ho, J.; Wilding, P. Micromachined analytical device: Microchips for semen testing. *J. Pharm. Biomed.* **1997**, *15*, 1443–1447. [CrossRef]

5. Jacobson, S.C.; Hergenröder, R.; Koutny, L.B.; Ramsey, J.M. Open channel electrochromatography on a microchip. *Anal. Chem.* **1994**, *66*, 2369–2373. [CrossRef]

6. Effenhauser, C.S.; Paulus, A.; Manz, A.; Widmer, H.M. High-speed separation of antisense oligonucleotides on a micromachined capillary electrophoresis device. *Anal. Chem.* **1994**, *66*, 2949–2953. [CrossRef]

7. Shoffner, M.A.; Cheng, J.; Hvichia, G.E.; Kricka, L.J.; Wilding, P. Chip PCR. I. Surface passivation of microfabricated silicon–glass chips for PCR. *Nucleic Acids Res.* **1996**, *24*, 375–379. [CrossRef] [PubMed]

8. Yu, L.; Huang, H.; Dong, X.; Wu, D.; Qin, J.; Lin, B. Simple, fast and high throughput single cell analysis on PDMS microfluidic chips. *Electrophoresis* **2008**, *29*, 5055–5060. [CrossRef] [PubMed]

9. Hong, T.F.; Ju, W.J.; Wu, M.C.; Tai, C.H.; Tsai, C.H.; Fu, L.M. Rapid prototyping of PMMA microfluidic chips utilizing a CO2 laser. *Microfluid. Nanofluid.* **2010**, *9*, 1125–1133. [CrossRef]

10. Fu, L.M.; Ju, W.J.; Yang, R.J.; Wang, Y.N. Rapid prototyping of glass-based microfluidic chips utilizing two-pass defocused CO_2 laser beam method. *Microfluid. Nanofluid.* **2013**, *14*, 479–487. [CrossRef]

11. Rodriguez, I.; Spicar-Mihalic, P.; Kuyper, C.L.; Fiorini, G.S.; Chiu, D.T. Rapid prototyping of glass microchannels. *Anal. Chim. Acta* **2003**, *496*, 205–215. [CrossRef]

12. Chen, Q.; Li, G.; Jin, Q.H.; Zhao, J.L.; Ren, Q.S.; Xu, Y.S. A rapid and low-cost procedure for fabrication of glass microfluidic devices. *J. Microelectromech. Syst.* **2007**, *16*, 1193–1200. [CrossRef]

13. Grétillat, M.A.; Paoletti, F.; Thiébaud, P.; Roth, S.; Koudelka-Hep, M.; De Rooij, N.F. A new fabrication method for borosilicate glass capillary tubes with lateral insets and outlets. *Sens. Actuators A Phys.* **1997**, *60*, 219–222. [CrossRef]

14. Bu, M.; Melvin, T.; Ensell, G.J.; Wilkinson, J.S.; Evans, A.G.R. A new masking technology for deep glass etching and its microfluidic application. *Sens. Actuators A Phys.* **2004**, *115*, 476–482. [CrossRef]

15. Mourzina, Y.; Steffen, A.; Offenhäusser, A. The evaporated metal masks for chemical glass etching for BioMEMS. *Microsyst. Technol.* **2005**, *11*, 135–140. [CrossRef]

16. Zhu, H.; Holl, M.; Ray, T.; Bhushan, S.; Meldrum, D.R. Characterization of deep wet etching of fused silica glass for single cell and optical sensor deposition. *J. Micromech. Microeng.* **2009**, *19*, 65013–65020. [CrossRef]

17. Ronggui, S.; Righini, G.C. Characterization of reactive ion etching of glass and its applications in integrated optics. *J. Vac. Sci. Technol.* **1991**, *9*, 2709–2712. [CrossRef]

18. Zeze, D.A.; Forrest, R.D.; Carey, J.D.; Cox, D.C.; Robertson, I.D.; Weiss, B.L.; Silva, S.R.P. Reactive ion etching of quartz and Pyrex for microelectronic applications. *J. Appl. Phys.* **2002**, *92*, 3624–3629. [CrossRef]

19. Ichiki, T.; Sugiyama, Y.; Ujiie, T.; Horiike, Y. Deep dry etching of borosilicate glass using fluorine-based high-density plasmas for microelectromechanical system fabrication. *J. Vac. Sci. Technol.* **2003**, *21*, 2188–2192. [CrossRef]

20. Park, J.H.; Lee, N.E.; Lee, J.; Park, J.S.; Park, H.D. Deep dry etching of borosilicate glass using SF_6 and SF_6/Ar inductively coupled plasmas. *Microelectron. Eng.* **2005**, *82*, 119–128. [CrossRef]

21. Leech, P.W. Reactive ion etching of quartz and silica-based glasses in CF_4/CHF_3 plasmas. *Vacuum* **1999**, *55*, 191–196. [CrossRef]

22. Yen, M.H.; Cheng, J.Y.; Wei, C.W.; Chuang, Y.C.; Young, T.H. Rapid cell-patterning and microfluidic chip fabrication by crack-free CO_2 laser ablation on glass. *J. Micromech. Microeng.* **2006**, *16*, 1143–1153. [CrossRef]

23. Sohn, I.B.; Lee, M.S.; Woo, J.S.; Lee, S.M.; Chung, J.Y. Fabrication of photonic devices directly written within glass using a femtosecond laser. *Opt. Express* **2005**, *13*, 4224–4229. [CrossRef] [PubMed]

24. Marcinkevičius, A.; Juodkazis, S.; Watanabe, M.; Miwa, M.; Matsuo, S.; Misawa, H.; Nishii, J. Femtosecond laser-assisted three-dimensional microfabrication in silica. *Opt. Lett.* **2001**, *26*, 277–279. [CrossRef]

25. Nikumb, S.; Chen, Q.; Li, C.; Reshef, H.; Zheng, H.Y.; Qiu, H.; Low, D. Precision glass machining, drilling and profile cutting by short pulse laser. *Thin Solid Films* **2005**, *477*, 216–221. [CrossRef]

26. Arif, M.; Rahman, M.; San, W.Y.; Doshi, N. An experimental approach to study the capability of end-milling for microcutting of glass. *Int. J. Adv. Manuf. Technol.* **2011**, *53*, 1063–1073. [CrossRef]

27. Arif, M.; Rahman, M.; San, W.Y. Ultraprecision ductile mode machining of glass by micromilling process. *J. Manuf. Process.* **2011**, *13*, 50–59. [CrossRef]

28. Slikkerveer, P.J.; Bouten, P.C.P.; De Haas, F.C.M. High quality mechanical etching of brittle materials by powder blasting. *Sens. Actuators A Phys.* **2000**, *85*, 296–303. [CrossRef]

29. Schlautmann, S.; Wensink, H.; Schasfoort, R.; Elwenspoek, M.; Van Der Berg, A. Powder-blasting technology as an alternative tool for microfabrication of capillary electrophoresis chips with integrated conductivity sensors. *J. Micromech. Microeng.* **2001**, *11*, 386. [CrossRef]

30. Plaza, J.A.; Lopez, M.J.; Moreno, A.; Duch, M.; Cane, C. Definition of high aspect ratio columns. *Sens. Actuators A-Phys.* **2003**, *105*, 305–310. [CrossRef]

31. Yan, B.H.; Wang, A.C.; Huang, C.Y.; Huang, F.Y. Study of precision micro-holes in borosilicate glass using micro EDM combined with micro ultrasonic vibration machining. *Int. J. Mach. Tools Manuf.* **2002**, *42*, 1105–1112. [CrossRef]

32. Dietrich, T.R.; Ehrfeld, W.; Lacher, M.; Krämer, M.; Speit, B. Fabrication technologies for microsystems utilizing photoetchable glass. *Microelectron. Eng.* **1996**, *30*, 497–504. [CrossRef]

33. Yan, J.; Zhou, T.; Yoshihara, N.; Kuriyagawa, T. Shape transferability and microscopic deformation of molding dies in aspherical glass lens molding press. *J. Manuf. Technol. Res.* **2009**, *1*, 85–102.

34. Zhou, T.; Yan, J.; Masuda, J.; Oowada, T.; Kuriyagawa, T. Investigation on shape transferability in ultraprecision glass molding press for microgrooves. *Precis. Eng.* **2011**, *35*, 214–220. [CrossRef]

35. Kobayashi, R.; Zhou, T.; Shimada, K.; Mizutani, M.; Kuriyagawa, T. Ultraprecision glass molding press for microgrooves with different pitch sizes. *Int. J. Automot. Technol.* **2013**, *7*, 678–685. [CrossRef]

36. Yi, A.Y.; Chen, Y.; Klocke, F.; Pongs, G.; Demmer, A.; Grewell, D.; Benatar, A. A high volume precision compression molding process of glass diffractive optics by use of a micromachined fused silica wafer mold and low Tg optical glass. *J. Micromech. Microeng.* **2006**, *16*, 2000–2005. [CrossRef]

37. Huang, C.Y.; Hsiao, W.T.; Huang, K.C.; Chang, K.S.; Chou, H.Y.; Chou, C.P. Fabrication of a double-sided micro-lens array by a glass molding technique. *J. Micromech. Microeng.* **2011**, *21*, 1–6. [CrossRef]

38. Saotome, Y.; Imai, K.; Sawanobori, N. Microformability of optical glasses for precision molding. *J. Mater. Process. Technol.* **2003**, *140*, 379–384. [CrossRef]

39. Chen, Q.; Chen, Q.; Maccioni, G. Fabrication of microfluidics structures on different glasses by simplified imprinting technique. *Curr. Appl. Phys.* **2013**, *13*, 256–261. [CrossRef]

40. Chen, Q.; Chen, Q.; Maccioni, G.; Sacco, A.; Ferrero, S.; Scaltrito, L. Fabrication of microstructures on glass by imprinting in conventional furnace for lab-on-chip application. *Microelectron. Eng.* **2012**, *95*, 90–101. [CrossRef]

41. Huang, C.Y.; Kuo, C.H.; Hsiao, W.T.; Huang, K.C.; Tseng, S.F.; Chou, C.P. Glass biochip fabrication by laser micromachining and glass-molding process. *J. Mater. Process. Technol.* **2012**, *212*, 633–639. [CrossRef]

42. Zhu, X.Y.; Wei, J.J.; Chen, L.X.; Liu, J.L.; Hei, L.F.; Li, C.M.; Zhang, Y. Anti-sticking Re-Ir coating for glass molding process. *Thin Solid Films* **2015**, *584*, 305–309. [CrossRef]

43. Lin, C.H.; Lee, G.B.; Lin, Y.H.; Chang, G.L. A fast prototyping process for fabrication of microfluidic systems on soda-lime glass. *J. Micromech. Miceoeng.* **2001**, *11*, 726–732. [CrossRef]

44. Takahashi, M.; Sugimoto, K.; Maeda, R. Nanoimprint of glass materials with glassy carbon molds fabricated by focused-ion-beam etching. *Jpn. J. Appl. Phys.* **2005**, *44*, 5600–5605. [CrossRef]

45. Lee, G.B.; Chen, S.H.; Huang, G.R.; Sung, W.C.; Lin, Y.H. Microfabricated plastic chips by hot embossing methods and their applications for dna separation and detection. *Sens. Actuators B Chem.* **2001**, *75*, 142–148. [CrossRef]

46. Castaño-Álvarez, M.; Ayuso, D.F.P.; Granda, M.G.; Fernández-Abedul, M.T.; García, J.R.; Costa-García, A. Critical points in the fabrication of microfluidic devices on glass substrates. *Sens. Actuators B Chem.* **2008**, *130*, 436–448. [CrossRef]

47. Evander, M.; Lenshof, A.; Laurell, T.; Nilsson, J. Acoustophoresis in wet-etched glass chips. *J. Anal. Chem.* **2008**, *80*, 5178–5185. [CrossRef] [PubMed]

48. Iliescu, C.; Taylor, H.; Avram, M.; Miao, J.; Franssila, S. A practical guide for the fabrication of microfluidic devices using glass and silicon. *Biomicrofluidics* **2012**, *6*, 016505. [CrossRef] [PubMed]

49. Qin, M.; Hou, S.; Wang, L.; Feng, X.; Wang, R.; Yang, Y.; Wang, C.; Yu, L.; Shao, B.; Qiao, M. Two methods for glass surface modification and their application in protein immobilization. *Colloids Surf. B Biointerfaces* **2007**, *60*, 243–249. [CrossRef] [PubMed]

Article

Highly Sensitive Label-Free Detection of Small Molecules with an Optofluidic Microbubble Resonator

Zihao Li [†], Chenggang Zhu [†], Zhihe Guo, Bowen Wang, Xiang Wu * and Yiyan Fei *

Department of Optical Science and Engineering, Shanghai Engineering Research Center of Ultra-Precision Optical Manufacturing, Key Laboratory of Micro and Nano Photonic Structures (Ministry of Education), Fudan University, Shanghai 200433, China; 15210720012@fudan.edu.cn (Z.L.); 16110720026@fudan.edu.cn (C.Z.); 17110720004@fudan.edu.cn (Z.G.); 16210720012@fudan.edu.cn (B.W.)
* Correspondence: wuxiang@fudan.edu.cn (X.W.); fyy@fudan.edu.cn (Y.F.); Tel.: +86-021-6564-2092 (Y.F.)
† These authors contributed equally to this work.

Received: 8 April 2018; Accepted: 29 May 2018; Published: 31 May 2018

Abstract: The detection of small molecules has increasingly attracted the attention of researchers because of its important physiological function. In this manuscript, we propose a novel optical sensor which uses an optofluidic microbubble resonator (OFMBR) for the highly sensitive detection of small molecules. This paper demonstrates the binding of the small molecule biotin to surface-immobilized streptavidin with a detection limit reduced to 0.41 pM. Furthermore, binding specificity of four additional small molecules to surface-immobilized streptavidin is shown. A label-free OFMBR-based optical sensor has great potential in small molecule detection and drug screening because of its high sensitivity, low detection limit, and minimal sample consumption.

Keywords: label-free sensor; optofluidic microbubble resonator; detection of small molecules

1. Introduction

Methods to detect small molecular analytes (<1000 Da) have found several, important uses in the bio-medical, food, and environmental fields [1–3]. In biomedical analyses, several small molecules such as steroids, thyroid hormones, and peptides derived from disease-specific proteins have been utilized as diagnostic markers [4,5]. Additionally, the detection of small molecules analytes could serve a vital role in the identification of bacterial pathogens in infectious diseases [6,7].

To detect small molecules, several electrical, mechanical, and optical sensors have been developed, such as the nanomechanical resonator sensor, nanowire sensor, and surface plasmon resonance sensor. Among them, optical detection methods stand out because of their advantages, such as greater sensitivity, electrical passiveness, and robustness. To date, the most widely used optical detection technology uses labels, such as radio- or fluorescent-labeling, for detection. Although popular and useful, the labeling step requires additional time and cost, and even worse, a labeled approach can significantly alter the activities of small molecules and lead to inaccurate conclusions. Various label-free optical techniques, such as surface plasmon resonance (SPR) [8], resonant waveguide grating (RWG) [9], resonant mirror (RM) [10,11], and high-Q optical microcavities [12–16] have been developed, but their sensitivities diminish with the size of the molecule, making it extremely challenging to detect small molecules. To date, the detection limit for small molecules with label-free optical detection technologies are in the range of µM to nM [17,18].

An optofluidic microbubble resonator (OFMBR) that supports high-Q whispering gallery modes (WGMs) is a promising candidate for the development of sensors [19–23] because of its unique hollow-core structure, capabilities of integration with microfluidics, high sensitivity, and excellent optical confinement of the WGMs [24,25]. Light couples into the microbubble resonator through the fiber taper and the WGMs shift in response to the change of the refractive index in the medium surrounding the resonator, which leads to additional applications of OFMBR in label-free biosensing. However, the OFMBR is typically exposed in the air, which is susceptible to environmental changes, greatly diminishing its performance capabilities in small molecule detection.

In this paper, we propose a packaged OFMBR sensor by fixing the ends of a microbubble and fiber taper on glass substrate with glue, and sealing the OFMBR inside a glass box. By immobilization of streptavidin on the inner surface of microbubble and injection of various concentrations of small molecule biotin, we obtained real-time binding curves of biotin to immobilized streptavidin, which gave us a binding affinity of 6.7×10^{14} M^{-1}. The detection limit of biotin to immobilized streptavidin on the packaged OFMBR sensor was determined to be 0.41 pM. In addition, we flew four small molecules over a streptavidin-immobilized OFMBR sensor; only magnolol bound to the surface-immobilized streptavidin, with binding affinity of 4.7×10^{10} M^{-1}. The packaged OFMBR sensor was demonstrated to be able to detect small molecules with high sensitivity and low detection limit, which has significant application potential in small molecule detection and drug screening.

2. Materials and Method

2.1. Materials

The 3-glycidoxypropyltrimethoxysilan (GOPTS) was from Sinopharm Chemical Regent Company (Shanghai, China). Bovine serum albumin (BSA) and phosphate buffered saline (PBS) were from Sigma-Aldrich (St. Louis, MO, USA). Streptavidin (SA) was from Life Technologies (Shanghai, China). Biotin was from Aladdin Industrial Corporation (Shanghai, China). Magnolol, medetomidine HCl, cetrimonium bromide, and reboxetine mesylat were from Selleck (Houston, TX, USA). The low refractive index polymer MY133 was from MY Polymers (Ness Ziona, Israel). The UV glue was from Thorlabs (Shanghai, China).

2.2. The Packaged OFMBR Sensor for Biomolecular Interaction Detection

Figure 1a shows the sketch of the OFMBR detection system, which consists of a tunable, continuous diode laser, with a tuning range from 765 nm to 781 nm (TLB6700, New Focus, San Diego, CA, USA), a polarization controller (EPC300, Connet Fiber Optics, Shanghai, China), a packaged OFMBR sensor, a photon detector (DET10C, Thorlabs, Newton, NJ, USA), and an oscilloscope at a working sampling rate of 500,000 S/s (TDS3012, Tektronix, Portland, OR, USA). Figure 1b shows the structure of a packaged OFMBR sensor. The OFMBR was on a glass substrate, with wall thickness about 2–4 µm, lying perpendicular to a fiber taper; optimal coupling was achieved by adjusting the gap between them. To fix the position and wrap the fiber, low refractive index polymer MY133 was on both ends of the fiber taper. To fix the OFMBR on the substrate, UV glue (NOA68, Thorlabs, Newton, NJ, USA) covered both ends of the microbubble. In addition, a cover glass was placed on top of the four piles on the glass substrate corners to seal the packaged sensor. The packaged OFMBR sensor was solidified in the air by its exposure to a UV lamp at 365 nm for 5 min. Instead of putting MY133 around the coupling region of OFMBR with fiber taper and on the ends of the fiber taper as described before [25], we put polymer MY133 only on the ends of the fiber taper in such a way that excited WGMs were not affected; as a result, the as-packaged OFMBR sensor was more sensitive than that described before [25].

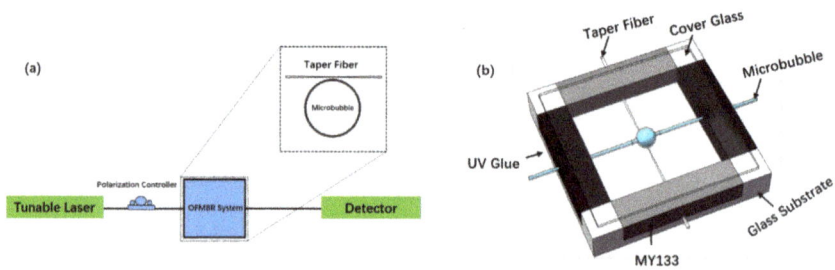

Figure 1. (**a**) Sketch of the detection system for the OFMBR sensor; (**b**) Structure of the packaged OFMBR sensor.

When biological molecules aggregated at the inner surface of the microbubble, they interacted with the evanescent part of the WGM field. Subsequently, a wavelength shift occurred as a result of the change of the radius or the refractive index of the microbubble. The resonance wavelength shift Δ_λ was proportional to the surface density of the small molecules $N_{AB}(t)$, which is expressed as [26],

$$\Delta_\lambda = \alpha_{ex} \lambda S \frac{2\pi \sqrt{n_2^2 - n_3^2}}{\varepsilon_0 \lambda^2} \frac{n_2}{n_3^2} \times N_{AB}(t) = \mu \times N_{AB}(t) \tag{1}$$

n_2, n_3 are the refractive index values of the OFMBR's wall (quartz) and the OFMBR's core (air or liquid), α_{ex} is the excess polarizability of the biomolecules, S is the sensitivity of refractive index. By following the wavelength shift variations with time, binding kinetics of biomolecule to the OFMBR inner surface was revealed.

2.3. Detection of Small Molecule Binding to Surface-Immobilized Streptavidin

Figure 2a schematically shows the functionalization of the inner surface of the microbubble with epoxy groups by using 3-glycidoxypropyltrimethoxysilan (GOPTS, Sinopharm Chemical Regent Company, Shanghai, China). The inner surface of OFMBR was first activated by 12% NaOH (Lingfeng Chemical Regent, Shanghai, China) for 1 h to obtain OH terminal groups. The microbubble was then filled with 1% GOPTS in 95% ethanol for 2 h, followed by washing with toluene, ethanol, and deionized water for 10 min each to remove residual unreacted GOPTS. The microbubble was finally dried in an oven at 120 °C for 2 h. The functionalized microbubble surface, having epoxy groups, formed covalent linkages with amine groups on proteins which were thus immobilized on the inner surface of microbubble with high efficiency.

Figure 2b shows experimental protocols for the detection of small molecule binding to surface-immobilized streptavidin. Protein streptavidin was first immobilized on the inner surface of the OFMBR by its incubation with streptavidin, at a concentration of 7.7 uM for 2 h, followed by 1 × PBS washing for 0.5 h. The inner surface was then blocked with 7.6 uM BSA at a flow rate of 2.5 µL/min for 2 h and followed by washing with 1 × PBS for 0.5 h.

Each binding curve of small molecule to surface-immobilized streptavidin included baseline, association phase, and dissociation phase. During baseline, 1 × PBS flew over surface-immobilized streptavidin at a flow rate of 2 µL/min for 6 min. During the association phase, small molecule solution flew over surface-immobilized streptavidin for 22 min at a flow rate of 2 µL/min. Dissociation phase included 1 × PBS flowing over the OFMBR at a flow rate of 2 µL/min for 8 min. Specific binding of small molecule to surface-immobilized streptavidin caused red-shift of the OFMBR WGMs, which was recorded from the beginning of the baseline to the end of the dissociation phase at a time resolution of 1 s. All binding curves were normalized against respective maximal resonance wavelength shift which does not affect reaction kinetic rate constants.

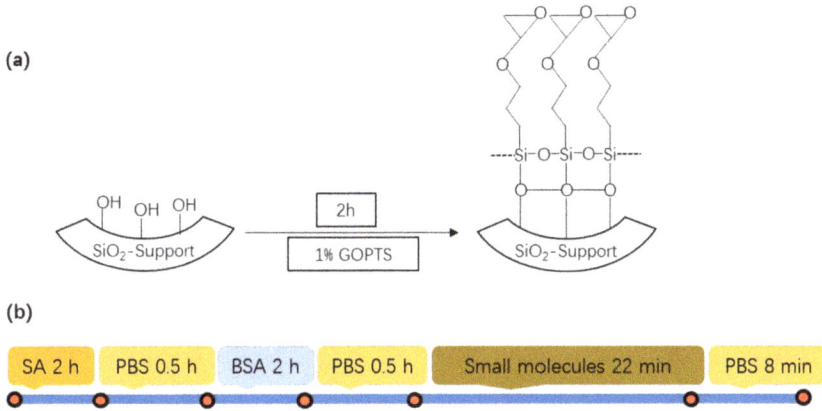

Figure 2. (**a**) Schematic illustration of surface functionalization of microbubble surface with epoxy groups; (**b**) Flow diagram of experimental processes for small biomolecule binding to surface-immobilized streptavidin.

3. Results

3.1. Transmission Spectra and Q Factors of OFMBR Sensor

An OFMBR sensor consists of a microbubble coupled with a fiber taper. The OFMBR sensor exhibits high sensitivity due to the presence of a significant part of the WGM field close to the inner surface of the microbubble. In addition, the OFMBR sensor has excellent fluidic capability because of its intrinsic hollow structure. However, the exposed OFMBR sensor is susceptible to environmental changes and usually an exposed sensor lasts less than 12 h. We thus propose to package the OFMBR sensor within a sealed glass box (as described in Section 2.2) to reduce the impact of the environment.

Figure 3a,b show transmission spectra of an OFMBR sensor with air inside before packaging and 24 h after packaging within a wavelength range of 770–780 nm, respectively. After packaging, WGMs remained excited. The decreased intensity of the transmission light after packaging resulted from the loss of scattering and absorption due to the coating of MY133 around the ends of the fiber taper. Figure 3c,d show fine scan of the transmission spectra of OFMBR sensor before and after packaging around 774.3 nm. Both spectra show the same four WGMs, indicating that the excited WGMs were all reserved after the packaging process. The typical Q factors were obtained by Lorenz fits of the WGMs, indicated by the red curves in Figure 3c,d and the Q values before and after packaging were approximately 3.7×10^5 and 3.0×10^5, respectively. The Q values before and after packaging were on the same order of magnitude, proving good maintenance of Q values after the packaging. The slight decrease of Q values was attributable to changes of coupling coefficiency between the microbubble and the fiber taper, caused by a slight movement of taper fiber during the packaging process. Figure 3e,f show resonant wavelength shift varying with time of an OFMBR sensor before and after packaging. The standard deviation of resonant wavelength shifts decreased from 2.6 pm to 0.15 pm after packaging, indicating that sensor packaging greatly decreased resonant wavelength drift. In addition, performance of the as-packaged OFMBR sensor was stable for months. Results indicate that OFMBR sensor packaging could provide lower noise and longer lift time, without sacrificing its performance.

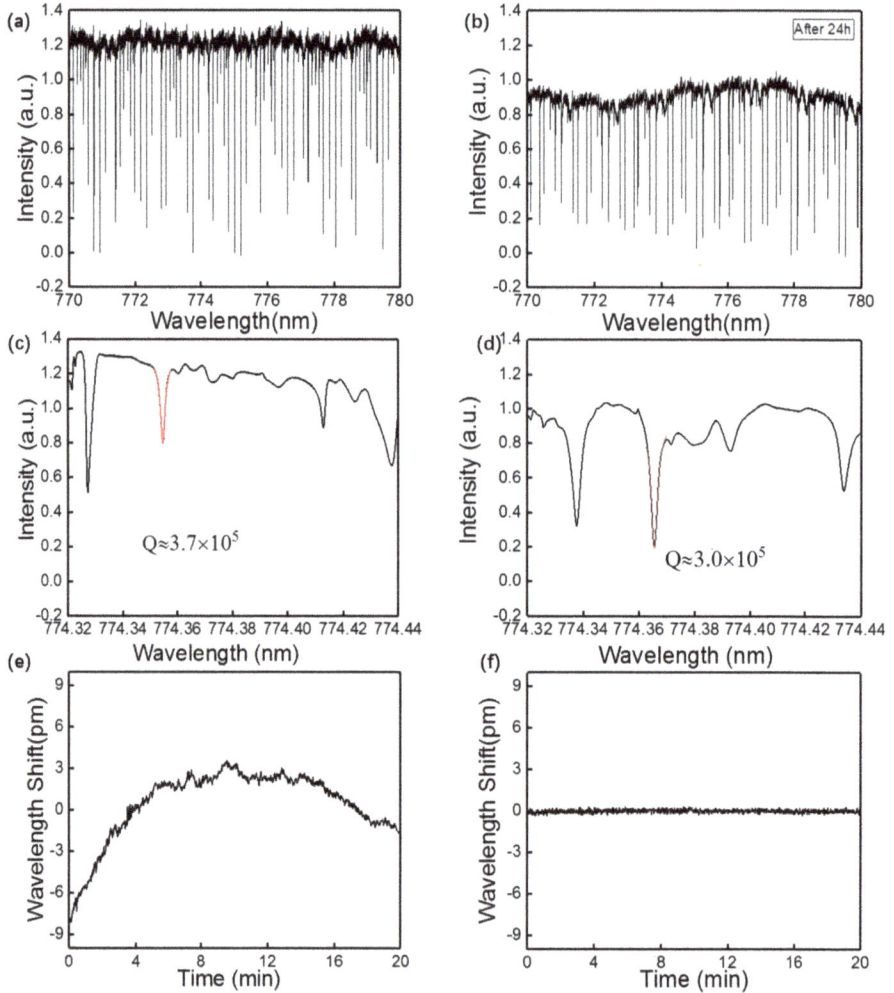

Figure 3. (**a**) Transmission spectra of an OFMBR before packaging in the wavelength range between 770 nm and 780 nm; (**b**) Transmission spectra of an OFMBR 24 h after packaging in the wavelength range between 770 nm and 780 nm; (**c**) Fine scan of the transmission spectra before packaging around 774.3 nm; (**d**) Fine scan of the transmission spectra after packaging around 774.3 nm; (**e**) Resonant wavelength shift of an OFMBR sensor varying with time before packaging; (**f**) Resonant wavelength shift of an OFMBR sensor varying with time after packaging.

3.2. Binding Kinetics of Biotin to Surface-Immobilized Streptavidin

Biotin is a small molecule with molecular weight of 224 Da, which binds strongly to streptavidin. We used biotin binding to surface-immobilized streptavidin as a model system to study the performance of a packaged OFMBR sensor on small molecule detection. We measured binding kinetics of biotin to surface-immobilized streptavidin at respective biotin concentrations of 205 pM, 410 pM, and 820 pM on three fresh OFMBR sensors; Figure 4a shows the three normalized real-time binding curves among them. Vertical lines indicate the start of the association and dissociation phases. Real-time binding curves contain information on kinetic rate constants, k_{on} (association rate) and

k_{off} (dissociation rate). For the monophasic molecular interaction with stoichiometry 1:1, the affinity constant k_a is the ratio of k_{on} to k_{off},

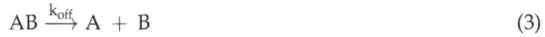

$$A + B \xrightarrow{k_{on}} AB \tag{2}$$

$$AB \xrightarrow{k_{off}} A + B \tag{3}$$

$$k_a = \frac{k_{on}}{k_{off}} \tag{4}$$

To extract binding kinetic rate constants, we used a Langmuir reaction model. As shown in Equations (5), small molecules are assumed to bind to surface-immobilized streptavidin at a rate proportional to the association rate k_{on}, small molecule concentration $[A]$, and the unbonded binding sites $N_{[AB]_{max}} - N_{[AB]}$ at time less than t_0. After the small molecule solution was replaced with a PBS buffer at time t_0, small molecule-protein complexes dissociated at a rate proportional to dissociation rate k_{off} and number of complexes per sensing area $N_{[AB]}$.

$$\begin{cases} \frac{dN_{[AB]}}{dt} = k_{on}[A]\left(N_{[AB]_{max}} - N_{[AB]}\right) - k_{off}N_{[AB]}, t \leq t_0, \\ \frac{dN_{[AB]}}{dt} = -k_{off}N_{[AB]}, t > t_0 \end{cases} \tag{5}$$

$N_{[AB]_{max}}$ is the maximal number of binding sites per sensing area. The number of small molecule complexes per unit sensing area $N_{[AB]}$ is calculated to be,

$$\begin{cases} N_{AB}(t) = \frac{k_{on}[A]N_{[AB]_{max}}}{k_{on}[A]+k_{off}}\left(1 - e^{-(k_{on}[A]+k_{off})t}\right), t \leq t_0, \\ N_{AB}(t) = \frac{k_{on}[A]N_{[AB]_{max}}}{k_{on}[A]+k_{off}}\left(1 - e^{-(k_{on}[A]+k_{off})t_0}\right)e^{-k_{off}(t-t_0)}, t > t_0, \end{cases} \tag{6}$$

We obtained reaction kinetic rate constants k_{on} and k_{off} by globally fitting three normalized binding curves in Figure 4a with Equations (6). As shown in Table 1, the binding affinity between small molecule biotin and surface-immobilized streptavidin was 6.7×10^{14} M^{-1}, which is close to the affinity range (10^{13}–10^{14} M^{-1}) reported by others [27,28]. Figure 4b shows the binding curve of surface-immobilized streptavidin to flowing biotin at a concentration of 0.41 pM, which is significantly lower than the detection limit achieved on other label-free optical sensing systems [12].

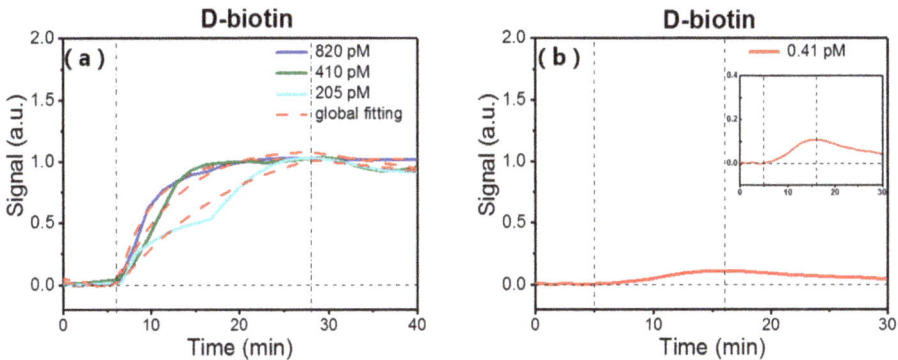

Figure 4. (a) Binding curves of surface-immobilized streptavidin with flowing biotin at respective concentrations of 205 pM, 410 pM, and 820 pM. Vertical lines indicate start of association and dissociation phases. Red dashed lines are global fitting results with the Langmuir reaction model; (b) Specific binding curve of surface-immobilized streptavidin with biotin at a concentration of 0.41 pM. Inset shows enlarged view of the binding curve.

Table 1. Kinetic constants of biotin and magnolol binding to immobilized streptavidin.

Small Molecule	k_{on} (min nM)$^{-1}$	k_{off} (min)$^{-1}$	k_a (M)$^{-1}$
Biotin	0.30	4.5×10^{-7}	6.7×10^{14}
Magnolol	0.29	6.2×10^{-3}	4.7×10^{10}

3.3. Specificity of Small Molecules Binding to Surface-Immobilized Streptavidin

To study binding specificity of small molecules to surface-immobilized streptavidin on the packaged OFMBR sensor, we flew four small molecules at respective concentrations of 205 pM sequentially over the surface-immobilized streptavidin. Three small molecules, medetomidine HCl, cetrimonium bromided, and reboxetine mesylat were randomly selected from our compound library. Magnolol, a binding ligand of streptavidin screened from 3375 compounds [29], was also included. Figure 5 show that the three randomly selected small molecules did not bind to the surface-immobilized streptavidin and magnolol bound specifically to streptavidin, indicating that the as-packaged OFMBR sensor functionalized with streptavidin binds specifically with small molecules. From binding curves between streptavidin and magnolol at concentrations of 100 pM, 500 pM, and 1000 pM, the binding affinity between them was found to be 4.7×10^{10} M^{-1}.

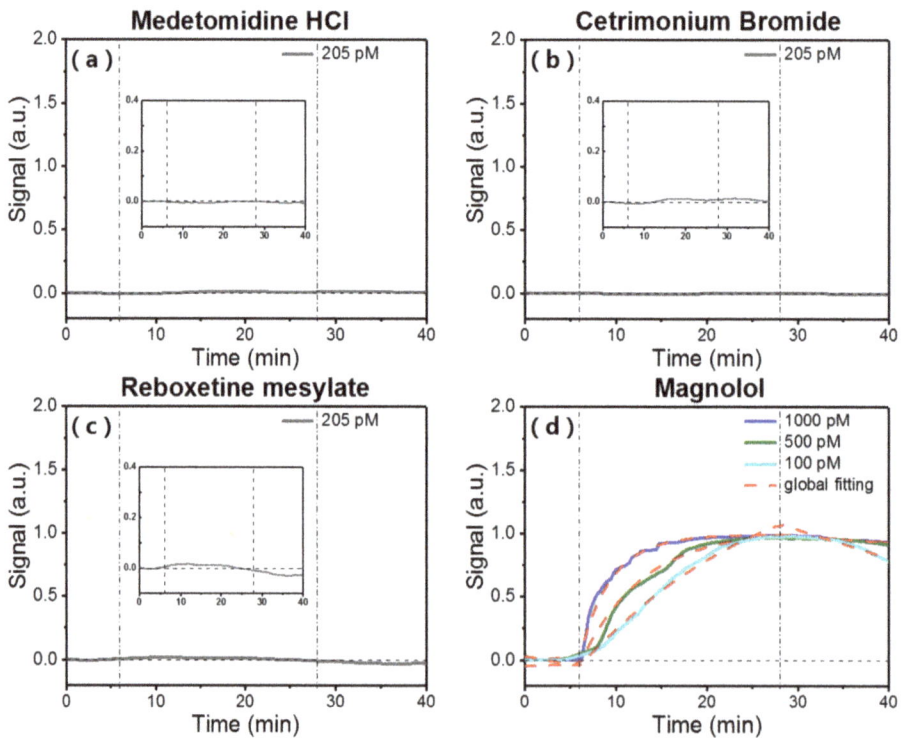

Figure 5. Real-time binding curves of small molecules (**a**) medetomidine HCl; (**b**) cetrimonium bromide; (**c**) reboxetine mesylate with surface-immobilized streptavidin on packaged OFMBR sensor at respective concentration of 205 pM (Insets show enlarged views of the binding curves); (**d**) Binding curves of surface-immobilized streptavidin with flowing magnolol at concentrations of 100 pM, 500 pM, and 1000 pM. Vertical lines are the starts of association and dissociation phases.

4. Discussion and Conclusions

The OFMBR sensor was demonstrated to be capable of detecting biomolecular interactions with high sensitivity, based on WGMs exited through a coupling of a microbubble resonator (OFMBR) and a fiber taper. Traditionally, a OFMBR sensor is exposed to the air and the sensor's performance is susceptible to environmental changes, especially dirt in the air. In this way, the exposed OFMBR sensor usually lasts less than 12 h. In addition, it is not convenient to carry the exposed OFMBR sensor. To improve the stability of an OFMBR sensor, we fixed the ends of the sensor with glue on the glass substrate and put the packaged sensor inside a glass box. The as-packaged OFMBR sensor displayed good stability, lasting for months. A significant difference between the package protocols of this work and the package protocols described before [25] is that we did not wrap the coupling position between the microbubble and the fiber taper with MY133; accordingly, the high radial order modes were reserved to provide high sensitivity.

Detection limit of the packaged OFMBR sensor can be further optimized by enhancing sensor sensitivity, increasing surface density of immobilized protein ligand, and reducing noise level. Thin shells and high radial modes are beneficial for improving sensor sensitivity. Optimization of surface functionalization and surface immobilization protocols is expected to maximize protein surface density. Noise level can be reduced by increasing the stability of the OFMBR sensor, keeping it at a constant temperature, and using a self-referencing sensing scheme, such as mode-splitting method [14].

In this paper, we demonstrate a novel optical sensor for label-free small molecule detection with low detection limit. With surface-immobilized streptavidin on the inner surface of a packaged OFMBR sensor, binding of small molecule biotin to streptavidin can be detected with a reduced detection limit at 0.41 pM. The specificity of small molecule binding on OFMBR sensor was demonstrated with four additional small molecules. Such a sensitive optical sensor will find wide applications in medical diagnoses, treatment of diseases, and drug screening.

Author Contributions: Z.L., Z.G., B.W., and X.W. designed and performed the experiments; C.Z. contributed samples and analyzed experimental data; Z.L. and Y.F. wrote and revised the paper.

Acknowledgments: This work was supported by the Special Project of National Key R&D Program of the Ministry of Science and Technology of China (2106YFC0201401), the National Natural Science Foundation of China (NSFC) (61505032, 61378080, 61327008).

Conflicts of Interest: The authors declare no conflicts of interest.

References

1. Zhang, J.; Yang, P.L.; Gray, N.S. Targeting cancer with small molecule kinase inhibitors. *Nat. Rev. Cancer* **2009**, *9*, 28. [CrossRef] [PubMed]
2. Wang, W.U.; Chen, C.; Lin, K.H.; Fang, Y.; Lieber, C.M. Label-free detection of small-molecule—Protein interactions by using nanowire nanosensors. *Proc. Natl. Acad. Sci. USA* **2005**, *102*, 3208–3212. [CrossRef] [PubMed]
3. Strausberg, R.L.; Schreiber, S.L. From knowing to controlling: A path from genomics to drugs using small molecule probes. *Science* **2003**, *300*, 294–295. [CrossRef] [PubMed]
4. Kenakin, T.; Christopoulos, A. Signalling bias in new drug discovery: Detection, quantification and therapeutic impact. *Nat. Rev. Drug Discov.* **2013**, *12*, 205. [CrossRef] [PubMed]
5. Holliger, P.; Hudson, P.J. Engineered antibody fragments and the rise of single domains. *Nat. Biotechnol.* **2005**, *23*, 1126. [CrossRef] [PubMed]
6. Gooding, J.J. Biosensor technology for detecting biological warfare agents: Recent progress and future trends. *Anal. Chim. Acta* **2006**, *559*, 137–151. [CrossRef]
7. Fitch, J.P.; Raber, E.; Imbro, D.R. Technology challenges in responding to biological or chemical attacks in the civilian sector. *Science* **2003**, *302*, 1350–1354. [CrossRef] [PubMed]
8. Anker, J.N.; Hall, W.P.; Lyandres, O.; Shah, N.C.; Zhao, J.; Van Duyne, R.P. Biosensing with plasmonic nanosensors. *Nat. Mater.* **2008**, *7*, 442–453. [CrossRef] [PubMed]

9. Fang, Y.; Ferrie, A.M.; Fontaine, N.H.; Mauro, J.; Balakrishnan, J. Resonant waveguide grating biosensor for living cell sensing. *Biophys. J.* **2006**, *91*, 1925–1940. [CrossRef] [PubMed]

10. Owen, V. Real-time optical immunosensors—A commercial reality. *Biosens. Bioelectron.* **1997**, *12*, i–ii. [CrossRef]

11. Daghestani, H.N.; Day, B.W. Theory and applications of surface plasmon resonance, resonant mirror, resonant waveguide grating, and dual polarization interferometry biosensors. *Sensors* **2010**, *10*, 9630–9646. [CrossRef] [PubMed]

12. Vollmer, F.; Yang, L. Review Label-free detection with high-Q microcavities: A review of biosensing mechanisms for integrated devices. *Nanophotonics* **2012**, *1*, 267–291. [CrossRef] [PubMed]

13. Soria, S.; Berneschi, S.; Brenci, M.; Cosi, F.; Nunzi Conti, G.; Pelli, S.; Righini, G.C. Optical microspherical resonators for biomedical sensing. *Sensors* **2011**, *11*, 785–805. [CrossRef] [PubMed]

14. Li, M.; Wu, X.; Liu, L.; Fan, X.; Xu, L. Self-referencing optofluidic ring resonator sensor for highly sensitive biomolecular detection. *Anal. Chem.* **2013**, *85*, 9328–9332. [CrossRef] [PubMed]

15. Zhang, X.; Liu, L.; Xu, L. Ultralow sensing limit in optofluidic micro-bottle resonator biosensor by self-referenced differential-mode detection scheme. *Appl. Phys. Lett.* **2014**, *104*, 033703. [CrossRef]

16. Ren, L.; Wu, X.; Li, M.; Zhang, X.; Liu, L.; Xu, L. Ultrasensitive label-free coupled optofluidic ring laser sensor. *Opt. Lett.* **2012**, *37*, 3873–3875. [CrossRef] [PubMed]

17. Arlett, J.L.; Myers, E.B.; Roukes, M.L. Comparative advantages of mechanical biosensors. *Nat. Nanotechnol.* **2011**, *6*, 203. [CrossRef] [PubMed]

18. Feng, C.; Dai, S.; Wang, L. Optical aptasensors for quantitative detection of small biomolecules: A review. *Biosens. Bioelectron.* **2014**, *59*, 64–74. [CrossRef] [PubMed]

19. Armani, A.M.; Kulkarni, R.P.; Fraser, S.E.; Flagan, R.C.; Vahala, K.J. Label-free, single-molecule detection with optical microcavities. *Science* **2007**, *317*, 783–787. [CrossRef] [PubMed]

20. Berneschi, S.; Farnesi, D.; Cosi, F.; Conti, G.N.; Pelli, S.; Righini, G.C.; Soria, S. High Q silica microbubble resonators fabricated by arc discharge. *Opt. Lett.* **2011**, *36*, 3521–3523. [CrossRef] [PubMed]

21. Cosci, A.; Quercioli, F.; Farnesi, D.; Berneschi, S.; Giannetti, A.; Cosi, F.; Barucci, A.; Conti, G.N.; Righini, G.; Pelli, S. Confocal reflectance microscopy for determination of microbubble resonator thickness. *Opt. Express* **2015**, *23*, 16693–16701. [CrossRef] [PubMed]

22. Farnesi, D.; Barucci, A.; Righini, G.C.; Conti, G.N.; Soria, S. Generation of hyper-parametric oscillations in silica microbubbles. *Opt. Lett.* **2015**, *40*, 4508–4511. [CrossRef] [PubMed]

23. Wang, H.T.; Wu, X. Optical manipulation in optofluidic microbubble resonators. *Sci. China Phys. Mech. Astron.* **2015**, *58*, 114206. [CrossRef]

24. Zhang, X.; Liu, T.; Jiang, J.; Liu, K.; Yu, Z.; Chen, W.; Liu, W. Micro-bubble-based wavelength division multiplex optical fluidic sensing. In Proceedings of the International Society for Optics and Photonics, Beijing, China, 9–11 October 2014.

25. Tang, T.; Wu, X.; Liu, L.; Xu, L. Packaged optofluidic microbubble resonators for optical sensing. *Appl. Opt.* **2016**, *55*, 395–399. [CrossRef] [PubMed]

26. Hu, H.; White, I.M.; Suter, J.D.; Dale, P.S.; Fan, X. Analysis of biomolecule detection with optofluidic ring resonator sensors. *Opt. Express* **2007**, *15*, 9139–9146.

27. Srisa-Art, M.; Dyson, E.C.; de Mello, A.J.; Edel, J.B. Monitoring of real-time streptavidin—Biotin binding kinetics using droplet microfluidics. *Anal. Chem.* **2008**, *80*, 7063–7067. [CrossRef] [PubMed]

28. Jung, L.S.; Nelson, K.E.; Stayton, P.S.; Campbell, C.T. Binding and dissociation kinetics of wild-type and mutant streptavidins on mixed biotin-containing alkylthiolate monolayers. *Langmuir* **2000**, *16*, 9421–9432. [CrossRef]

29. Zhu, C.; Zhu, X.; Landry, J.P.; Cui, Z.; Li, Q.; Dang, Y.; Mi, L.; Zheng, F.; Fei, Y. Developing an efficient and general strategy for immobilization of small molecules onto microarrays using isocyanate chemistry. *Sensors* **2016**, *16*, 378. [CrossRef] [PubMed]

micromachines

MDPI

Review

A Review of the Precision Glass Molding of Chalcogenide Glass (ChG) for Infrared Optics

Tianfeng Zhou [1,*], Zhanchen Zhu [2], Xiaohua Liu [2], Zhiqiang Liang [1] and Xibin Wang [1]

[1] Key Laboratory of Fundamental Science for Advanced Machining, Beijing Institute of Technology, Beijing 100081, China; liangzhiqiang@bit.edu.cn (Z.L.); cutting0@bit.edu.cn (X.W.)
[2] School of Mechanical Engineering, Beijing Institute of Technology, Beijing 100081, China; zhuzhanchen@163.com (Z.Z.); liuxh89@126.com (X.L.)
*   Correspondence: zhoutf@bit.edu.cn; Tel.: +86-010-6891-2716

Received: 10 June 2018; Accepted: 22 June 2018; Published: 2 July 2018

Abstract: Chalcogenide glass (ChG) is increasingly demanded in infrared optical systems owing to its excellent infrared optical properties. ChG infrared optics including ChG aspherical and freeform optics are mainly fabricated using the single point diamond turning (SPDT) technique, which is characterized by high cost and low efficiency. This paper presents an overview of the ChG infrared optics fabrication technique through precision glass molding (PGM). It introduces the thermo-mechanical properties of ChG and models the elastic-viscoplasticity constitutive of ChG. The forming accuracy and surface defects of the formed ChG are discussed, and the countermeasures to improve the optics quality are also reviewed. Moreover, the latest advancements in ChG precision molding are detailed, including the aspherical lens molding process, the ChG freeform optics molding process, and some new improvements in PGM.

Keywords: chalcogenide glass; infrared optics; precision glass molding; aspherical lens; freeform optics

1. Introduction

1.1. Introduction of ChG Infrared Optics

1.1.1. Characteristics of ChG

Infrared optical systems, including thermal imaging devices and night vision cameras, are gaining more attention in the optical fields [1,2]. Crystalline infrared materials such as single-crystal germanium (Ge) and zinc selenide (ZnSe) have been used and primarily fabricated by single point diamond turning (SPDT) and chemical vapour deposition (CVD) in past decades, though they are rare and expensive [3–5]. Chalcogenide glass (ChG), which is a kind of artificial material mainly made of chalcogens (S, Se, and Te) and some other elements, shows much wider transmission wave band from near to far infrared wavelength compared with the oxide glasses and has more excellent properties of athermalization and achromatism compared with the crystalline infrared materials [6,7]. ChG infrared optics could be formed in mass production using the precision glass molding (PGM) process with low-cost compared to the silicon (Si) and germanium (Ge) optics machined by the diamond turning or grinding. Hence, ChG is now deemed as an alternative for crystalline infrared materials for various infrared optics [8].

ChG can be divided into sulfide glass, selenide glass, and telluride glass according to the elemental composition. The As–S and Ge–S glass are the first sulfide glasses to be studied [9,10]. As–S glass has the advantages of good infrared transmittance, high refractive index, low sound speed, and high quality factor. However, its shortcomings, such as high intrinsic loss, poor chemical stability, and so on, hinder its application in the optical industry. In order to improve the comprehensive properties of

As–S glass, the preparation and properties of Ge–As–S glass have been studied. The results show that the density, hardness, softening temperature, and chemical stability of glass increase and the expansion coefficient decreases with the increase of element Ge content. The transmittance range of Ge–As–S glass is 0.6–11 µm [11].

The preparation of selenide glass is easier than sulfide glass due to the higher rate of chemical reaction between Se and other elements, as well as lower pressure when Se melts [12–15]. The Ge–As–Se glass and Ge–Sb–Se glass are the most suitable selenide glasses for infrared optical system [2,16], because Ge–As–Se glass has intrinsic optical stability and a wide temperature range of glass forming process. It has a great transmittance of far infrared wave, the range of which is 0.8–15 µm. Besides, the value of the third-order optical nonlinearity of Ge–As–Se glass is the largest among all ChGs.

Telluride glass has a wider infrared transmittance than sulfide glass and selenide glass. The infrared transmission spectrum of telluride glass can be extended above 20 µm [17–19]. However, the glass forming ability of Te is weak, because it has stronger metallicity compared with S and Se. The telluride glass has low glass transition point, poor thermal stability, and mechanical strength, which limits its applications in far infrared optics [20,21].

1.1.2. Application of ChG Infrared System

ChG has been widely used in various infrared optical systems due to its excellent optical properties [22–25]. ChG infrared night vision system uses infrared light to enable human eyes to observe scenes at night by converting the invisible infrared image of the scene into a visually sensible image. ChG has a lower temperature dependency of the refractive index compared with single-crystal germanium (Ge) and zinc selenide (ZnSe). The night vision system designed with ChG can realize the temperature self-adaptation control and make the system image useable in a large temperature range [26].

The ChG infrared night vision system consists of an infrared optical imaging device and a photoelectric conversion device, as shown in Figure 1. The ChG lens is responsible for transmitting infrared radiation of the targets. The photoelectric conversion device converts the infrared radiation into the visible image for human eyes, and then the image is projected on the display screen after denoising, reshaping, and amplifying [27]. The image quality is related to the shape accuracy, surface roughness, and infrared transmittance of ChG lens.

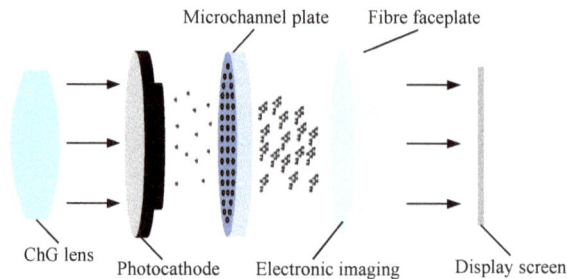

Figure 1. Schematic diagram of night vision system.

ChG is also extensively applied in infrared thermal imaging technology [28,29]. Figure 2 shows the infrared thermal imaging system. The infrared detector and ChG lens are used to accept the infrared radiation energy distribution of measured target, which reflects in the photosensitive element of infrared detector to obtain infrared thermography. The infrared thermal imaging system can transform the invisible infrared radiation of the object into the visible thermal image, and different colors in the thermal image represent the different temperatures corresponding to the heat distribution of the measured object.

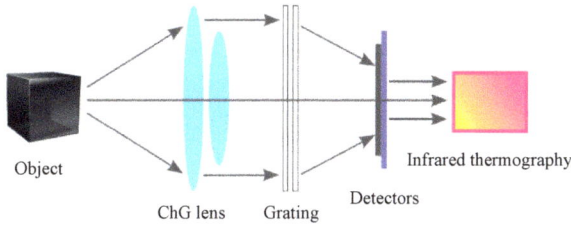

Figure 2. Schematic diagram of thermal imaging system [30].

1.1.3. Classification of ChG Infrared Optics

ChG infrared optics/lenses are usually designed with complex geometric shapes to obtain superior optical performance, including the ChG aspherical lens and ChG freeform optics. ChG aspherical lens can significantly improve the image quality and infrared optical properties. As shown in Figure 3, it can change the light path to reduce the number of infrared optical elements, thereby simplifying the structure of infrared system and reducing the volume of the infrared system [31,32].

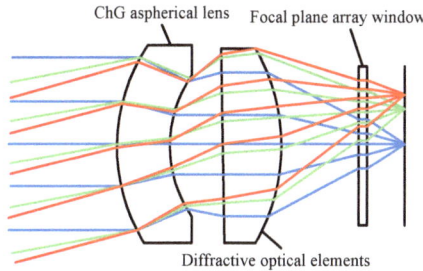

Figure 3. The optical layout of ChG aspherical lens.

ChG freeform optics are generally defined as an infrared element with microlens array or microstructure array on the surface (shown in Figure 4). The surface microstructures of the infrared optical elements can improve the infrared thermal imaging quality and eliminate light aberration, and the microstructure array is widely used to simplify the infrared system and meet specific requirements. For example, the detection capability of infrared detectors can be improved by coupling ChG microlens array with infrared detector. At the same time, ChG microlens array has a cold shield effect. The incident angle of infrared radiation is limited to ensure that each photosensitive element of the infrared detector has the same incident angle and background radiation, and then the noise can be reduced [33].

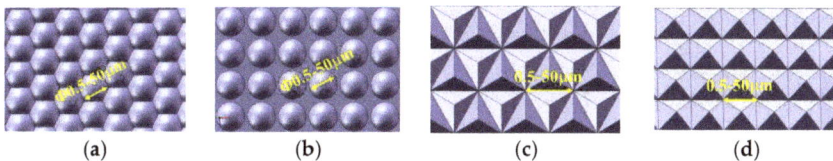

Figure 4. Element shapes of the ChG freeform optics: (**a**) adjacent microlens arrays, (**b**) distributed microlens arrays, (**c**) triangular pyramid arrays, and (**d**) rectangular pyramid arrays [34].

1.2. Methods of ChG Optics Manufacturing

In order to meet the application requirements of ChG optics, various fabrication techniques have been developed. The conventional methods for manufacturing ChG aspherical lens are single point diamond turning and polishing. They are relatively mature and can obtain ultraprecision surface accuracy. However, they are time-consuming processes and are very ill-suited to mass production with high accuracy [35].

There are also some processes to produce ChG freeform optics. Based on energy assisted machining, the focused ion beam machining and laser processing can fabricate surface microstructures directly on optical components. However, these two methods are complex in processing, high in cost, and poor in microstructures uniformity [36–38]. The ChG freeform optics can also be processed by photolithography, but the shape and size of the microstructures are limited [39]. Using traditional material removal machining techniques such as single point diamond turning or ultraprecision grinding, complex surface microstructures can be fabricated. The surface microstructures have ultraprecision shape accuracy. However, these methods still have many problems such as high cost, poor efficiency, and so on. They are not capable enough to meet the needs of the market [40]. PGM can solve the problems of the processing technology mentioned above, and it is one of the most promising technologies with which to manufacture ChG optics.

1.3. Precision Glass Molding of ChG

PGM was earliest used to process optical aspherical lens, replacing traditional Polymethyl Methacrylate (PMMA) resin optics to achieve better optical characteristics and stability [41]. It has high processing efficiency and can achieve mass production to reduce processing cost with good formation accuracy of optical components [42,43]. Compared with single-crystal germanium, ChG is an amorphous material without a fixed melting point. The viscosity of ChG gradually decreases during heating, and it is suitable for ChG optics to be fabricated by PGM [44]. From the viewpoint of fabrication cost and process time, the PGM is undoubtedly a better approach for producing precision ChG aspherical lens and ChG freeform optics. PGM of ChG were first reported by Zhang et al. at the French company Umicore IR Glass S.A. in 2003 [44]. They produced ChG lenses by using $Ge_{22}As_{20}Se_{58}$ and $Ge_{20}Sb_{15}Se_{65}$. The maximum shape errors of their molded lenses were approximately 0.3 μm and 2 μm, respectively. In the same year, Zhang et al., at the University of Rennes I [3], reported a similar result that the shape error of the molded $Ge_{22}As_{20}Se_{58}$ lens was approximately 0.4 μm. In 2006, Curatu et al. [29] designed and fabricated an athermalization lens over the entire operating temperature range (−40 °C~+80 °C). Cha et al. [45] studied the effect of temperature on the molding process of $Ge_{10}As_{40}Se_{50}$ in 2010. They suggested that the ChG should be molded at a temperature higher than softening temperature to prevent breakage. In 2011, Liu et al. [46] conducted PGM numerical simulations for $Ge_{33}As_{12}Se_{55}$ to investigate the variations of its thermos-mechanical properties. Ju et al. [35] characterized the moldability of ChG and selected the preferential Ge/Sb ratio in the Ge–Sb–Se based ChG system to fabricate lenses in 2014. Zhou et al. [47] evaluated the stress relaxation behavior of As_2S_3 and calculated its refractive index change in 2017.

ChG molding process can be divided into four stages: heating, pressing, annealing, and cooling, as shown in Figure 5. In a PGM cycle, the molds and ChG are heated simultaneously. The ChG is heated above softening point and fully softened. Then, the upper mold is moved down to compress the ChG preforms, and the stress in ChG is relaxed by holding the pressing load without further deformation of ChG for a short time. After that, the ChG is annealed at a slow cooling rate. Finally, the formed ChG optics are cooled to ambient temperature and released from the molds. In this way, the geometric shapes of the molds are replicated to the ChG surfaces.

Though PGM for optical glasses is relatively established, it is still in early stages for ChGs. Directly using the constitutive model of oxide glass to describe the thermo-mechanical behavior of ChG could be problematic, because the molding temperature for oxide glass is between the yielding point and the softening point, while the molding temperature is above the softening point for ChG [45,46]. In contrast to oxide glass, most ChG has a greater saturated vapor pressure, which enables trace gases to release during the molding process. If these gases cannot escape, the lens shape and surface quality will be severely impaired [48].

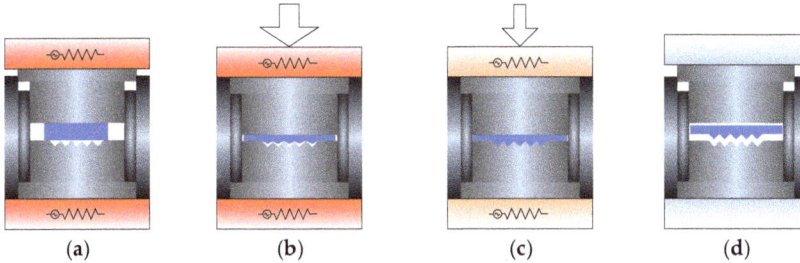

Figure 5. Four stages of a PGM cycle: (**a**) heating, (**b**) pressing, (**c**) annealing, and (**d**) cooling.

2. Modeling and Simulation of ChG Molding

2.1. Modeling of Elastic-Viscoplasticity Constitutive of ChG

2.1.1. Thermo-Mechanical Behavior Test of ChG

Finite element method can solve issues when some variables are difficult to measure in experiments. However, a reliable simulation model requires an accurate constitutive of materials. The thermo-mechanical behavior is the basis for modelling constitutive of the ChG. The thermo-mechanical behavior test of the ChG is mainly applied in the pressing stage. The true strain and stress of the ChG are calculated by measuring the displacement curve of the upper mold. Figure 6 [49] shows the displacement w of the upper mold under different pressing forces in which the w is defined as $w(t) = y(t) - y(0)$, and $y(t)$ is the position of the upper mold.

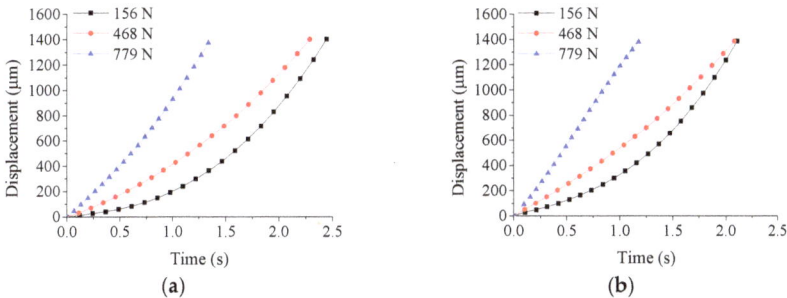

Figure 6. The upper mold displacement w during compression test under different pressing forces at the following molding temperatures: (**a**) 382 °C and (**b**) 392 °C.

The true strain ε_t, also named as logarithmic strain or Hencky strain [50], considering as an incremental strain, can be derived by Equation (1):

$$\varepsilon(t) = \int_{z_0}^{z_t} \frac{dz}{z} = \ln(1 - \frac{w(t)}{z_0}) \tag{1}$$

in which z_0, z_t are the initial and instantaneous heights of the cylindrical specimen. Both stress and strain should be negative in compression but are taken as absolute values in the following figures for convenience. The changes of strain with time during deformation under different pressing forces are shown in Figure 7 [49].

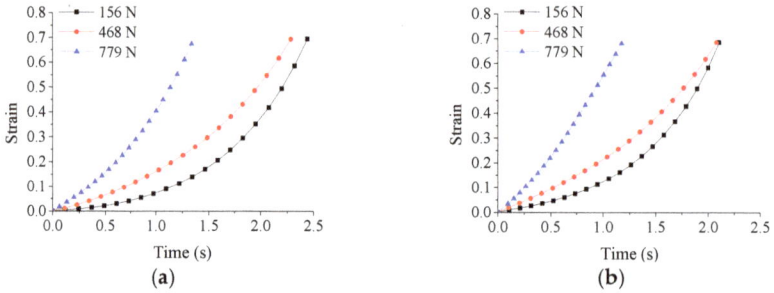

Figure 7. The variations of strain with time under different pressing forces at the following molding temperatures: (**a**) 382 °C and (**b**) 392 °C.

The true stress $\sigma(t)$ can be derived by Equation (2):

$$\sigma(t) = \frac{F}{\pi \rho_0^2} e^{\varepsilon(t)} \tag{2}$$

in which F is the molding force, and ρ_0 is the radius of ChG preform. The changes of stress with time under different pressing forces are shown in Figure 8 [49]. It is noted that the stress decreases more rapidly under higher pressing force.

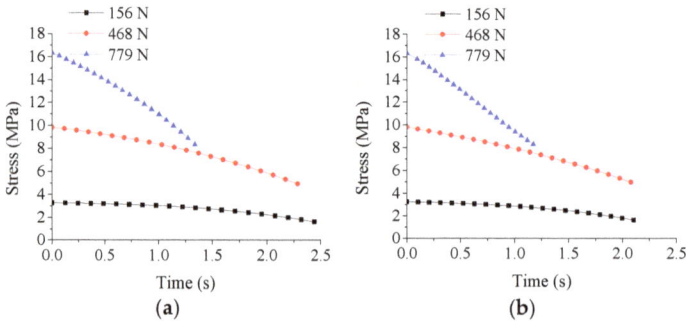

Figure 8. The variations of stress with time under different pressing forces at the following molding temperatures: (**a**) 382 °C and (**b**) 392 °C.

2.1.2. Elastic-Viscoplasticity Modeling

At the molding temperature, ChG exhibits elastic-viscoplastic behavior, which is a combination of viscoelastic and viscoplastic mechanisms. As ChG is an isotropic material, the elastic-viscoplastic behavior of which can be simplified as a combination of spring, dashpot, and slider, representing elasticity, viscosity, and plasticity, respectively. The constitutive relations of these three elements are, respectively, listed as below:

$$\sigma_e = E\varepsilon_e \tag{3}$$

$$\sigma_v = \eta \dot{\varepsilon}_v \tag{4}$$

$$\sigma_p = \sigma_f \text{sign}(\dot{\varepsilon}_p) \tag{5}$$

in which σ_e, σ_v, and σ_p are the elastic stress, viscous stress, and plastic stess, respectively; σ_f is the yield stress of the friction element; ε_e, ε_v, and ε_p are corresponding strains; E and η are elastic modulus and viscosity; and sign$(\dot{\varepsilon}_p)$ is sign function, being equal to 1, 0, and −1 when the plastic strain rate $\dot{\varepsilon}_p$ is positive, zero, and negative.

In order to describe all deformation mechanisms of the ChG, Schofield–Scott Blair model is used [51], as shown in Figure 9. The instantaneous elastic and plastic deformation, delayed elastic and plastic deformation, and viscous flow coexist in this model. The constitutive equation based on the Schofield-Scott Blair model is

$$
\begin{cases}
\frac{\eta_1}{E_2}\dot{\varepsilon} + \varepsilon = \frac{\eta_1}{E_1 E_2}\dot{\sigma} + \frac{E_1 + E_2}{E_1 E_2}\sigma, & \sigma < \sigma_f \\
\frac{\eta_1}{E_2}\ddot{\varepsilon} + \dot{\varepsilon} = \frac{\eta_1}{E_1 E_2}\ddot{\sigma} + \frac{1}{E_2}\left(1 + \frac{E_2}{E_1} + \frac{\eta_1}{\eta_2}\right)\dot{\sigma} + \frac{1}{\eta_2}\sigma - \frac{\sigma_f}{\eta_2}, & \sigma \geq \sigma_f
\end{cases}
\tag{6}
$$

in which E_1, E_2 are elastic moduli and η_1, η_2 represent the viscosities of the elastic-viscoplastic material. As the ChG preform is compressed at a uniform temperature, the pressing stage can be treated as isothermal process without heat transfer [52].

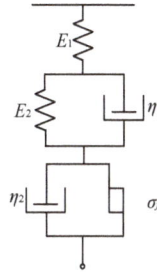

Figure 9. The Schofield–Scott Blair constitutive model.

Considering the small strain scenario, the first order Taylor expansion of Equation (2) leads to

$$
\varepsilon(t) = C\sigma(t) - 1
\tag{7}
$$

in which $C = \pi \rho_0^2 / F$ is a constant. Substituting Equation (7) into Equation (6), and considering the cases of $\sigma \geq \sigma_f$ results in [49]:

$$
\begin{aligned}
\sigma(t) = \sigma_f + C\eta_2 + & \{[(E_1\eta_1 + E_1\eta_2 + E_2\eta_2 + C_1)(\sigma_0 - \sigma_f) - CC_1\eta_2 - CE_1\eta_1\eta_2 + 2\eta_1\eta_2\dot{\sigma}_0 \\
-C_3 - C_2]e^{-\frac{(E_1\eta_1 + E_1\eta_2 + E_2\eta_2 - C_1)t}{2\eta_1\eta_2}} & \}/2C_1 - \{[(E_1\eta_1 + E_1\eta_2 + E_2\eta_2 - C_1)(\sigma_0 - \sigma_f) + CC_1\eta_2 - \\
CE_1\eta_1\eta_2 + 2\eta_1\eta_2\dot{\sigma}_0 - C_3 - C_2]e^{-\frac{(E_1\eta_1 + E_1\eta_2 + E_2\eta_2 + C_1)t}{2\eta_1\eta_2}} & \}/2C_1 , \quad \sigma \geq \sigma_f
\end{aligned}
\tag{8}
$$

in which $C_1 = \sqrt{E_1^2\eta_1^2 + 2E_1^2\eta_1\eta_2 + E_1^2\eta_2^2 - 2E_1 E_2\eta_1\eta_2 + 2E_1 E_2\eta_2^2 + E_2^2\eta_2^2}$, $C_2 = CE_2\eta_2^2$, $C_3 = CE_1\eta_2^2$, and σ_0 is the initial stress when the upper mold just contacts the ChG surface and $\dot{\sigma}_0$ is the stress rate at $t = 0$. The values of σ_0 and $\dot{\sigma}_0$ can be obtained from Figure 8, and they can be utilized in Equation (8) to determine the constitutive model parameters from compression tests mentioned above.

Considering for the large strain, the constitutive equation is

$$
\frac{\eta_1}{E_1 E_2}\ddot{\sigma} + \frac{1}{E_2}\left(1 + \frac{E_2}{E_1} + \frac{\eta_1}{\eta_2}\right)\dot{\sigma} + \frac{1}{\eta_2}\sigma - \frac{1}{\sigma} + \frac{\eta_1}{E_2\sigma^2} - \frac{\sigma_f}{\eta_2} = 0, \quad \sigma \geq \sigma_f
\tag{9}
$$

2.2. Simulation of Molding Process of ChG Infrared Optics

2.2.1. Simulation of Molding for Aspherical Lens

Based on the thermo-mechanical behavior test of ChG and Schofield-Scott Blair model, the two-dimensional (2D) axisymmetric simulation model of aspherical lens is created. In the model, three blocks are assembled from top to bottom representing upper mold, ChG preform, and lower mold, respectively, as shown in Figure 10 [49]. The lower mold is fixed, and a constant pressing force is applied on the upper mold to press the ChG with initial field temperature for the entire model of 25 °C. The distribution of equivalent stress in course of PGM is shown in Figure 11 [49].

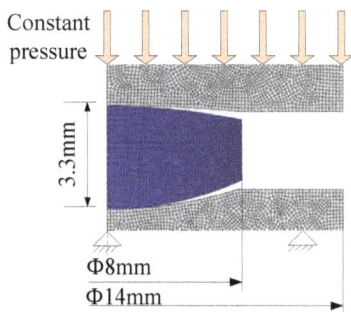

Figure 10. A half simulation model of ChG aspherical lens molding [49].

The contact area gradually increases from point contact, and the distribution of stress is initially localized near the contact region, as shown in Figure 11a. The stresses are significantly large at this early stage of aspherical lens molding. With pressing going on, the stress decreases when the contact area becomes larger, as shown in Figure 11b. Therefore, the material flow and stress changes in PGM are obtained to predict the forming accuracy of aspherical lens.

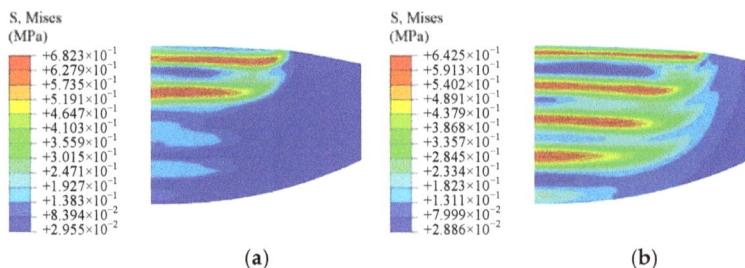

Figure 11. Stress distribution of ChG aspherical lens molding: (**a**) early stage and (**b**) later stage.

2.2.2. Simulation of Molding for Freeform Optics

The molding conditions for freeform optics at different process parameters are optimized. The two-dimensional simulation model is shown in Figure 12. The flat upper mold moves downward to press the softened ChG. The lower mold with microstructures is fixed. In order to avoid the stress convergence in the contact region and save remeshing time at the sharp corners during the molding simulation, the sharp corners of the microstructures are rounded with a small radius of 0.5 μm. As the ChG preform and the molds are pressed at a uniform temperature, the pressing stage can be treated as isothermal process without heat transfer.

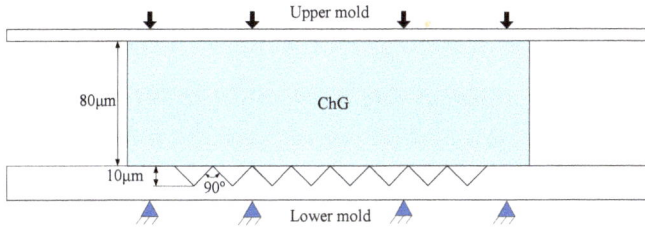

Figure 12. A two-dimensional simulation model of microstructures molding [30].

Two representative points are selected in the simulation model of the ChG, as shown in Figure 13. The point A is at the interface between the ChG and the top of the mold microstructures. Point B is at the interface between the ChG and the mold surface without microstructures.

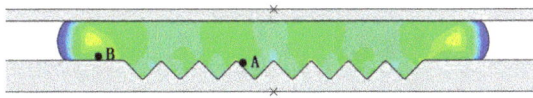

Figure 13. Two representative points of microstructures simulation [30].

Figure 14 illustrates the equivalent stress changes of point A and point B at different temperatures, pressing velocities, and friction coefficients. When the temperature of the ChG rises continuously, the internal stress of ChG decreases gradually. The stresses of point A and point B increase gradually with the increase of pressing velocity and friction coefficient.

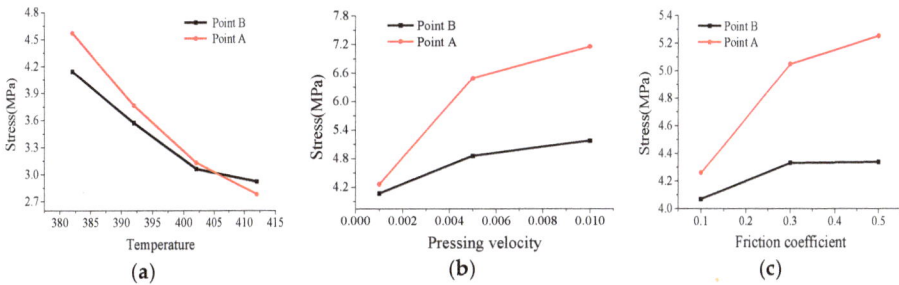

Figure 14. Effect of process parameters on stresses of ChG: (**a**) temperature, (**b**) pressing velocity, and (**c**) friction coefficient [30].

3. Molding Process of ChG for Aspherical Lens

3.1. ChG Molding Condition Optimization

The ChG molding parameters can be partly determined by the FEM simulation results, and the cylindrical molding tests can be conducted to explore the other molding conditions. Most ChGs have a greater saturated vapor pressure, which enables the release of trace gases during the molding process. When these gases cannot escape, microdimples will be formed on the ChG pillars. The maximum peak-to-valley height difference of microdimples is 1.562 μm, and the shape and surface quality of lens are severely impaired (in Figure 15). Therefore, the gas release and gas escape must be controlled [53].

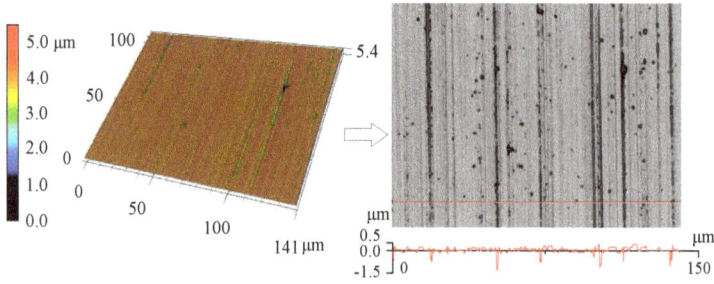

Figure 15. The surface morphology and contour of a formed ChG pillar.

For gas generation, the solubility of gas has been studied, and it can be expressed using A. Sieverts' square root law [54]:

$$S = k\sqrt{P}e^{-\Delta H/(RT)} \tag{10}$$

in which S is the dissolved concentration, k is a constant, P is the gas partial pressure, ΔH is the dissolution heat, R is the gas constant, and T is the temperature.

The movements of the molecules become more intense as the temperature increases, leading to more reactions, smaller gas solubility, and more gas generation. Pressure is another contributor that affects gas solubility. The gas density grows with the increasing pressure, so the free molecular motion decreases, and the gas increasingly dissolves, as expressed by Equation (10). According to Dalton's law of partial pressure, the total pressure is equal to all partial pressures of the mixed gas [55]. Meanwhile, based on Newton's third law [56], the total pressure increases with the increase of the pressing force.

The area ratios of the microdimples increase approximately logarithmically with the increase of molding temperature, as shown in Figure 16. By extrapolating the trend line in Figure 16, the temperature for preventing microdimple generating is approximately 379.8 °C. However, this temperature is too low and could lead to surface scratching. Therefore, it is inferred that the optimum temperature would be in the range of 380–382 °C [53].

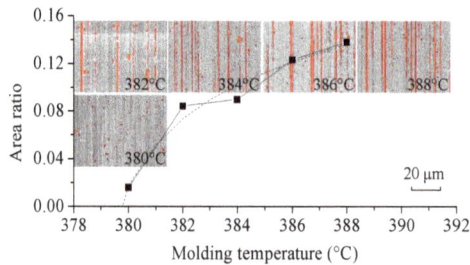

Figure 16. Area ratios of the microdimples of the formed ChG pillars at different molding temperatures under a pressing force of 1362 N.

As shown in Figure 17, the area ratios of the microdimples decrease with the increase of the pressing force at a molding temperature of 382 °C. They decrease slowly from 1362 N to 2723 N and then continue to decrease rapidly from 2723 N to 4085 N. Therefore, if the contact pressure is larger than the saturated vapor pressure of the ChG, the gas will no longer be generated.

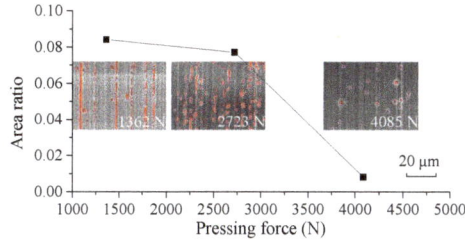

Figure 17. Area ratios of the microdimples of the formed ChG pillars under different pressing forces at 382 °C.

Due to the surface microstructures of the mold and glass, the enclosed spaces result in a lower local pressure. The pressure of the enclosed spaces is lower than the saturated vapor pressure of ChG, which intensifies the evaporation of selenide gases, leading to incomplete reproduction and the formation of microdimples. Simultaneously, when the contact surfaces are smoother, the gas escapes more completely, which can result in a better surface quality [53]. Figure 18 shows the area ratios of the microdimples with different mold surface roughnesses. It can be concluded that smoother mold surfaces lead to fewer microdimples, which verifies the effect of the contact surface roughness on surface quality of molded ChG microstructures.

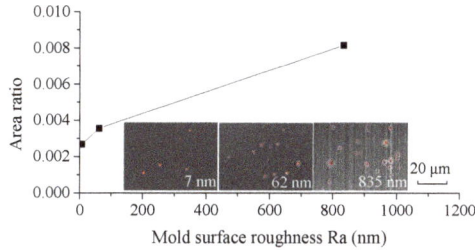

Figure 18. Area ratios of the microdimples of the formed ChG pillars with different mold surface roughness values under the pressure force of 4085 N at 382 °C.

When the ChG surface is curved with a smaller radius, the gas escape tends to be easier, as shown in Figure 19, and the area of contact surface is smaller when the pressing displacement stays the same, which means the pressure is larger under the same force with the same displacement. Hence, smaller radius ChG surface can reduce microdimples. It can be seen from Figure 20 that the area ratios of the microdimples of the formed pillars decrease, or are even completely eliminated, with the decrease of the glass curvature radius.

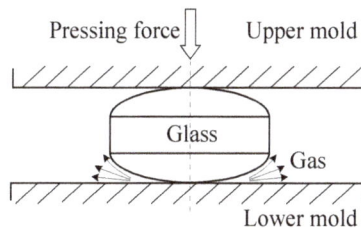

Figure 19. Schematic diagram of gas escape mode during the spherical glass molding process.

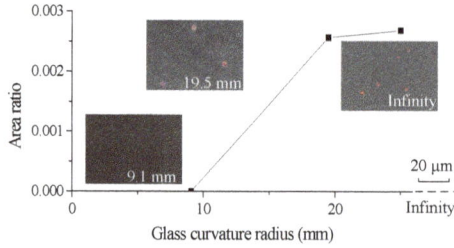

Figure 20. Area ratios of the microdimples of the formed ChG pillars with different glass surface curvature radii under 4085 N at a molding temperature of 382 °C.

3.2. Forming Accuracy and Surface Quality Control

The forming accuracy of aspherical lens impacts the optical properties of infrared optics, and the aspherical lens molding experiments are carried out to obtain high forming accuracy and surface quality. The precision aspherical molds and a molded ChG aspherical lens are shown in Figure 21. The formed lenses have an excellent surface finish of Ra 8 nm, and the surface profiles of formed lenses are consistent with the designed values.

Figure 21. Photographs of aspherical molds and aspherical lens: (**a**) upper mold, (**b**) lower mold, (**c**) upper surface of aspherical lens, and (**d**) lower surface of aspherical lens.

To demonstrate the conformity and replication fidelity of the formed lenses, the profile errors between molds and formed lenses are compared, as shown in Figure 22. Both the upper and lower finish surfaces of formed lenses are better than the molds, because the formed lens does not fill into the micro cavities caused by surface asperities of molds. However, higher molding temperature leads to better conformation between mold and glass surface, and larger roughness of the molded lens, indicating that the surface finish of the molded lens can be improved at lower molding temperature. Besides, the radius of curvature of the formed lens is less than the molds due to the shrinkage of glass in annealing and cooling stages. There is a shrinkage-inducing depression in the lower surface center of formed lens, where the cooling rate is larger than other places on this lens. Additionally, this shrinkage becomes greater with the increase of molding temperature.

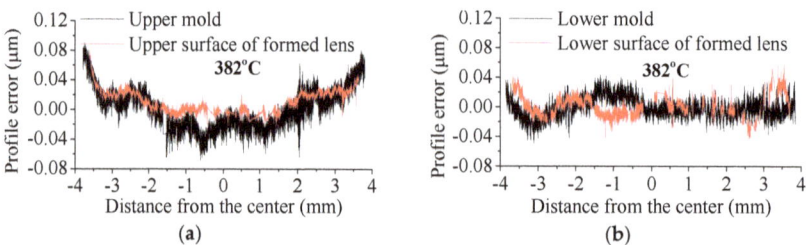

Figure 22. *Cont.*

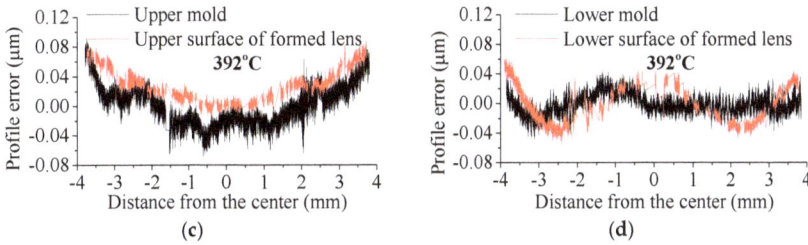

Figure 22. Evaluations of surface topography comparison of profile error: (**a**) between upper mold and lens formed at 382 °C, (**b**) between lower mold and lens formed at 382 °C, (**c**) between upper mold and lens formed at 392 °C, and (**d**) between lower mold and lens formed at 392 °C.

Hence, the surface roughness of the formed aspherical lens can be improved by using a lower forming temperature. When replicating the specific micromorphology on the mold surface, the forming accuracy can be improved by increasing the mold temperature. Due to the shrinkage, there is also a difference between the profile of the formed aspherical lens and the mold. In the mold profile design, the shrinkage should be compensated.

4. Molding Process of ChG for Freeform Optics

4.1. ChG Molding for Microlens Array

Although silicon carbide (SiC) and tungsten carbide (WC) are preferable mold materials owing to their high hardness and toughness under molding conditions, it is quite difficult to generate microlens array and microstructures on these materials [57,58]. In order to extend the PGM to creating microlens array and microstructures on the ChG surface, the mold should be machined with high accuracy and efficiency. Electroless nickel phosphorus (Ni-P) has emerged as an outstanding hard coating material because of its great hardness, outstanding corrosion resistance, and antiwear property [59–61]. Therefore, Ni-P plating has been used as mold to fabricate microlens array and microstructures for PGM [62,63].

In the microlens arrays mold preparation, the Ni-P layer is coated on the substrate through electro-less plating method, and the upper and lower molds are processed by diamond cutting, using a round diamond tool. Then, microlens arrays are generated on the lower mold by using a high-speed diamond-ball nose-end milling tool with an included fillet radius of 500 µm and a shank diameter of 6 mm (in Figure 23).

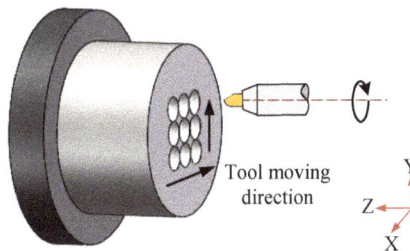

Figure 23. Single point diamond cutting process of ChG microlens arrays.

Figure 24a is the microscopic image of microlens arrays on the Ni-P mold. The diameter of the microlens is 80 µm, and the height of the microlens is 1 µm. Figure 24b is the microscopic image of microlens arrays on the ChG. To demonstrate the conformity and replication fidelity of microlens arrays, the microlens arrays profiles of mold and ChG are compared, and the profile error is analyzed by the least square method (shown in Figure 25). The peak-to-valley height error is 0.121 μm, and the average error of the height is 0.034 μm, which meet design requirements.

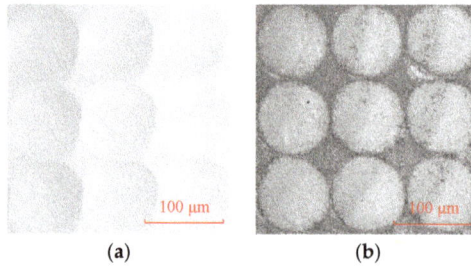

(a)   (b)

Figure 24. Microscope images of microlens arrays: (**a**) microlens arrays on the mold and (**b**) microlens arrays on the ChG.

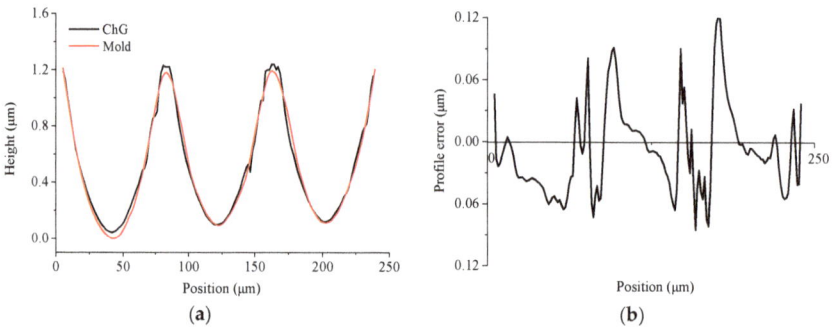

(a)   (b)

Figure 25. The profile curves of microlens arrays: (**a**) microlens arrays on the mold and the ChG, and (**b**) profile error of microlens arrays.

4.2. ChG Molding for Microstructures

Before the microstructures ChG molding, mirror mold surfaces are generated on the Ni-P plating layer by diamond cutting using a round diamond tool. Moreover, microstructures are generated using a triangle diamond tool with an included angle of 90° by feeding the diamond tool on the lower mold with three-axle linkage. Therefore, microstructures with more complicated geometric are formed with high accuracy, as shown in Figure 26.

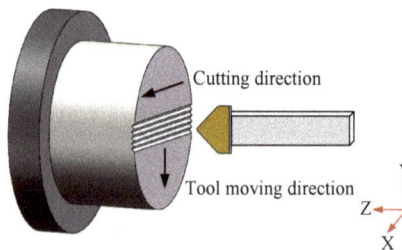

Figure 26. Single point diamond turning process of ChG microstructures.

Microscopic observations of the ChG surface and the mold surface are performed using a confocal laser scanning microscope, and the microscopic observations of the microstructures are shown in

Figure 27. The contour curves of the microstructures on the ChG surface and mold surface are extracted to analyze the forming quality, as shown in Figure 28. The average error of the height is 0.36 µm, and the peak-to-valley height error is 0.97 µm. It can be seen from the contour curve that the top of the microgroove is sharp, although the forming quality of bottom is slightly worse compared with that of the top. Therefore, PGM can achieve high replication accuracy of the ChG microstructures. Moreover, the minimum feature size of ChG microstructures is limited by the size of microstructures on molds, and the ChG molding can produce optics with feature sizes from millimetres to micrometres to nanometres.

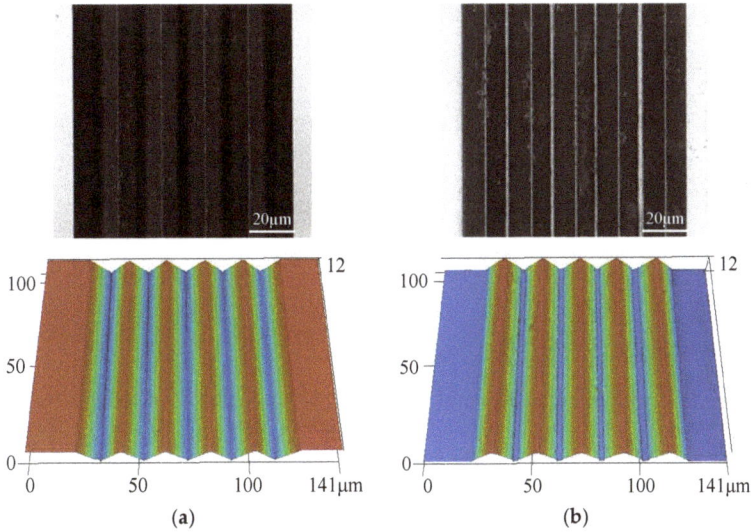

Figure 27. Microscope images of microstructures on (**a**) mold surface and (**b**) ChG surface.

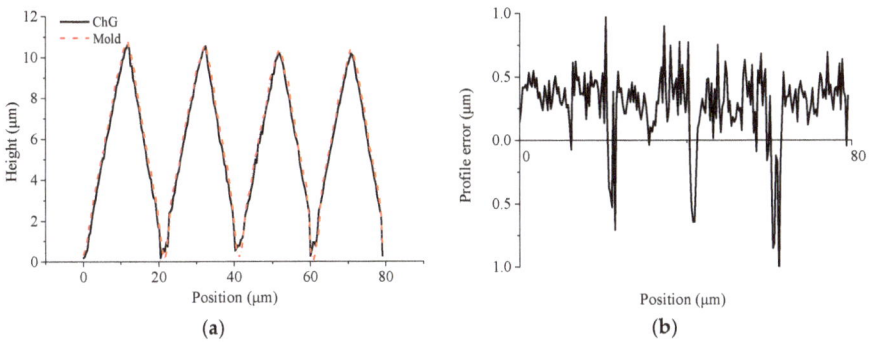

Figure 28. The profile curves of microstructures: (**a**) microstructures on the mold and the ChG, and (**b**) profile error of microstructures.

Raman spectroscopy detections of ChG preform and formed ChG made into 2 mm sheets are carried out to study the influence of molding process on infrared transmission characteristics of ChG, as shown in Figure 29. Infrared transmittances of the ChG formed at various molding temperatures are almost identical to ChG preform, and the infrared transmission wavelength band of them are 0.8–16 µm. Though the transmittance of ChG is low at the wavelength of 6 um due to the absorption band of oxide impurities in ChG, it covers three atmospheric windows of 1–3 µm, 3–5 µm, and 8–12 µm, which are generally used in infrared systems. Therefore, ChG infrared optics fabricated by PGM meet the requirements of infrared optical systems.

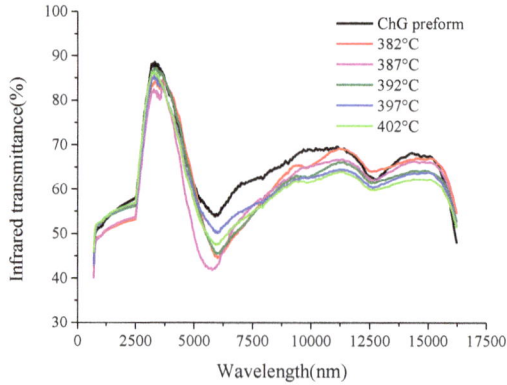

Figure 29. Infrared transmittance of the formed ChG at different molding temperatures.

5. Innovations of ChG Molding

5.1. Localized Rapid Heating Process for ChG Molding

PGM can achieve mass production of ChG optics with a good forming accuracy. However, conventional PGM is a bulk heating process that usually requires a long thermal cycle. The molding assembly and ChG are heated and cooled together, which often cause large thermal expansion in both the molds and ChG infrared optics. A localized rapid heating process is developed to effectively heat only bottom surface of the ChG. Graphene film coated on the mold surface only elevates the temperature of area where microlens array and microstructures need to be replicated, which can reduce the undesired thermal expansion of the mold and ChGs [64,65].

Localized rapid heating glass molding assembly is shown in Figure 30. A specially designed polymer base plate is fixed on the lower mold, placing under the graphene-coated, fused silica wafer and a thermocouple mounted on the fused silica wafer. Two copper electrodes are sandwiched on the front and back edges of the polymer base plate. The lower mold is moved up by a linear drive until the top surface of the glass touched the bottom of the tungsten carbide. A direct current (60 V/0.84 A) is applied on the copper electrodes during heating stage. The surface of the fused silica wafer is heated up rapidly. Nitrogen flow is introduced into the chamber to accelerate the cooling during annealing and cooling stages. Afterwards, the load is released and the formed ChG is removed from the chamber. The molded surface of the ChG using localized rapid heating and the mold surface are shown in Figure 31. It can be confirmed the ChG fills precisely into microstructures on mold by localized rapid heating process.

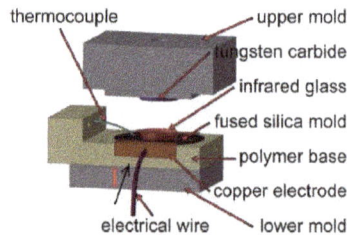

Figure 30. The localized rapid heating molding assembly [65].

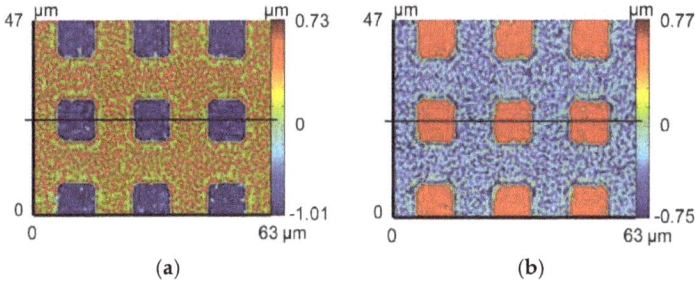

Figure 31. Microscope images of microstructures on (**a**) mold surface and (**b**) molded ChG surface after localized rapid heating process [65].

5.2. Contactless Molding of ChG Microlens Array

Most of ChGs have a high content of unstable volatile elements, and these elements tend to diffuse out from performs at high temperature and deposit on the surface of the molds, which induce severe interfacial adhesion problems and make it more difficult for molding process. Therefore, gas-assisted contactless molding process is developed to form microlens array avoiding interfacial adhesion problems [66].

The gas-assisted contactless molding processes machine is shown in Figure 32. The IR heater is placed on the top of the fused silica chamber with programming control for the molding temperature. The position of the temperature sensor is located outside the chamber but directly contacts the quartz tube near the sample. The stainless steel plate with arrayed through holes is used as a mold. The heating temperature is gradually increased to approach the softening point to allow the glass plate to be driven downward by the N$_2$ gas pressure, so that the ChG flows into the arrayed through holes and forms the microlens array.

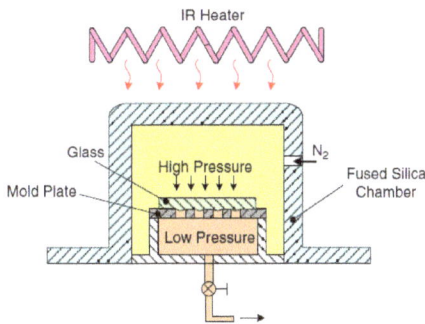

Figure 32. Schematic diagram of gas pressure-assisted glass molding machines [66].

The SEM images of ChG microlens array are shown in Figure 33. Hence, the contactless gas-assisted molding system can avoid contact-induced glass sticking and gas bubble problems. Besides, the surface profile of the molded lenses can be precisely controlled by changing the applied gas pressure, molding temperature, and time duration.

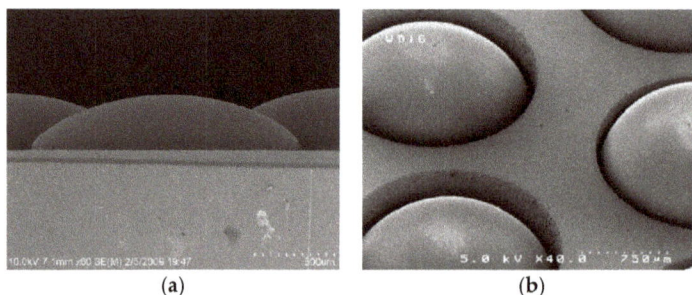

Figure 33. SEM images of ChG microlens array: (**a**) the cross-sectional image of the molded ChG microlens array and (**b**) the appearance of molded ChG microlens array [66].

6. Outlook

In this article, we have overviewed recent progresses in infrared optics fabrication by PGM. To date, some part of them, like elastic-viscoplasticity constitutive model and ChG molding surface defect reduction, are relatively mature. However, several hot topics, such as atomic diffusion, interfacial adhesion, and anti-reflection film technology, still need further study [67–69].

(1)   Atomic diffusion

Atomic diffusion will come up between the surfaces of ChG and mold in molding process. The motion of atoms is violent, and the diffusion phenomenon is serious at high temperature, which will lead to the change of optical property of ChG and thermal mechanical property of mold material. The diffusion degree of atoms is related to the surface energy and surface quality of the mold materials [70]. Therefore, the influence of the mold materials on atomic diffusion is necessarily studied to choose the suitable mold material.

(2)   Interfacial adhesion

Interfacial adhesion has been a major challenge in PGM. It results in the surface quality deterioration of the mold and shortens the service life of the mold, which escalates the overall cost and decreases the repeatability. Protective coatings, such as diamond-like carbon (DLC), platinum-iridium (Pt-Ir), rhenium-iridum (Re-Ir), and so on have been employed to prevent interfacial adhesion [71,72]. These coatings have low friction coefficient and antisticking property. However, the alternating stress during repeated thermal cycles in PGM will cause coatings to peel off. It is urgent to study the mechanism of interfacial adhesion and develop suitable techniques to prevent interfacial adhesion.

(3)   Anti-reflection film technology

The elements of sulfur S, selenium Se, and tellurium Te in ChG are easily influenced by environment. Meanwhile, the ChG has high refractive index and large reflection loss. The refractive index of ChG will reduce after PGM [73]. In order to improve the infrared optical performance of the ChG, anti-reflective film technology needs to be further explored to effectively reduce the infrared reflection loss of ChG and improve the infrared optical transmittance. Hence, an anti-reflective coating on the ChG infrared optics should be applied after molding, and the maximum utility of ChG infrared optics can be achieved in the infrared systems.

Author Contributions: T.Z. and Z.Z. developed the structure of the paper; T.Z., X.L., Z.L. and X.W. revised the paper; Z.Z. wrote the paper.

Funding: This work was financially supported by the National Key Basic Research Program of China (No. 2015CB059900) and the National Natural Science Foundation of China (No. 51775046). The authors would also like to acknowledge the support from the Fok Ying-Tong Education Foundation for Young Teachers in the Higher Education Institutions of China (No. 151052).

Micromachines **2018**, *9*, 337

Conflicts of Interest: The authors declare no conflicts of interest.

References

1. Tang, B.; Yang, Y.; Fan, Y.; Zhang, L. Barium gallogermanate glass ceramics for infrared applications. *J. Mater. Sci. Technol.* **2010**, *26*, 558–563. [CrossRef]
2. Bureau, B.; Zhang, X.; Smektala, F.; Adam, J.; Troles, J.; Ma, H.; Boussard-Plèdel, C.; Lucas, J.; Lucas, P.; Coq, D.; et al. Recent advances in chalcogenide glasses. *J. Non-Cryst. Solids* **2004**, *345*, 276–283. [CrossRef]
3. Zhang, X.; Ma, H.; Lucas, J. Applications of chalcogenide glass bulks and fibres. *J. Optoelectron. Adv. Mater.* **2003**, *5*, 1327–1333.
4. Wang, Z.; Li, J.; Qin, H.; Zhang, Y.; Zhang, F.; Su, Y. Research of forming characteristic of precision glass molding. *Int. Symp. Adv. Opt. Manuf. Test. Technol. Adv. Opt. Manuf. Technol.* **2016**, *9683*, 96831G.
5. Ju, H.; Du, H.; Kang, H.; Kim, J.; Kim, H. Development of chalcogenide glass with thermal stability for molded infrared lens. *Proc. SPIE Int. Soc. Opt. Eng.* **2014**, *8982*, 395–406.
6. Yu, H. *Infrared Optical Materials*; National Defense Industry Press: Beijing, China, 2007.
7. Luo, S.; Huang, F.; Zhan, D.; Wang, M. Development of chalcogenide glasses for infrared thermal imaging system. *Laser Infrared* **2010**, *40*, 9–13.
8. Mehta, N.; Kumar, A. Recent advances in chalcogenide glasses for multifunctional applications in fiber optics. *Recent Pat. Mater. Sci.* **2013**, *6*, 59–67. [CrossRef]
9. Tsuchihashi, S.; Kawamoto, Y.; Adachi, K. Some physico-chemical properties of glasses in the system As-S. *J. Ceram. Assoc. Jpn.* **1968**, *76*, 101–106. [CrossRef]
10. Tsuchihashi, S.; Kawamoto, Y. Properties and structure of glasses in the system As-S. *J. Non-Cryst. Solids* **1971**, *5*, 286–305. [CrossRef]
11. Wang, Y.; Zu, C.; Zhao, H.; He, K.; Zhao, H.; Chen, J.; Han, B. Study on long-wavelength infrared glasses. *J. Funct. Mater.* **2010**, *41*, 196–200.
12. Zhang, X.; Calvez, L.; Seznec, V.; Ma, H.; Danto, S.; Houizot, P.; Boussard-Plédel, C.; Lucas, J. Infrared transmitting glasses and glass-ceramics. *J. Non-Cryst. Solids* **2006**, *352*, 2411–2415. [CrossRef]
13. Zhao, D.; Zhang, X.; Wang, H.; Zeng, H.; Ma, H.; Adam, J.; Chen, G. Thermal properties of chalcogenide glasses in the GeSe$_2$-As$_2$Se$_3$-CdSe system. *J. Non-Cryst. Solids* **2008**, *354*, 1281–1284. [CrossRef]
14. Zhang, X.; Ma, H.; Lucas, J. Evaluation of glass fibers from the Ga-Ge-Sb-Se system for infrared applications. *Opt. Mater.* **2004**, *25*, 85–89. [CrossRef]
15. Wang, Y.; Chen, F.; Wang, G.; Shen, X.; Zhou, Y.; Li, J.; Dai, S. Investigation of GeSe$_2$-In$_2$Se$_3$-AgI chalcogenide glasses. *Acta Photonica Sin.* **2012**, *41*, 718–722. [CrossRef]
16. Zakery, A.; Elliott, S. Optical properties and applications of chalcogenide glasses: A review. *J. Non-Cryst. Solids* **2003**, *330*, 1–12. [CrossRef]
17. Wilhelm, A.; Boussard-Pledel, C.; Coulombier, Q.; Lucas, J.; Bureau, B.; Lucas, P. Development of far-infrared-transmitting Te based glasses suitable for carbon dioxide detection and space optics. *Adv. Mater.* **2007**, *19*, 3796–3800. [CrossRef]
18. Wang, G.; Nie, Q.; Wang, X.; Shen, X.; Chen, F.; Xu, T.; Dai, S.; Zhang, X. New far-infrared transmitting Te-based chalcogenide glasses. *J. Appl. Phys.* **2011**, *110*, 043536. [CrossRef]
19. Wang, G.; Li, C.; Nie, Q.; Pan, Z.; Li, M.; Xu, Y.; Wang, H.; Shi, D. Thermal stability and far infrared transmitting property of GeTe$_4$-AsTe$_3$-AgI glasses and glass-ceramics. *J. Non-Cryst. Solids* **2017**, *463*, 80–84. [CrossRef]
20. Cheng, C.; Wang, X.; Xu, T.; Zhu, Q.; Liao, F.; Sun, L.; Pan, Z.; Liu, S.; Dai, S.; Shen, X.; et al. Research on properties of far infrared Ge-Ga-Te-Ag chalcogenide glasses. *Acta Photonica Sin.* **2015**, *44*, 1116001. [CrossRef]
21. Xu, H.; Nie, Q.; Wang, X.; He, Y.; Wang, G.; Dai, S.; Xu, T.; Zhang, P.; Zhang, X.; Bruno, B. Optical properties of the Ge-Ga-Te-Cu far-infrared-transmitting chalcogenide glasses. *J. Optoelectron. Laser* **2013**, *24*, 93–98.
22. Kulakova, N.; Nasyrov, A.; Nesmelova, I. Current trends in creating optical systems for the IR region. *J. Opt. Technol.* **2010**, *77*, 324–330. [CrossRef]
23. Song, B.; Zhang, Y.; Wang, Q.; Dai, S.; Xu, T.; Nie, Q.; Wang, X.; Shen, X.; Wu, L.; Lin, C. Optical properties measurement of infrared chalcogenide glasses and analysis on its influencing factors. *Infrared Laser Eng.* **2012**, *41*, 1442–1447.

24. Xue, J.; Man, X.; Gong, Y.; Zhao, X. Preparation, characteristic and application of chalcogenide glasses. *Optoelectron. Technol. Inf.* **2003**, *16*, 28–31.

25. Gai, X.; Han, T.; Prasad, A.; Madden, S.; Choi, D.; Wang, R.; Bulla, D.; Luther-Davies, B. Progress in optical waveguides fabricated from chalcogenide glasses. *Opt. Express* **2010**, *18*, 26635–26646. [CrossRef] [PubMed]

26. Chen, G.; Zhang, X. Development of fine molded chalcogenide glasses for IR night vision. *Bull. Chin. Ceram. Soc.* **2004**, *23*, 3–7.

27. Lu, Y.; Song, B.; Xu, T.; Dai, S.; Nie, Q.; Shen, X.; Lin, C.; Zhang, P. Design of refractive-diffractive night vision system based on chalcogenide glass. *Laser Optoelectron. Prog.* **2013**, *50*, 168–174.

28. Shi, G.; Zhang, X.; Wang, L.; He, F.; Zhang, J. Application of the new chalcogenide glass in design of low cost thermal imaging systems. *Infrared Laser Eng.* **2011**, *40*, 615–619.

29. Curatu, G.; Binklev, B.; Tinch, D.; Curatu, C. Using molded chalcogenide glass technology to reduce cost in a compact wide-angle thermal imaging lens. *Proc. SPIE* **2006**, *6206*, 62062M.

30. Zhou, T. Precision molding of microstructures on chalcogenide glass for infrared optics. In *Micro and Nano Fabrication Technology*; Springer: Berlin/Heidelberg, Germany, 2018.

31. Curatu, G. Design and fabrication of low-cost thermal imaging optics using precision chalcogenide glass molding. *Proc. SPIE Int. Soc. Opt. Eng.* **2008**, *7060*, 706008.

32. Cha, D.; Kim, H.; Hwang, Y.; Jeong, J.; Kim, J. Fabrication of molded chalcogenide-glass lens for thermal imaging applications. *Appl. Opt.* **2012**, *51*, 5649–5656. [CrossRef] [PubMed]

33. Eisenberg, N.; Klebanov, M.; Lyubin, V.; Manevich, M.; Noach, S. Infrared microlens arrays based on chalcogenide photoresist, fabricated by thermal reflow process. *J. Optoelectron. Adv. Mater.* **2000**, *2*, 147–152.

34. Zhou, T.; Liu, X.; Liang, Z.; Liu, Y.; Xie, J.; Wang, X. Recent advancements in optical microstructure fabrication through glass molding process. *Front. Mech. Eng.* **2017**, *12*, 46–65. [CrossRef]

35. Ju, H.; Jang, Y.; Du, H.; Kim, J.; Kim, H. Chalcogenide glass with good thermal stability for the application of molded infrared lens. *Proc. SPIE Int. Soc. Opt. Eng.* **2014**, *9253*, 925310.

36. Naessens, K.; Ottevaere, H.; Van Daele, P.; Baets, R. Flexible fabrication of microlenses in polymer layers with excimer laser ablation. *Appl. Surf. Sci.* **2003**, *208*, 159–164. [CrossRef]

37. Hisakuni, H.; Tanaka, K. Optical fabrication of microlenses in chalcogenide glasses. *Opt. Lett.* **1995**, *20*, 958–960. [CrossRef] [PubMed]

38. Chiu, C.; Lee, Y. Fabricating of aspheric micro-lens array by excimer laser micromachining. *Opt. Lasers Eng.* **2011**, *49*, 1232–1237. [CrossRef]

39. Manevich, M.; Klebanov, M.; Lyubin, V.; Varshal, J.; Broder, J.; Eisenberg, N. Gap micro-lithography for chalcogenide micro-lens array fabrication. *Chalcogenide Lett.* **2008**, *5*, 61–64.

40. Davies, M.; Evans, C.; Bergner, B. Application of precision diamond machining to the manufacture of microphotonics components. *Proc. SPIE Int. Soc. Opt. Eng.* **2002**, *5183*, 94–108.

41. Lin, C.; Fang, Y.; Yang, P. Optical film with microstructures array for slim-type backlight applications. *Opt. Int. J. Light Electron Opt.* **2011**, *122*, 1169–1173. [CrossRef]

42. Zhou, T.; Yan, J.; Masuda, J.; Oowada, T.; Kuriyagawa, T. Investigation on shape transferability in ultraprecision glass molding press for microgrooves. *Precis. Eng.* **2011**, *35*, 214–220. [CrossRef]

43. Xie, J.; Zhou, T.; Liu, Y.; Kuriyagawa, T.; Wang, X. Mechanism study on microgroove forming by ultrasonic vibration assisted hot pressing. *Precis. Eng.* **2016**, *46*, 270–277. [CrossRef]

44. Zhang, X.; Guimond, Y.; Bellec, Y. Production of complex chalcogenide glass optics by molding for thermal imaging. *J. Non-Cryst. Solids* **2003**, *327*, 519–523. [CrossRef]

45. Cha, D.; Kim, H.; Park, H.; Hwang, Y.; Kim, J.; Hong, J.; Lee, K. Effect of temperature on the molding of chalcogenide glass lenses for infrared imaging applications. *Appl. Opt.* **2010**, *49*, 1607–1613. [CrossRef] [PubMed]

46. Liu, G.; Shen, P.; Jin, N. Viscoelastic properties of chalcogenide glasses and the simulation of their molding processes. *Phys. Procedia* **2011**, *19*, 422–425. [CrossRef]

47. Zhou, J.; Yu, J.; Lee, L.; Shen, L.; Allen, Y. Stress relaxation and refractive index change of As_2S_3 in compression molding. *Int. J. Appl. Glass Sci.* **2017**, *8*, 255–265. [CrossRef]

48. Hilton, A. *Chalcogenide Glasses for Infrared Optics*; McGraw-Hill Companies: New York, NY, USA, 2010.

49. Zhou, T.; Zhou, Q.; Xie, J.; Liu, X.; Wang, X.; Ruan, H. Elastic-viscoplasticity modeling of the thermo-mechanical behavior of chalcogenide glass for aspheric lens molding. *Int. J. Appl. Glass Sci.* **2018**, *9*, 252–262. [CrossRef]

50. Hencky, H. Über die Form des Elastizitätsgesetzes bei ideal elastischen Stoffen. *Z. Tech. Phys.* **1928**, *9*, 215–220.

51. Liu, L.; Wang, G.; Chen, J.; Yang, S. Creep experiment and rheological model of deep saturated rock. *Trans. Nonferr. Met. Soc. China* **2013**, *23*, 478–483. [CrossRef]

52. Zhou, T.; Yan, J.; Kuriyagawa, T. Evaluating the viscoelastic properties of glass above transition temperature for numerical modeling of lens molding process. *Proc. SPIE Int. Soc. Opt. Eng.* **2007**, *6624*, 662403.

53. Zhou, T.; Zhou, Q.; Xie, J.; Liu, X.; Wang, X.; Ruan, H. Surface defect analysis on formed chalcogenide glass $Ge_{22}Se_{58}As_{20}$ lenses after the molding process. *Appl. Opt.* **2017**, *56*, 8394–8402. [CrossRef] [PubMed]

54. Gupta, C. *Chemical Metallurgy: Principles and Practice*; John Wiley & Sons: Hoboken, NJ, USA, 2006.

55. Silberberg, M. *Chemistry: The Molecular Nature of Matter and Change*; McGraw-Hill Companies, Inc.: New York, NY, USA, 2000.

56. Kelley, J.; Leventhal, J. *Newtonian Physics Problems in Classical and Quantum Mechanics*; Springer International Publishing: Berlin/Heidelberg, Germany, 2017.

57. Yan, J.; Oowada, T.; Zhou, T.; Kuriyagawa, T. Precision machining of microstructures on electroless-plated NiP surface for molding glass components. *J. Mater. Process. Technol.* **2009**, *209*, 4802–4808. [CrossRef]

58. Muhammad, A.; Mustafizur, R.; Wong, Y. A study on the effect of tool-edge radius on critical machining characteristics in ultra-precision milling of tungsten carbide. *Int. J. Adv. Manuf. Technol.* **2013**, *67*, 1257–1265.

59. Hitchiner, M.; Wilks, J. Factors affecting chemical wear during machining. *Wear* **1984**, *93*, 63–80. [CrossRef]

60. Casstevens, J.; Daugherty, C. Diamond turning optical surfaces on electroless nickel. *Precis. Mach. Opt.* **1978**, *159*, 109–113.

61. Krishnan, K.; John, S.; Srinivasan, K.; Praveen, J.; Ganesan, M.; Kavimani, P. An overall aspect of electroless Ni-P depositions—A review article. *Metall. Mater. Trans. A* **2006**, *37*, 1917–1926. [CrossRef]

62. Liu, Y.; Zhao, W.; Zhou, T.; Liu, X.; Wang, X. Microgroove machining on crystalline nickel phosphide plating by single-point diamond cutting. *Int. J. Adv. Manuf. Technol.* **2017**, *91*, 477–484. [CrossRef]

63. Liu, X.; Zhou, T.; Pang, S.; Xie, J.; Wang, X. Burr formation mechanism of ultraprecision cutting for microgrooves on nickel phosphide in consideration of the diamond tool edge radius. *Int. J. Adv. Manuf. Technol.* **2017**, *94*, 3929–3935. [CrossRef]

64. Xie, P.; He, P.; Yen, Y.; Kwang, J.; Daniel, G.; Chang, L.; Liao, W.; Allen, Y.; Lee, L. Rapid hot embossing of polymer microstructures using carbide-bonded graphene coating on silicon stampers. *Surf. Coat. Technol.* **2014**, *258*, 174–180. [CrossRef]

65. Li, H.; He, P.; Yu, J.; Lee, L.; Allen, Y. Localized rapid heating process for precision chalcogenide glass molding. *Opt. Lasers Eng.* **2015**, *73*, 62–68. [CrossRef]

66. Ma, K.; Chien, H.; Huang, S.; Fu, W.; Chao, C. Contactless molding of arrayed chalcogenide glass lenses. *J. Non-Cryst. Solids* **2011**, *357*, 2484–2488. [CrossRef]

67. Saiz, E.; Cannon, R.; Tomsia, A. High-temperature wetting and the work of adhesion in metal/oxide systems. *Annu. Rev. Mater. Res.* **2008**, *38*, 197–226. [CrossRef]

68. Rieser, D.; Spieß, G.; Manns, P. Investigations on glass-to-mold sticking in the hot forming process. *J. Non-Cryst. Solids* **2008**, *354*, 1393–1397. [CrossRef]

69. Wang, J.; Fan, J.; Zhang, Y.; Wang, G.; Wang, W.; Chan, K. Diffusion bonding of a Zr-based metallic glass in its supercooled liquid region. *Intermetallics* **2014**, *46*, 236–242. [CrossRef]

70. Monfared, A.; Liu, W.; Zhang, L. On the adhesion between metallic glass and dies during thermoplastic forming. *J. Alloys Compd.* **2017**, *711*, 235–242. [CrossRef]

71. Masuda, J.; Yan, J.; Zhou, T.; Kuriyagawa, T.; Fukase, Y. Thermally induced atomic diffusion at the interface between release agent coating and mould substrate in a glass moulding press. *J. Phys. D Appl. Phys.* **2011**, *44*, 215302. [CrossRef]

72. Zhu, X.; Wei, J.; Chen, L.; Liu, J.; Hei, L.; Li, C.; Zhang, Y. Anti-sticking Re-Ir coating for glass molding process. *Thin Solid Films* **2015**, *584*, 305–309. [CrossRef]

73. Zhang, L.; Zhou, W.; Naples, N.; Allen, Y. Investigation of index change in compression molding of $As_{40}Se_{50}S_{10}$ chalcogenide glass. *Appl. Opt.* **2018**, *57*, 4245–4252. [CrossRef] [PubMed]

micromachines

MDPI

Article

2D Optical Gratings Based on Hexagonal Voids on Transparent Elastomeric Substrate

Valentina Piccolo [1], Andrea Chiappini [2,*], Cristina Armellini [2], Mario Barozzi [3], Anna Lukowiak [4], Pier-John A. Sazio [5], Alessandro Vaccari [6], Maurizio Ferrari [2,7] and Daniele Zonta [2,8]

[1] DICAM-University of Trento, Via Mesiano 77, 38123 Trento, Italy; valentina.piccolo@unitn.it
[2] IFN-CNR CSMFO Lab & FBK CMM, Via alla Cascata 56/C, 38123 Trento, Italy; cristina.armellini@unitn.it
[3] CMM-MNF, Fondazione Bruno Kessler, Via Sommarive 18, 38123 Trento (Povo), Italy; barozzi@fbk.eu
[4] Institute of Low Temperature and Structure Research PAS, 50-422 Wroclaw, Poland; a.lukowiak@intibs.pl
[5] ORC, University of Southampton, University Road, Southampton SO17 1BJ, UK; P.A.Sazio@soton.ac.uk
[6] CMM-ARES, Fondazione Bruno Kessler, Via Sommarive 18, 38123 Trento (Povo), Italy; vaccari@fbk.eu
[7] Enrico Fermi Centre, Piazza del Viminale 1, 00184 Roma, Italy; maurizio.ferrari@ifn.cnr.it
[8] Department of Civil and Environmental Engineering, University of Strathclyde, Montrose Street, 75, Glasgow G1 1XJ, UK; daniele.zonta@strath.ac.uk
* Correspondence: andrea.chiappini@ifn.cnr.it; Tel.: +39-0461-314920

Received: 29 June 2018; Accepted: 9 July 2018; Published: 10 July 2018

Abstract: A chromatic vectorial strain sensor constituted by hexagonal voids on transparent elastomeric substrate has been successfully fabricated via soft colloidal lithography. Initially a highly ordered 1.6 microns polystyrene spheres monolayer colloidal crystal has been realized by wedge-shaped cell method and used as a suitable mold to replicate the periodic structure on a polydimethylsiloxane sheet. The replicated 2D array is characterized by high periodicity and regularity over a large area, as evidenced by morphological and optical properties obtained by means of SEM, absorption and reflectance spectroscopy. In particular, the optical features of the nanostructured elastomer have been investigated in respect to uniaxial deformation up to 10% of its initial length, demonstrating a linear, tunable and reversible response, with a sensitivity of 4.5 ± 0.1 nm/%. Finally, it has been demonstrated that the specific geometrical configuration allows determining simultaneously the vectorial strain-stress information in the x and y directions.

Keywords: micro/nano patterning; 2D colloidal crystal; soft colloidal lithography; strain microsensor; vectorial strain gauge

1. Introduction

Among the different fabrication techniques that allow obtaining micro/nanostructured surfaces, colloidal lithography is attracting big interest due to low cost, time efficiency, simplicity, and the possibility to pattern over a large surface area [1].

This bottom-up approach exploits the self-assembly of hard dielectric micro and nano spheres such as silica or polystyrene (PS) in order to fabricate two dimensional arrays. In recent literature, 2D colloidal crystals have been realized by self-assembly under electrophoresis deposition [2], Langmuir–Blodgett deposition [3], spin coating [4] and capillary forces [5]. Considering this last approach, Sun et al. [6] have demonstrated that the use of wedge-shaped cell allows obtaining large domains 2D colloidal crystals, with centimeter size, taking advantage on the capillary forces and drying front formed in the cell.

In this contest it is worth mentioning that 2D colloidal crystals are interesting and promising systems for micro and nanopatterning due to their periodicity and specific size [7]. In micro and nano patterning field, colloidal crystals can be employed as lithographic masks or as molds for the production of micro and nanostructures for light trapping applications [8] or for the realization of SERS (Surface Enhanced Raman Spectroscopy) substrates [9]. Furthermore, they can act as masters by means of soft lithography in order to produce hexagonally arrayed structures.

These types of systems can be employed for the realization of responsive materials able to measure physical quantities such as magnetic fields [10], and temperature [11], or detect different chemicals, [12–14] including important analytes such as glucose [15], creatinine [16] and nerve gas agents [17].

Focusing the attention on mechanical parameters (i.e., strain), periodic polymeric photonic materials demonstrated sensitivity to deformation, in particular different configurations have been employed such as opal-type photonic crystals infiltrated with elastomeric materials [18,19]; 1D grating based on buckled thin film with periodic sinusoidal patterns on a transparent elastomeric substrate [20]; 1D array of gold nanoparticles on flexible substrate [21] and double sided 1D orthogonal polydimethylsiloxane (PDMS) gratings [22].

In particular, the realization of surface stress-based sensors has become fundamental in several fields in order to detect acoustic waves and forces on different structures such as spacecrafts, submarines, buildings or bridges.

Recently Guo et al. [22] have demonstrated that a double sided 1D orthogonal polydimethylsiloxane grating can be used as a vector mechanical sensor, able to detect mechanical parameters and giving information about their direction and strength.

In this work we have developed a strain/stress vector sensor based on hexagonal voids on a transparent elastomeric substrate: due to the specific geometric configuration, the application of a horizontal strain induces an opposite movement of the diffraction spots created by a white light impinging on the structure. The relative displacement of these spots can be investigated to estimate the vectorial strain/stress information and to characterize the applied strain in both the x and y directions. This structure paves the way for the development of low cost vector strain sensor systems.

2. Materials and Methods

2.1. Materials

The PS latex beads were delivered by Thermo Scientific (Waltham, MA, USA), the PDMS Sylgard 184 by Dow Corning (Midland, MI, USA) and all the chemicals (Absolute ethanol, Clorothrimethylsilane and Dimethylformamide), used as received, by Aldrich (St. Louis, MO, USA).

2.2. PS Colloidal Particles and Substrate Preparation

Monodisperse latex particles 1.6 microns in diameter and size distribution of 0.021 μm, 1.3% CV were purchased from Thermo Scientific and used as received at standard concentration of 1 wt% suspension in water. The v-SiO$_2$ substrates were cleaned firstly by brushing with neutral glassware detergent and then by ethanol. Finally, they were treated in an ozone cleaner for 30 min.

2.3. Assembly of the PS 2D Template

The PS spheres monolayer was used as a template for the fabrication of the PDMS grating and was deposited on v-SiO$_2$ by means of the wedge-shaped cell method. This growth method allows the deposition of large domains two-dimensional colloidal crystals that self-organize by controlling the drying front of evaporation when the constituting particles are confined within two slides holding at an angle of about 2°. After the infiltration of 125 μL of PS suspension, the cell was maintained at room temperature (RT) and relative humidity (RH) of 40% for 1 day. Due to the evaporation of the solvent in the suspension, the latex beads crystallized in an ordered hexagonal structure.

2.4. Functionalization and Infiltration of the PS 2D Template

The 2D PDMS grating was obtained by infiltrating the template with the elastomer. Before the infiltration, to facilitate the following peeling off from the glassy substrate, the PS monolayer was functionalized by silanization with clorotrimethylsilane in a Petri dish for 90 min. As a second step a mixture of a 10:1 base:curing Sylgard 184 elastomer was poured on the functionalized template and thermally cured for 4 h at 65 °C. Finally, the PDMS with embedded PS spheres was gently removed from the glass substrate by peeling off the elastomer.

2.5. PS Particles Chemical Etching

The last step of the fabrication protocol was the etching of the PS particles in the elastomeric matrix that was performed by immersion of the PDMS slab in dimethylformamide for 90 min. Dimethylformamide is a solvent for PS and a non-solvent for PDMS, hence it provides a selective etching of the latex beads allowing the formation of an inverse replica of the template based on hexagonal voids in elastomeric matrix. After the etching process the sample was rinsed in water and blown with nitrogen.

2.6. Sample Characterization

Morphological investigation of the samples has been carried out by means of scanning electron microscopy (SEM) measurements using a SEM JEOL JSM 7401-F FEG (Akishima, Tokyo, Japan). Transmittance measurements have been performed using a double beam VIS-NIR spectrophotometer Varian Cary 5000 (Palo Alto, CA, USA) in the range between 1000 and 2500 nm. The spectra of the samples were obtained illuminating the whole sample with a white light (halogen lamp) and collecting the diffracted light using a fiber-optic UV-Vis spectrometer Ocean Optics USB 4000 (Edinburgh, UK) as shown in Figure 1. Measurement of wavelength shift has been performed analyzing the displacement of two different diffraction spots, under the application of a horizontal strain. For both spots, this effect has been investigated by means of the wavelength shift of the 1st order of the transmittance diffraction, keeping the detection angle fixed at a specific value in order to have an initial spectrum centered in the visible region.

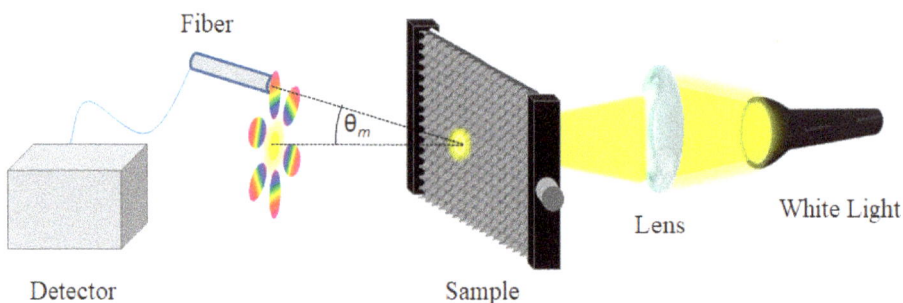

Figure 1. Sketch of the experimental set-up for the 2D diffraction grating measurements.

3. Results and Discussion

The first step, as shown in Figure 2a, concerned the realization of two-dimensional assembly of PS colloidal particles later used as a mold; Figure 3a reports a typical optical image of an ordered 2D colloidal crystal obtained by wedge-shaped cell method, where we can notice the presence of large areas of ordered domains (about 100 × 70 μm with few punctual defects). In Figure 3b, three transmission dips at λ = 1954 nm, 1598 nm, 1480 nm can be distinguished at normal incidence ($\theta = 0°$) and are the result of the excitation of the photonic eigenmodes of the periodic dielectric structure due to its coupling with the incident light as proposed by Sun et al. [6], which, confirm the high optical quality. Furthermore, we can notice a decrease in the transmittance values attributed to an increase in the scattered radiation for lower wavelength affecting the collection of the zero order transmitted signal.

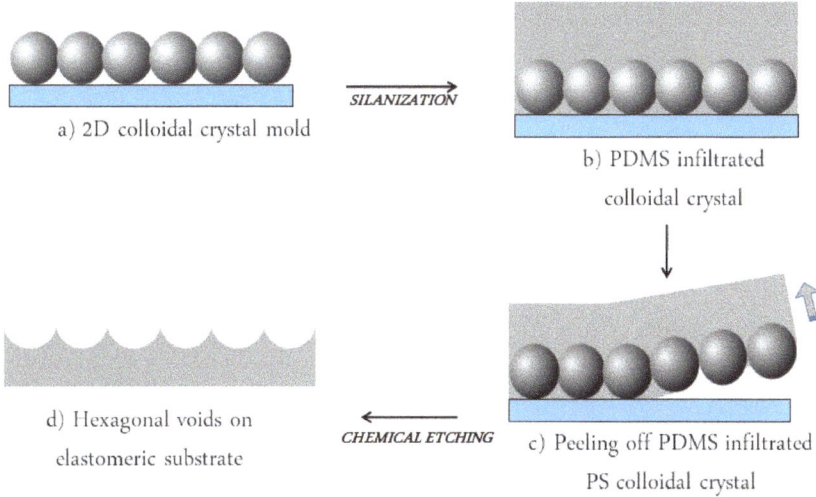

a) 2D colloidal crystal mold

SILANIZATION

b) PDMS infiltrated colloidal crystal

d) Hexagonal voids on elastomeric substrate

CHEMICAL ETCHING

c) Peeling off PDMS infiltrated PS colloidal crystal

Figure 2. Schematic illustration of the experimental approach employed for the realization of 2D PDMS replica patterns (**a**) formation of 2D colloidal crystal by means of wedge-shaped cell (**b**) functionalization and infiltration of PDMS by capillary force; (**c**) peeling off PDMS infiltrated PS colloidal crystal; (**d**) chemical etching of the PS spheres.

Moreover, the similar results acquired on different points, reported in Figure 3b, indicate the good homogeneity of the 2D colloidal crystal mold.

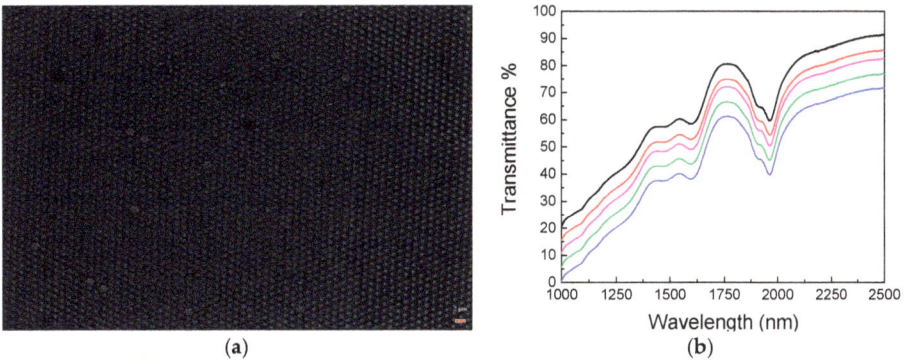

(a)

(b)

Figure 3. (**a**) Optical microscopy image of a typical area of the 2D colloidal crystals self-assembled using a wedge-shaped cell (scale bar of 2 μm). (**b**) Transmittance spectrum obtained on a 2D colloidal crystal deposited on a v-SiO$_2$ substrate. The individual spectra are offset vertically by 5% for clarity (the black spectrum is the original one).

Following the procedure described in Figure 2, a hexagonal voids regular structure (sketched in Figure 4a), has been fabricated via soft colloidal lithography. Figure 4b,c shows SEM-images of the resulting patterned polymeric structure where the reciprocal morphology of the PS mold has been successfully obtained. From a morphological point of view the hexagonal voids structure presents a periodicity of about 1600 nm with a depth of the voids of about 460 nm.

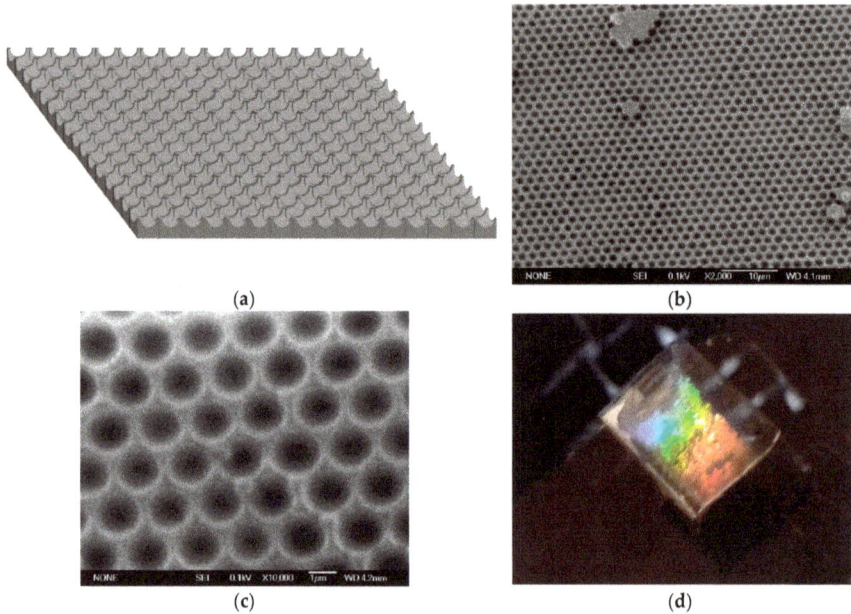

Figure 4. (**a**) Sketch of the concave structure obtained via soft lithography (not in scale) (**b**) SEM surface image of PDMS inverted colloidal crystal. (**c**) detail of the ordered hexagonal array. (**d**) Photograph of hexagonal voids on transparent elastomeric substrate.

Furthermore, as shown in Figure 4d, the hexagonal voids structure, that can be seen as a 2D grating, presents an iridescent color that is attributed to the high order over a large area. In this case the morphology of the periodic hexagonal pattern satisfies the diffraction features that can be expressed through the simple law of diffraction (Equation (1)).

$$n \cdot \sin(\theta_m) - n_i \cdot \sin(\theta_i) = \frac{m \cdot \lambda}{d} \tag{1}$$

where θ_i is the incident angle while θ_m corresponds to the mth diffraction order angle; n_i and n are the refractive indices of the incident medium and of the medium where the diffracted orders propagate respectively; λ represents the wavelength of the incident light; and d is the period of the grating.

From an optical point of view illuminating the grating by white light, and collecting the diffraction projected on a screen, we can clearly notice the presence of a chromatic hexagonal pattern (see Figure 5) due to the arrangement of the semispherical voids.

In order to verify the optical response of the system to mechanical deformation, the structure has been mounted on a linear stage and a deformation in the horizontal direction was applied.

As evidenced in Figure 5b, the application of a horizontal strain produces a change in the diffraction pattern. In this case it is worth mentioning that the movement of the first-order diffraction spot (see points 1 and 6 as labeled in the inset) is attributed to the variation of the grating period as a function of the strain, as predicted by the multi-slit Fraunhofer diffraction theory.

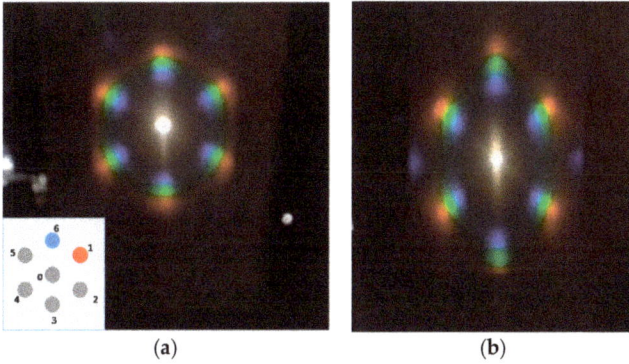

Figure 5. Strain induced diffraction spot movements: (**a**) Optical diffraction pattern without strain; inset: labelling of the investigated spots (1 and 6); (**b**) optical diffraction with a strain (ε) ε = 10% along the horizontal direction.

In particular comparing Figure 5a,b, focusing the attention on spot number 1, we can observe its movement towards the center (0), while if we consider spot number 6 we can notice that it moved away from the zero order. This effect can be attributed to an increase in the diffraction pitch in the parallel direction of the strain, and a consequent decrease (contraction) in the opposite side.

These features have been investigated by means of reflectance measurements detecting the wavelength shift of the 1st order of the transmittance diffraction, maintaining fixed the detector and applying a different strain to the grating.

Analyzing Figure 6a, related to spot number 1, we can notice that the first order of the diffraction peak presents a noticeable red-shift when increasing the applied strain. The diffraction peak wavelength passes from 510 to 553 nm for a uniaxial deformation of the structure up to 10% of its initial length. On the other hand, for spot 6 we have observed a decrease in wavelength of the diffraction peak from 575 to 551 nm. The images shown in Figure 5 and the difference in the peak wavelength shifts indicate that the strain induces an elliptical modification of the voids. Evidently, Figure 5 is suggestive of the fact that the grating's sensitivity is much higher against longitudinal geometrical changes then transversal ones. Indeed, the former are due to the imposed strain while the latter are due to Poisson's effect. Clearly, the initially circular semi-voids become elongated ellipses in the direction of the applied strain.

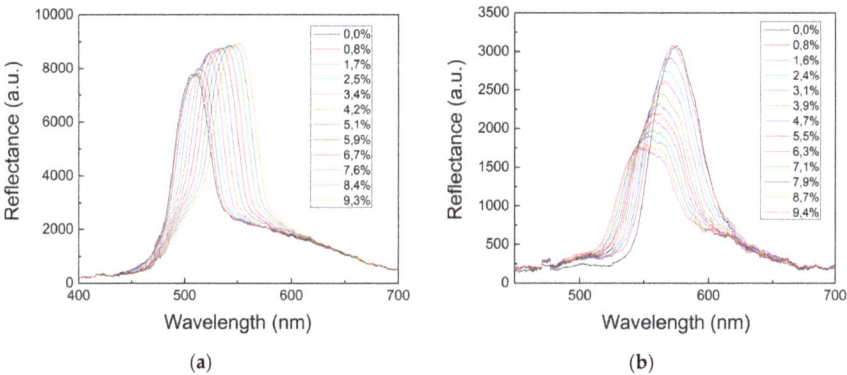

Figure 6. Reflectance spectra collected considering: (**a**) spot 1 as a function of the applied strain; (**b**) spot 6 as a function of the applied strain.

Moreover, we can see an increase of the intensity of the transmitted diffracted light at longer wavelength. To explain this effect, we have devised a simple model of its optical response. According to this model, the grating optical behavior is assimilable to two 2D arrays of secondary sources, both having the periodicity of the hexagonal semi-voids structure. The two arrays however, are half shifted in the grating plane, because one of them corresponds to rays emerging from the semi-void tops and the other corresponds to rays emerging from the semi-void bottoms. The two kinds of rays have an inherent optical path length difference due to the difference in the top and bottom substrate height. After calculations, the resulting intensity pattern for the downstream interference formula is thus depending from the primary beam wavelength, and in such a way that at an increase of its value necessarily implies an increase in the revealed intensity of a given secondary maximum [23].

Now in order to determine the sensitivity of the 2D grating as strain sensor we have analyzed the variation in wavelength of the diffraction peak as a function of the applied strain. In Figure 7 we report the variation of the peak positions of the diffracted light (a) for spot 1 and (b) spot 6 in respect to the % applied strain. First of all, we can notice a linear behavior, moreover we can determine a sensitivity equal to 4.5 ± 0.1 nm/% and 2.5 ± 0.1 nm/% for spot 6 and 1 respectively. These results, if compared with those reported in the literature for mechanochromic systems, permit to include the developed structure among the most sensitive as strain sensors [21].

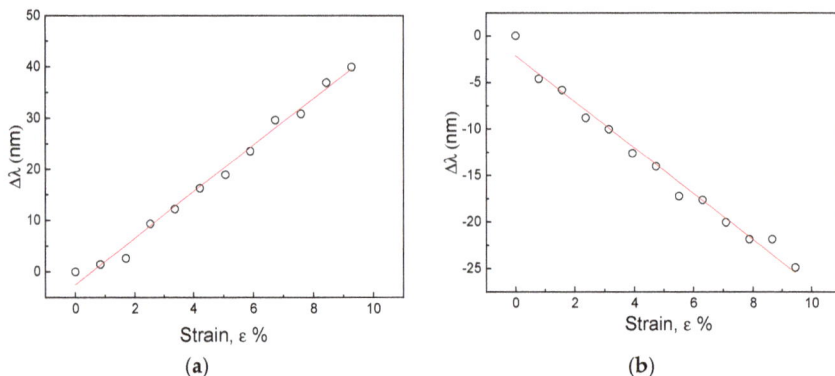

Figure 7. Experimental relationship between the peak position of the diffracted light (**a**) for spot 1 and (**b**) spot 6 in respect of the strain as a result of the elongation tests, (error bars are hidden by the circle points).

4. Conclusions

A chromatic strain sensor based on hexagonal voids on a transparent PDMS elastomeric substrate has been realized via soft colloidal lithography. The fabricated 2D grating can be employed for the development of a low cost and innovative sensor able to determine simultaneously the vectorial strain-stress information in the x and y directions.

Moreover, we have demonstrated that the sensor exhibits a tunable and reversible response under the application of a mechanical strain.

Optical reflection measurements have evidenced a linear behavior under the application of a horizontal strain up to 10% of its original length. The sensitivity of 4.5 ± 0.1 nm/%, when compared with mechanochromic photonic systems already present in literature, permits to classify the structure developed among the most sensitive strain sensors, paving the way for its applications in several fields such as smart sensing, mechanical sensing, and strain imaging.

Author Contributions: A.C., C.A., A.V. conceived and designed the experiments; A.C., V.P. performed the experiments; V.P. analyzed the data; C.A. realized the 2D grating; M.B. performed morphological analysis of the samples; A.C., V.P., A.L., P.J.A.S., D.Z. and M.F. wrote and revised the paper.

Acknowledgments: The research activity is performed in the framework of Centro Fermi MiFo project.

Conflicts of Interest: The authors declare no conflicts of interest.

References

1. Weiler, M.; Pacholski, C. Soft colloidal lithography. *RSC Adv.* **2017**, *7*, 10688–10691. [CrossRef]
2. Liao, C.-H.; Hung, P.-S.; Cheng, Y.; Wu, P.-W. Combination of microspheres and sol-gel electrophoresis for the formation of large-area ordered macroporous SiO₂. *Electrochem. Commun.* **2017**, *85*, 6–10. [CrossRef]
3. Askar, K.; Phillips, B.M.; Fang, Y.; Choi, B.; Gozubenli, N.; Jiang, P.; Jiang, B. Self-assembled self-cleaning broadband anti-reflection coatings. *Colloids Surf. A* **2013**, *439*, 84–100. [CrossRef]
4. Luo, C.-L.; Yang, R.-X.; Yan, W.-G.; Zhao, J.; Yang, G.-W.; Jia, G.-Z. Rapid fabrication of large area binary polystyrene colloidal crystals. *Superlattices Microstruct.* **2016**, *95*, 33–37. [CrossRef]
5. Ye, X.; Qi, L. Two-dimensionally patterned nanostructures based on monolayer colloidal crystals: Controllable fabrication, assembly, and applications. *Nano Today* **2011**, *6*, 608–631. [CrossRef]
6. Sun, J.; Tang, C.J.; Zhan, P.; Han, Z.L.; Cao, Z.S.; Wang, Z.L. Fabrication of Centimeter-Sized Single-Domain Two-Dimensional Colloidal Crystals in a Wedge-Shaped Cell under Capillary Forces. *Langmuir* **2010**, *26*, 7859–7864. [CrossRef] [PubMed]
7. Fournier, A.C.; Cumming, H.; McGrath, K.M. Assembly of two- and three-dimensionally patterned silicate materials using responsive soft templates. *Dalton Trans.* **2010**, *39*, 6524–6531. [CrossRef] [PubMed]
8. Kohoutek, T.; Parchine, M.; Bardosova, M.; Fudouzi, H.; Pemble, M. Large-area flexible colloidal photonic crystal film stickers for light trapping applications. *Opt. Mater. Express* **2018**, *8*, 960–967. [CrossRef]
9. Guddala, S.; Kamanoor, S.A.; Chiappini, A.; Ferrari, M.; Desai, N.R. Experimental investigation of photonic band gap influence on enhancement of Raman-scattering in metal-dielectric colloidal crystals. *J. Appl. Phys.* **2012**, *112*, 084303. [CrossRef]
10. Wang, W.; Fan, X.; Li, F.; Qiu, J.; Umair, M.M.; Ren, W.; Ju, B.; Zhang, S.; Tang, B. Magnetochromic Photonic Hydrogel for an Alternating Magnetic Field-Responsive Color Display. *Adv. Opt. Mater.* **2018**, *6*, 1701093. [CrossRef]
11. Yu, B.; Song, Q.; Cong, H.; Xu, X.; Han, D.; Geng, Z.; Zhang, X.; Usman, M. A smart thermo- and pH-responsive microfiltration membrane based on three-dimensional inverse colloidal crystals. *Sci. Rep.* **2017**, *7*, 12112. [CrossRef] [PubMed]
12. Cai, Z.; Kwak, D.H.; Punihaole, D.; Hong, Z.; Velankar, S.S.; Liu, X.; Asher, S.A. A Photonic Crystal Protein Hydrogel Sensor for Candida albicans. *Angew. Chem. Int. Ed.* **2015**, *54*, 13036–13040. [CrossRef] [PubMed]
13. Zhang, J.-T.; Cai, Z.; Kwak, D.H.; Liu, X.; Asher, S.A. Two-dimensional photonic crystal sensors for visual detection of lectin concanavalin A. *Anal. Chem.* **2014**, *86*, 9036–9041. [CrossRef] [PubMed]
14. Zhang, J.-T.; Chao, X.; Liu, X.; Asher, S.A. Two-dimensional array Debye ring diffraction protein recognition sensing. *Chem. Commun.* **2013**, *49*, 6337–6339. [CrossRef] [PubMed]
15. Chen, C.; Dong, Z.-Q.; Shen, J.-H.; Chen, H.-W.; Zhu, Y.-H.; Zhu, Z.-G. 2D Photonic Crystal Hydrogel Sensor for Tear Glucose Monitoring. *ACS Omega* **2018**, *3*, 3211–3217. [CrossRef]
16. Xu, D.; Zhu, W.; Jiang, Y.; Li, X.; Li, W.; Cui, J.; Yin, J.; Li, G. Rational design of molecularly imprinted photonic films assisted by chemometrics. *J. Mater. Chem.* **2012**, *22*, 16572–16581. [CrossRef]
17. Souder, B.; Prashant, P.; Seo, S.S. Hafnium polystyrene composite particles for the detection of organophosphate compound. *Soft Mater.* **2013**, *11*, 40–44. [CrossRef]
18. Fudouzi, H.; Tsuchiya, K.; Todoroki, S.-I.; Hyakutake, T.; Nitta, H.; Nishizaki, I.; Tanaka, Y.; Ohya, T. Smart photonic coating for civil engineering field: For a future inspection technology on concrete bridge. *Proc. SPIE* **2017**, *10168*, 1016820. [CrossRef]
19. Piccolo, V.; Chiappini, A.; Vaccari, A.; Calà Lesina, A.; Ferrari, M.; Deseri, L.; Perry, M.; Zonta, D. Finite difference analysis and experimental validation of 3D photonic crystals for structural health monitoring. *Proc. SPIE* **2017**, *10168*, 101681E. [CrossRef]

20. Yu, C.; O'Brien, K.; Zhang, Y.H.; Yu, H.; Jiang, H. Tunable optical gratings based on buckled nanoscale thin films on transparent elastomeric substrates. *Appl. Phys. Lett.* **2010**, *96*, 041111. [CrossRef]

21. Minati, L.; Chiappini, A.; Armellini, C.; Carpentiero, A.; Maniglio, D.; Vaccari, A.; Zur, L.; Lukowiak, A.; Ferrari, M.; Speranza, G. Gold nanoparticles 1D array as mechanochromic strain sensor. *Mater. Chem. Phys.* **2017**, *192*, 94–99. [CrossRef]

22. Guo, H.; Tang, J.; Qian, K.; Tsoukalas, D.; Zhao, M.; Yang, J.; Zhang, B.; Chou, X.; Liu, J.; Xue, C.; Zhang, W. Vectorial strain gauge method using single flexible orthogonal polydimethylsiloxane gratings. *Sci. Rep.* **2016**, *6*, 23606. [CrossRef] [PubMed]

23. Born, M.; Wolf, E. *Principles of Optics*, 4th ed.; Pergamon Press: Oxford, UK, 1970; pp. 401–414, ISBN 0072561912.

micromachines

MDPI

Review

Compound Glass Microsphere Resonator Devices

Jibo Yu [1], Elfed Lewis [2], Gerald Farrell [3] and Pengfei Wang [1,4,*]

[1]   Key Laboratory of In-fiber Integrated Optics of the Ministry of Education, College of Science, Harbin Engineering University, Harbin 150001, China; yu20131164@hrbeu.edu.cn

[2]   Optical Fibre Sensors Research Centre, Department of Electronic and Computer Engineering, University of Limerick, Limerick V94 T9PX, Ireland; Elfed.Lewis@ul.ie

[3]   Photonics Research Centre, Dublin Institute of Technology, Kevin Street, 8 Dublin D08 NF82, Ireland; gerald.farrell@dit.ie

[4]   Key Laboratory of Optoelectronic Devices and Systems of Ministry of Education and Guangdong Province, College of Optoelectronic Engineering, Shenzhen University, Shenzhen 518060, China

*   Correspondence: pengfei.wang@dit.ie; Tel.: +86-755-2653-1591

Received: 14 June 2018; Accepted: 17 July 2018; Published: 19 July 2018

Abstract: In recent years, compound glass microsphere resonator devices have attracted increasing interest and have been widely used in sensing, microsphere lasers, and nonlinear optics. Compared with traditional silica resonators, compound glass microsphere resonators have many significant and attractive properties, such as high-Q factor, an ability to achieve high rare earth ion, wide infrared transmittance, and low phonon energy. This review provides a summary and a critical assessment of the fabrication and the optical characterization of compound glasses and the related fabrication and applications of compound glass microsphere resonators.

Keywords: compound glass; microsphere; resonator; lasing; sensing

---

1. Introduction

Over the past few decades, research interest in microsphere resonators has grown rapidly. For a microsphere resonator, pump light is coupled into the microsphere through a tapered optical fiber or via free space. The coupled light signal is totally internally reflected and contained within the microsphere cavity to provide a 'whispering-gallery mode' (WGM) light resonance. Because of its extremely high-Q and small mode volume, microsphere resonators have many important roles in both active and passive photonic devices, such as in optical feedback, non-linear optics, low threshold lasers, dispersion managed optical systems, and energy storage [1–7].

Most current microsphere resonators are fabricated by melting the tip of a fused silica optical fiber [8], but is also possible to fabricate microsphere resonator from compound glass materials other than silica. Different host materials have different physical-chemical properties when in the form of compound glass, and many optical phenomena are limited by the properties of the material, therefore microspheres made from compound glass can behave differently in different practical applications. For example, glass materials with a high nonlinear coefficient can be used in wavelength conversion, optical switching and signal regeneration; high rare-earth ion doped glass materials have wide applications in near Near-infrared (NIR) and Mid-infrared (MIR) lasers, and some glasses are sensitive to temperature, light, and greenhouse gases. Such materials can be used to fabricate many different compound glass microsphere sensors [9–13].

In this paper, the progress of compound glass based microsphere-resonator devices over the past few decades is discussed. In the first section, the properties of the various compound glass materials are introduced, where the glass materials are divided into conventional glass and heavy metal glass types. In the second section, the fabrication methods for conventional silica microspheres and compound glass microspheres are reviewed. Finally, the applications of compound glass microsphere in microcavity lasing, nonlinear optical phenomena, and optical sensing are discussed, and the characterization of some compound glass microspheres with high-Q resonance are also presented.

2. Glass Materials

Oxide glasses have good chemical stability and excellent mechanical properties. Compared with metals, oxide glass materials have lower cost and typically possess an immunity to electromagnetic interference. Therefore, practical applications have focused on oxide glasses. Oxide glasses are generally divided into conventional oxide glasses and heavy metal oxide glasses. Conventional oxide glasses usually include silicate, germanate, and phosphate glasses, while heavy metal oxides usually comprise lead silicate, tellurite, and bismuth glasses. While the former have better thermal and chemical stability, the latter involve relatively simple fabrication processes as they can be processed at lower temperatures.

Heavy metal oxide glasses are predominantly used as a MIR material as they have excellent prospects for future developments in this wavelength domain e.g., Gas Sensing. They are principally based on PbO, TeO, and BiO. Heavy metal oxide glasses have many attractive optical properties, including high density, high refractive index, and excellent infrared transmission, for which the infrared wavelength band offers a broad range of applications. Compared with conventional oxide glasses, they have lower phonon energy and a broader infrared transmission range [14]. The low phonon energy in glasses reduces the relaxation rate of multiple phonons and therefore increases the probability of radiative transitions [15–17].

Heavy metal oxide glasses have better physical-chemical characteristics and preparation processes than chalcogenide glasses, while chalcogenide glasses have lower phonon energy and probability of multiple phonon relaxation compared with oxide glasses; chalcogenide glasses are also very suitable as host materials for rare earth doping. In particular, chalcogenide glasses have a wide infrared transmission window, and have received much attention as a material for MIR emission [18–22].

The physical-chemical properties of conventional oxide glasses, heavy metal oxide glasses, and chalcogenide glasses are introduced in detail in the following sub-sections, and are pinned to their related application areas where appropriate.

2.1. Conventional Oxide Glass

2.1.1. Silicate Glass

The main components of silicate glasses tend to be SiO_2-Al_2O_3-R_2O, SiO_2-P_2O_5-R_2O, SiO_2-B_2O_3-R_2O, SiO_2-GeO_2-R_2O, or a mix of the above, with a high SiO_2 content and relatively a low R_2O content. The glass often exhibits a small expansion coefficient, small dispersion, as well as excellent chemical and thermal stability. Silicate glasses are often used as optical glasses, in solar panels, liquid crystal display substrates, and heat collectors and silicate glasses have also been used in blue-violet LEDs to provide a new lighting sources [23]. In addition, it is worth noting that for the application of this glass in solar panels, high energy photon cutting can be achieved by rare-earth ions doping, which can significantly improve solar cell conversion efficiency [5,24].

The silicon-oxygen tetrahedron network of the silicate glass provides the excellent macroscopic mechanical and chemical stability of the material. When the silicate glass is melted in the fabrication process, the resulting loss of glass is relatively small, but a high melting temperature is required.

At present, active and passive optical fibers based on quartz materials are widely used, especially in high-power fiber lasers [25,26]. Quartz optical fibers exhibit low transmission loss, have a high thermal damage threshold, high mechanical strength and a high resistance to bending. However, quartz glass also has significant shortcomings. The silicon oxygen network of the quartz glass cannot provide enough non-bridging oxygens, which makes it very easy for rare earth ions to produce clusters, which leads to undesirable fluorescence quenching. In practical applications, the background loss of the optical fiber needs to be controlled to a very low value, and the introduction of other metal ions should be avoided as much as possible. Therefore, there is a pressing need to find a new way to increase the solubility of rare earth ions in silica glasses.

2.1.2. Phosphate Glass

Doping of rare earth ions in the phosphate glass can result in the fabricated phosphate fiber having the attractive characteristics of short length, small volume, high energy conversion efficiency, and low cost for fiber lasers, arousing the interest of many researchers [27–31]. Phosphate glasses have a higher rare earth ion solubility compared with silicate glasses, up to 10^{26} ions/m^3, so they can be doped with high concentrations of rare earth ions to obtain higher gains [32,33]. Furthermore, rare earth doped phosphate glass fibers can achieve the required gain within a few centimeters, avoiding the disadvantages of unwanted nonlinear phenomena associated with the long lengths of quartz fibers.

The basic structural unit of P$_2$O$_5$ glass is a phosphorus-oxygen tetrahedron [PO$_4$], where the phosphorus atom bonds to each oxygen atom with a double bond [34]. However, all the basic structural polyhedrons of tellurite glass and silicate glasses are connected by bridging oxygens, while the phosphorus tetrahedrons with double bonds lead to asymmetry in the glass structure of P$_2$O$_5$, resulting in a low viscosity, a large thermal expansion coefficient, and a poor chemical stability. In spite of these disadvantages, phosphate glasses still possesses many advantageous characteristics, such as a long fluorescence lifetime, large stimulated emission cross section, large gain coefficient, moderate phonon energy, and low fluorescence quenching.

2.2. Low-Phonon-Energy Oxide Glass

2.2.1. Germanate and Germanosilicate Glass

Germanate glass is a heavy metal oxide glass, with a wide infrared transmission window (~6 μm) and low phonon energy (~850 cm^{-1}), making it an ideal candidate material in for use in the MIR wavelength region [35]. Compared with fluoride and chalcogenide glasses, germanate glasses have good thermal stability, a simple fabrication process, and superior mechanical properties, leading to greater robustness and stability [36].

Conventional silicate fibers have excellent chemical stability and mechanical properties and their good plasticity enables fabrication into a variety of shapes, such as rods, plates, and optical fiber. This type of glass is therefore currently used in a wide variety of optical materials. However, the silicate glass material has a high phonon energy, resulting in an increase in the probability of non-radiative transitions, which limits the application of silicate glasses in photonics. Germanate glass have a lower phonon energy, which is crucial to increasing the radiative transition rate and probability of infrared transmission of rare earth ions and thus is a better material for obtaining high luminosity in the infrared band. However, high purity germanate glasses are extremely expensive, so only a limited amount of research has been conducted on this material [37].

Both silicon and germanium are in the periodic table group IVA and the outermost electron layer structure is in the form of ns^{2}np^2. The same main group elements have many similar chemical and physical properties. Silicon and germanium exist in the form of tetrahedral structure [SiO$_4$] and [GeO$_4$], thus the partial replacement of silicon oxide with germanium oxide in silicate glasses not only retains the excellent physical-chemical properties of the silicate glass, but also reduces the viscosity and fusion temperature of the glass, and improves the solubility of rare earth ions.

Germanosilicate glasses have received a lot of attention because they make up for the lack of development and utilization of germanate and silicate glasses in the optical field, which has resulted in germanium silicate glass having many applications in optical fiber communications, military detection and lasers.

2.2.2. Tellurite Glass

Early in 1952, Stanworth J. had studied the formation and structure of tellurite glasses [38]. However, the TeO_2 raw material was expensive, and hence tellurite glasses were considered to be of low practical value and had not been pursued as a candidate optical material until 1994. In this year, Wang. J. S of Rutgers University studied the tellurite glass as an optical material and found that it had a high rare earth ion solubility and constituted a new type of glass that could be used in optical fiber devices. Shortly afterwards, the Nd^{3+} doped tellurite fiber was fabricated and used to demonstrate single-mode laser output [39]. In 1997, Japan's NTT company successfully prepared an erbium doped fiber that could be used for broadband amplifiers, and quickly inspired many scientists to research tellurite glasses [40,41].

The phonon energy in tellurite glasses is low, generally in the range 650–800 cm^{-1}. The low non-radiative transition rate enhances the luminescence of the glass in the infrared. In addition, the maximum phonon energy of the tellurite glass system is the closest to that of fluoride glass, and the tellurite glass is most likely to replace fluoride glass as a host material, making it capable of forming a laser output in mid-infrared wavelength band.

Tellurite glass has a higher rare earth ion solubility than silica, which can be attributed to the fact that the rare earth ions in the tellurite glass can replace the position of the network modifier, which weakens the clustering phenomenon in the tellurite glass. Compared with fluoride glasses, they have good chemical stability, thermal stability, and mechanical properties and can also be fabricated using a relatively simple process. The melting temperature of tellurite glasses is generally around 800 °C. Tellurite glass has a high refractive index (1.8–2.3) compared to fluoride (~1.4), germanate glass (~1.6), and quartz (~1.45) and different structural units, such as $[TeO_4]$, $[TeO_3]$, and $[TeO_3{}^{+1}]$ [14]: this useful diversity of structural units can provide a more coordinated field environment for rare earth ions, thus the quenching phenomenon in tellurite glasses is greatly reduced.

2.2.3. Bismuth Glass

The electronic layer structure of Bi is $[Xe]4f^{14}5d^{10}6s^26p^3$, and this element has been widely studied. Since the electrons of the p orbital are easily involved in chemical bonding, Bi is also known as "The Wonder Metal". There are three distinct properties of the Bi element. Firstly, Bi has many valence states, such as 0, +1, +2, +3, and +5, so there are many states of matter, and the electrons of Bi element in the 6p, 6s, and 5d orbitals are very sensitive to the surrounding environment. Secondly, Bi has a strong cluster phenomenon, which is widely found in molten Lewis acids, molecular crystals, or porous zeolite solids. Thirdly, Bi shows a strong spin-orbit coupling effect, which allows them to act as optically active centers in different host materials. Bi doped materials exhibit a rich luminescent characteristic, which makes them different from the conventional active centers such as lanthanides and transition metals.

Bismuth glass has many advantages including low phonon energy (circa 600 cm^{-1}), high refractive index (circa 1.9), a wide infrared transmission range (0.4–6.5 μm), strong corrosion resistance, good solubility of rare earth ions, relatively good chemical stability, and low material cost, which results in many applications in photonics.

On one hand, bismuth glass is considered an innovative host material, and it is capable of providing efficient up conversion to red and green light output. On the other hand, bismuth glass is one of the most promising gain media materials currently used for rare earth doped fiber amplifiers. Many reports have shown that the host materials of bismuth glass have good performance in high capacity and ultra-wideband communications [42–44].

2.2.4. Lead Silicate Glass

Heavy metal silicate glasses have attracted wide attention in the field of photonic crystal fibers, because the addition of heavy metal ions that can increase the luminescent properties of the glass and change its nonlinear coefficient [44]. The nonlinear coefficient n_2 of lead silicate glasses is more than 20 times that of quartz, and it has the advantages of low melting temperature, moderate phonon energy (955 cm^{-1}), high rare earth ion doping ability, and a large emission cross section, making it an ideal choice for manufacturing photonic crystal fibers.

Lead silicate glass is a mature optical host material, which has been applied in lasers, military, construction, medical and other areas, as well as its application as a gain medium in laser and amplifier systems. Lead silicate glass has a high damage threshold, better resistance to crystallization, and greater mechanical strength. When added to conventional quartz fibers, Pb loosens the network structure of quartz, which accepts a high concentration of dopant ions and results in a suitably low fiber drawing temperature and could result in higher power laser outputs [45].

2.3. Chalcogenide Glass

Research on the optical properties of chalcogenide glass started nearly 60 years ago [46]. Chalcogenide glasses are composed of heavy elements joined by covalent bonds, resulting in unique optical properties making it highly suitable as a material for use in the MIR region, nonlinear optics, and optical waveguides. The emission of the chalcogenide glass red shifts to the visible or MIR region of the spectrum because the interatomic bond energy in chalcogenide glass is weaker than that in the oxide glass case. As the constituent atoms are heavier, the energy of the bond energy is very low, which means chalcogenide glass is transparent in the MIR region, and the low phonon energy (550 cm^{-1}) makes it an excellent host material for rare earth doping [47].

In general, the infrared transmission of chalcogenide glass extends up to 11 μm, with selenides reaching up to 15 μm and tellurite exceeding 20 μm. However, the physical properties of the glass transition temperature, hardness, strength, and durability usually degrade with weaker valence bonds, narrowing the long wave transparency band.

The low transition temperature means that precision glass forming offers a viable solution for the manufacture of low cost optical components such as those used in thermal imaging [48]. Chalcogenide glasses have a high refractive index, up to 2 to 3. According to Miller's formula [49], the higher the refractive index of a material, the higher its nonlinear coefficient n_2. As a consequence, the third-order Kerr effect of chalcogenide is several thousand times higher than that of silica [50–52], which means that chalcogenide glasses are considered excellent media for all-optical signal processing [43].

Chalcogenide glasses are also photosensitive. When exposed near the energy band, the chemical bond energy changes [53], and similar changes occur under heating and exposure to X-rays and electron beams. The inherent transparent window of chalcogenide glass is mainly in the molecular fingerprint region of 2–25 μm, making chalcogenide glass suitable for use in MIR optical fiber transmission, optical sensing, and as a waveguide material in optical communication.

Chalcogenide glass fibers were first reported in 1980 [54], when it was found that the limitation of high transmission loss was mainly due to impurity absorption. The relatively high loss of chalcogenides is still a significant problem, limiting its use to short lengths of fiber. Chalcogenide glasses require a high degree of purification during processing regardless of the final application. The high purity chalcogenide glass materials which do not include distillation in their processing often include oxygen, carbon, and hydrogen impurities [55,56], and these result in strong absorption peaks which occur within the 1.4–14.9 μm band.

There are many ways to reduce the impurities in chalcogenide glasses, which include: removing surface oxides in vacuum; chemical distillation using oxygen sorbents; treatment with thorium halide or active chlorine; evaporation through a porous silica frit; dynamic pyrolysis and purification of chalcogenide by high temperature oxidation [57–61]. These methods can reduce the impurity content to 10–5%, which significantly increases the infrared transparency.

3. Fabrication of Compound Glass Microspheres

At present, the principal method used for making microsphere resonators is based on melting of the glass materials. The optical loss factor of organic materials is generally large, and it is difficult to obtain a high-Q microsphere resonator, thus most microspheres are made from glass. The melting method uses the surface tension of molten glass to fabricate microspheres. In the case of a quartz fiber microspheres, fabrication usually involves the use of a CO_2 laser or heating furnace in their preparation. For other compound glass materials, there are slight differences in the preparation process. In this section, the manufacturing methods of traditional silica microspheres and compound glass microspheres are discussed.

3.1. Fabrication of Silica Microspheres

Low-cost standard single-mode fiber is widely used to produce high-Q silica microsphere resonators. The optical fiber generally used in the fabrication process is SMF-28 single-mode fiber (Corning), and its transmission loss in the telecommunication window is less than 0.2 dB/km. The diameters of the core and cladding are 8.2 µm and 125 µm, respectively.

The schematic diagram of the experimental setup for making a microsphere resonator is shown in Figure 1. The main instrument used in the experiment is a precision three-dimensional (3D) translation stage, a continuous CO_2 laser with a wavelength of 10.6 µm and a ZnSe lens for focusing. The experimental step of fabricating the silica microsphere resonator can be divided into three stages. In the first step, the coating layer at the end of the single-mode is removed, the fiber is mounted vertically on the 3D translation stage, and a weight is hung at the end of the fiber. Using a ZnSe lens to focus the laser beam on the single-mode fiber, the fiber absorbs light, resulting in a temperature rise. The glass softens and gradually turns into a tapered fiber under the influence of the weight. The heating is terminated when the waist diameter of the tapered fiber reaches around 100 µm. In the second step, the tapered fiber is accurately cleaved at the waist region to obtain a half tapered fiber. In the third step, using a ZnSe lens once more to focus the laser beam on the end of the half tapered fiber, the silica microsphere is formed at the fiber end due to the surface tension acting on the molten glass. The microscope image of a silica microspheres fabricated in this manner is shown in Figure 2.

In addition to employing the method referred to above, another commonly method for making silica microsphere is based on arc discharge [62,63]. In the literature [63], a fabrication process for silica microspheres using a Fitel S182PM fusion splicer was introduced. Firstly, the coating layer of the SMF28 fiber was removed, then the electrode of the splice was used to discharge the tip of the fiber, and the silica microsphere was formed due to the surface tension. In the experiment, the arc power was 110 a.u., the premelting time was 240 ms, and the arc duration was 2000 ms.

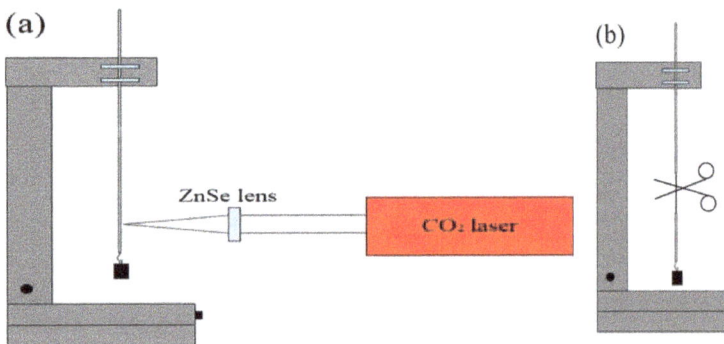

Figure 1. *Cont.*

(c)

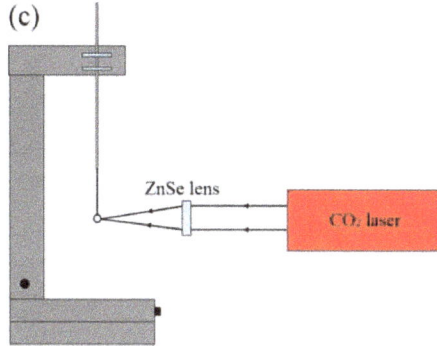

Figure 1. Schematic diagram of the experimental setup for making a silica microsphere. (**a**) A ZnSe lens is used to focus a CO_2 laser beam on the silica fiber; (**b**) the waist region of tapered fiber is cleaved; (**c**) a silica microsphere is obtained by focusing a CO_2 laser beam on the end of the cleaved tapered fiber.

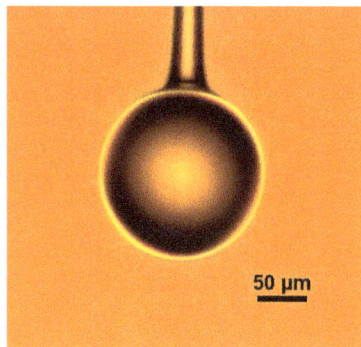

50 μm

Figure 2. Microscope image of silica microsphere made with CO_2 laser.

3.2. Fabrication of Lead Silicate Microspheres

In this section, a method of fabricating a lead silicate microspheres is introduced [64], and is similar to the method that uses a CO_2 laser to fabricate a silica optical microsphere as described in the previous section. The glass fiber is softened using a heat source and then the microspheres are formed due to surface tension of the glass fiber.

The schematic diagram of the experimental setup for making a lead silicate microsphere is shown in Figure 3. First of all, the lead silicate glass fiber is tapered to make the diameter of the lead silicate glass microsphere much smaller than the outside diameter of the lead silicate glass fiber. Then, the tapered section is placed in a resistive microheater with a Ω-shaped opening and heated to circa 500 °C. The microheater is moved back and forth along the fiber, while both ends of the fiber are carefully drawn using a computer-controlled translation stage. In this way, a tapered fiber of uniform waist diameter (d < 10 μm) and transition region are obtained [65]. The tapered fiber is then cut at the middle of the uniform waist. The newly formed half tapered fiber is then positioned in close proximity to the microheater and heated to 900 °C, which is higher than the softening point of the glass and thus the fused lead silicate glass fiber is melted to form the microsphere due to the surface tension of the lead silicate glass fiber.

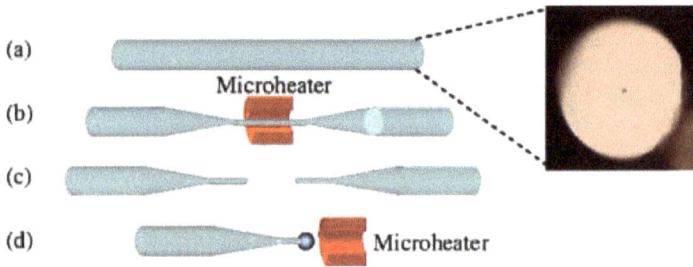

Figure 3. Schematic diagram of the experimental setup used for making a lead silicate microsphere. (**a**) Cross-section of lead silicate fiber; (**b**) a fiber is tapered by using a microheater; (**c**) the uniform waist of the tapered fiber is cleaved; (**d**) a lead silicate microsphere is formed. Reprinted/Adapted with permission from [64].

3.3. Fabrication of Germanium Microspheres

An experimental method for fabricating germanium microspheres using a germanium glass optical fiber is described in [66]. The material of the glass cladding is borosilicate, and the core is germanium. The core and the cladding diameters are typically 15 μm and 150 μm [67], respectively.

In order to obtain a germanium microsphere, the first step was to heat the core of the germanium with the fabrication method outlined for the quartz microsphere mentioned above, focusing the CO_2 laser beam on the core of the germanium glass. It is worth noting that when the temperature was between the melting point (938 °C) of the core and the softening point (1260 °C) of the cladding, a germanium microsphere would be formed inside the fiber cladding due to the lower cladding viscosity, which results in excellent smoothness for the glass surface.

The diameter of the resulting microsphere was 40–50 μm, and Figure 4a shows the microscope image of the microsphere embedded in a borosilicate cladding. The majority of the glass cladding was then etched away using a 48% HF solution, followed by immersion of the microsphere in a mixture of ammonium fluoride buffer and HF (20:1) for 5–10 min. Finally, the residual HF solution on the surface of the microsphere was cleaned with deionized water. The microscope image of the resulting germanium microsphere is shown in Figure 4b.

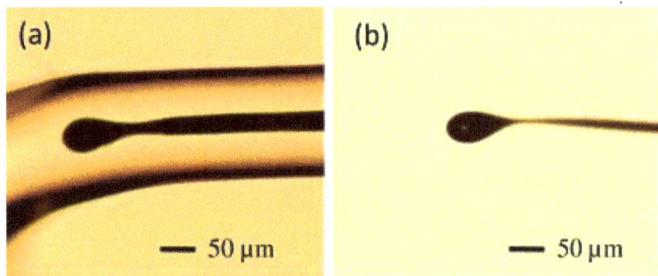

Figure 4. Microscope image of (**a**) a germanium microsphere in borosilicate cladding and (**b**) a fabricated germanium microsphere. Reprinted/Adapted with permission from [66].

3.4. Fabrication of Chalcogenide Microspheres

There have been many reports on the fabrication methods for chalcogenide glass microspheres [22,68–71]. One of them is a relatively simple method involving the use of a ceramic microheater to pull apart the molten chalcogenide glass [70]. This is possible due to the low softening temperature (100–400 °C) and transition temperature (100–300 °C) of chalcogenide glasses.

The chalcogenide glass fiber used in the experiment was a commercial step multimode fiber provided by Oxford Electronics. The core and cladding materials are As_2S_3 and As_xS_{1-x}, with diameters of 180 μm and 275 μm, respectively. The experimental step for fabricating the chalcogenide glass microsphere is shown in Figure 5. At first, the ceramic microheater was heated to about 200 °C, and a chalcogenide glass fiber was moved in close proximity to the microheater. Then, the end of the chalcogenide fiber was placed in contact with the outer wall of the ceramic microheater and as a result, the end of the glass fiber softened. Next, the fiber was pulled at the speed of 0.1–1 m/s until the end of the glass fiber was broken, resulting in a long half tapered microfiber formed at the chalcogenide fiber end. In the final step, the tip of the tapered microfiber was moved close to ceramic microheater, at which time the temperature of the ceramic microheater was raised to 500 °C. Due to the surface tension of the chalcogenide glass, the end of the microfiber melted into a microsphere, thus resulting in a chalcogenide glass microsphere resonator.

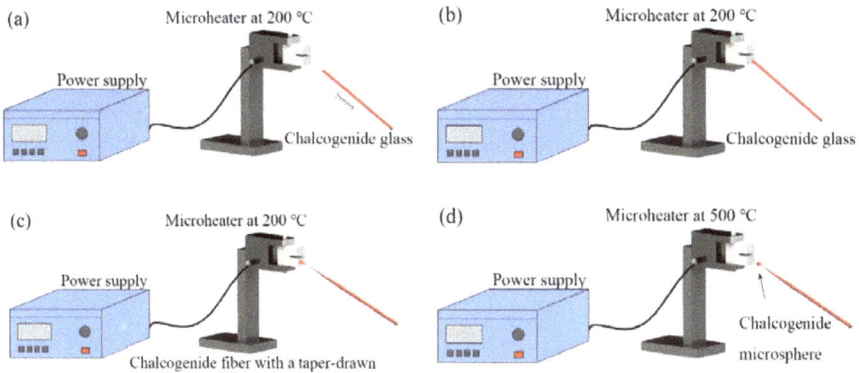

Figure 5. Experimental step for fabricating a chalcogenide microsphere. (**a**) Chalcogenide glass is moved to a microheater at 200 °C; (**b**) the chalcogenide glass is attached on the surface of microheater; (**c**) the chalcogenide glass is drawn to a half tapered fiber; (**d**) a chalcogenide microsphere is formed.

Figure 6 shows a microscope image of the chalcogenide microsphere fabricated by the above method: the surface of chalcogenide microsphere appears smooth and uniform.

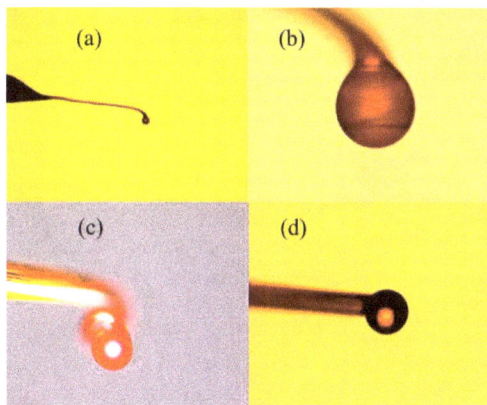

Figure 6. (**a**) Microscope image of the chalcogenide glass microsphere sample with different diameters; (**b**) 53 μm; (**c**) 98 μm; (**d**) 109 μm. Reprinted/Adapted with permission from [70].

3.5. Fabrication of Microspheres by the Powder Floating Method

The previously mentioned methods for fabricating compound glass microspheres are only capable of producing one microsphere at a time, and the size of the microsphere is determined by the size of the glass optical fiber. Moreover, the method of fabricating the microsphere from the optical fiber introduces additional heating steps during the fabrication process, and this will contaminate the glass material. This is extremely important for chalcogenide glass introduced in the previous section. Therefore, a powder floating method has been proposed [72,73], which is not only capable of producing microspheres of different sizes at the same time, but also effectively avoids the introduction of impurities.

The schematic diagram of fabrication of microspheres by the powder floating method is shown in Figure 7. During the fabrication process, a sample of compound glass material was ground into powder form, and the powders were allowed to pass through a sieve with openings on the order of tens of micrometers, depending on the required size of the microsphere. Then, the glass powders were dropped from the upper inlet of the furnace. Due to the inert gas (usually argon) flowing in the furnace, the microspheres were suspended in the furnace, with the temperature of the melting furnace (up to 1100 °C) set by the nature of the added glass material. In order to make the glass material melt fully in the furnace, the heating time of the powder in the furnace was increased by increasing the length of the heating zone of the furnace. Finally the molten powder forms glass microspheres due to the surface tension, and the smoothness of the microspheres is ensured by long term heating within the furnace. The advantages of the powder floating method is that the glass microspheres with a size distribution in a certain predetermined range can be fabricated in one batch, which greatly improves the efficiency for fabricating microspheres.

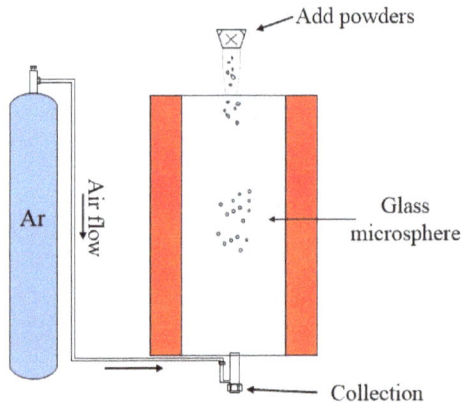

Figure 7. Schematic diagram of fabrication of microsphere by powder floating method.

4. Application and Characterization of Compound Glass Microspheres

A wide variety of applications based on compound glass microsphere resonators have been reported, which include fiber microcavity lasers, nonlinear optics, optical sensors, and quantum optics. Before introducing the application of the compound glass microspheres, the coupling methods of the microsphere resonators are introduced due to their significant role in experimental realization of these devices. There are two main methods for coupling incident light into the microsphere resonators. The first involves the use of tapered fiber coupling. Most tapered fibers are fabricated in silica fiber, which have lower refractive index when compared with most compound microspheres and tend to excite higher order modes in the microspheres [1,74]. The second method involves prism coupling, the input and output coupling being achieved using total internal reflection in a prism whose surface is located close to the microsphere [75–78]. The main advantage of this is it facilitates greater structural stability.

In this section, compound glass microcavity lasers and Stimulated Brillouin Scattering (SBS) in tellurite glass microspheres are initially reviewed, followed by an introduction to the application of chalcogenide glass-based microspheres in temperature sensing. Finally, high-Q compound glass microsphere resonators are characterized.

4.1. Compound Glass Microspheres Lasers

4.1.1. Silica Microsphere Laser

As early as 2003, the California Institute of Technology proposed a microcavity laser based on a silica microsphere [79], by coating Erbium-doped gels on silica microspheres to obtain a microcavity laser with a low threshold of 28 μW.

Before the experiment, a gel solution of tetraethoxysilane (TEOS) and $ErNO_3 \cdot 5H_2O$ was prepared in an acidic mixture. The silica microsphere was then immersed in the gel solution, and was immediately exposed to CO_2 laser radiation in order to cure the gel on the microsphere surface. The resulting microsphere had a diameter of 50–80 μm, including the gel layer thickness of 1–10 μm.

A single-frequency laser source with a wavelength of 980 nm was coupled to the doped microsphere using a silica tapered fiber with a diameter of 1.6 μm. The WGM transmitted in the microsphere provided the pump energy for the population inversion in the Erbium ions, and generated a single-mode laser output near the wavelength of 1540 nm. The output spectrum is included in Figure 8, which also shows the relationship between the laser output power of silica glass microsphere and the absorbed power; Figure 8 shown that the laser threshold value is 28 μW, and the absorbed power and laser output power are linear.

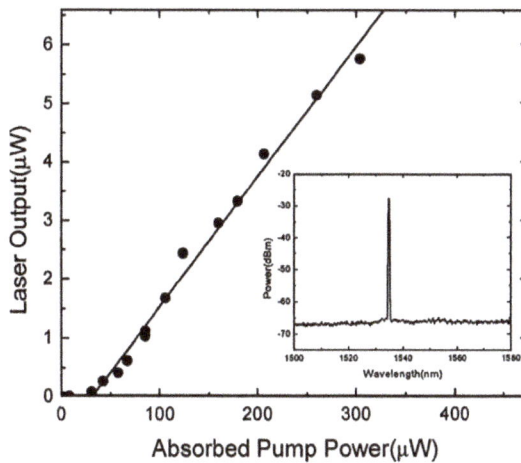

Figure 8. Relationship between the absorbed pump power and laser output; inset: spectrum of the laser output. Reprinted/Adapted with permission from [79].

4.1.2. Phosphate Microsphere Lasers

Phosphate glasses have a higher rare earth ion doping solubility compared to silica. A number of early studies were conducted on the doping of rare earth ions in phosphate glass microspheres. These include Er^{3+}, Er^{3+}-Yb^{3+} and Er^{3+}-Yb^{3+}-Cr^{3+} [28,29,31], all of which were successfully made to operate as a microcavity laser. The Er^{3+}-Yb^{3+} doped phosphate glass microsphere mentioned in [31] had the lowest laser threshold.

In the case of the microsphere with Er^{3+}-Yb^{3+} doped into the phosphate glass microsphere, the diameters of the resulting microsphere and silica tapered fiber were 57 μm and 1.8 μm, respectively. The pump laser was a semiconductor laser with a wavelength of 977 nm. When the excitation power was lower than the laser threshold, the fluorescence spectrum of the Er ions doped microsphere resembled that shown in Figure 9, where the free spectral range (FSR) values at the wavelength region of 1040 nm and 1550 nm are 4 nm and 9 nm, respectively.

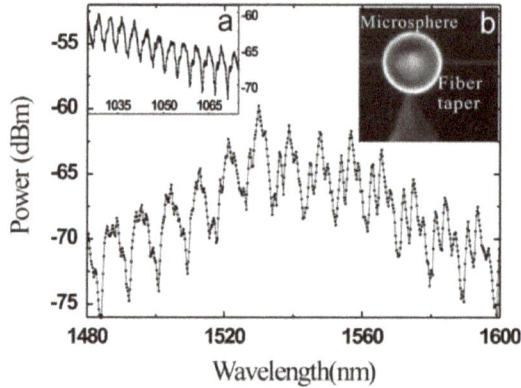

Figure 9. Fluorescence spectra of Er ions doped microsphere. (**a**) The spectrum in the band of 1040 nm; (**b**) the image of the microsphere and tapered fiber. Reprinted/Adapted with permission from [31].

The relationship between the laser output power of the Er^{3+}-Yb^{3+} doped microsphere and the absorbed power was obtained by increasing the pump power continuously in a range beyond the laser threshold. As shown in Figure 10, the laser threshold at a wavelength of 1032 nm and 1563 nm are 32 μW and 30 μW, respectively.

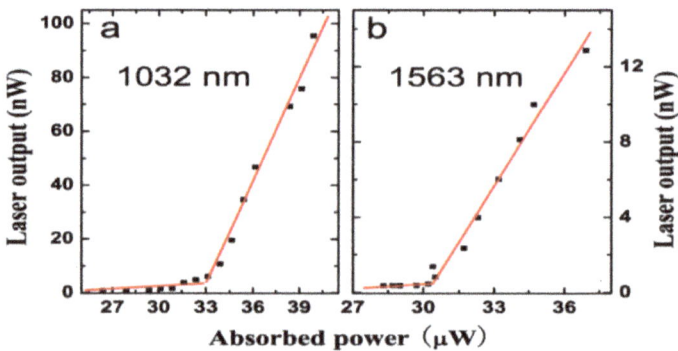

Figure 10. (**a**) Relationship between absorbed power and laser output, the lasing wavelength is 1032 nm; (**b**) the lasing wavelength is 1563 nm. Reprinted/Adapted with permission from [31].

4.1.3. Germanate Microsphere Laser

Traditional rare earth ion doped glass microspheres have a specific spectral region in the wavelength range of 1100–1400 nm, where no laser output has been achieved until recently. Bi-doped germanate glass microspheres have been effectively used to compensate for this gap in spectral coverage [37]. Compared to the conventional silica fiber (less than 0.1 mol%), the doping concentration of Bi ions in germanate glass is higher (3.5 mol%), and the melting temperature of the germanate is lower, which is more conducive to the fabrication of Bi doped germanate glass. In addition, the gain coefficient of Bi in germanate glasses is much higher than that in silica, mainly because the phonon relaxation rate in germanate glass (circa 600 cm^{-1}) is lower.

Microcavity lasers based on Bi-doped germanate glasses have been successfully fabricated and are described in the literature [37]. In this experiment, the diameter of the silica tapered coupling fiber and the microsphere were 3 μm and 80 μm, respectively. A laser light source with a wavelength of 808 nm was chosen as the pump. The laser output spectrum of the Bi-doped germanate could be observed on an optical spectrum analyzer (OSA), and is shown in the Figure 11. Figure 11a shows a spectrum of the WGMs when the pump light is coupled into the Bi-doped germanate glass microsphere; the full width half maximum (FWHM) is 0.0052 nm at 1310.05 nm, resulting in a Q value of the Bi-doped glass microsphere up to 2.5×10^5. When the pump absorption power is reaches 215 μW, the output of the single-mode laser at the wavelength of 1305.8 nm was observed, as shown by the red line in Figure 11b. The blue line in Figure 11b is the fluorescence spectrum of the Bi-doped germanate glass.

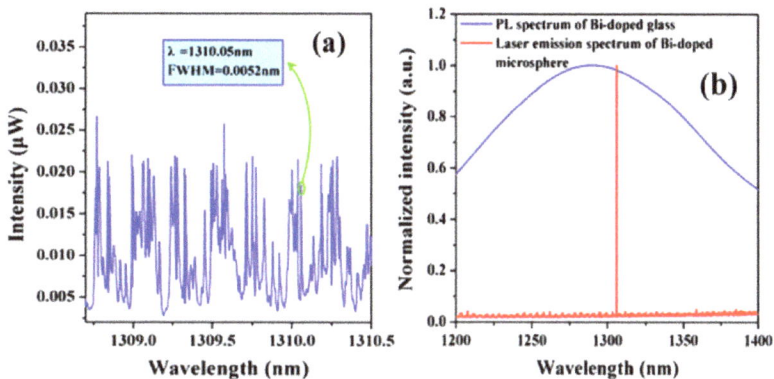

Figure 11. (**a**) Spectrum of the WGMs when the pump light is coupled into the microsphere; (**b**) red line: spectrum of the Bi-doped microsphere laser output; blue line: fluorescence spectrum of the Bi-doped glass. Reprinted/Adapted with permission from [37].

4.1.4. Tellurite Microsphere Laser

Tellurite glass has emerged as a promising material for use in the MIR wavelength region [80]. In this section, a Tm-Ho co-doped tellurite microcavity laser in the MIR region is introduced [81], with the underlying principle of operation being the transition of the $^3H_6$ to $^3H_4$ energy level in Tm^{3+} ions. This was achieved using a pump with a wavelength of 808 nm, resulting in the transition from $^3H_4$ to $^3F_4$ energy level in some Tm^{3+} ions, so that the energy in the Tm^{3+} ions could be partially transferred to the adjacent Ho^{3+} ions, prompting the transition of the ions from $^5I_8$ to $^5I_7$. Finally, the spontaneous emission of Tm^{3+} ions from $^5I_7$ to $^5I_8$ produces a fluorescence emission in the wavelength range 1.8–2.2 μm. The fluorescence spectra of Tm^{3+} doped tellurite glass and Tm-Ho co-doped tellurite glass are shown in Figure 12.

In this experiment, the diameters of the silica tapered fiber and the tellurite glass microsphere were 1.38 μm and 59.52 μm, respectively. A laser diode with a wavelength of 808 nm was used to pump the microsphere laser. When the pump power was lower than the threshold value, the fluorescence spectrum was observed using the OSA, and the resulting spectrum is shown in Figure 13a. Each peak of the fluorescence spectrum represents a mode transmitted in the microsphere and the observed FSR of the WGM was 12.79 nm. The black line in Figure 11a represents the fluorescence spectrum of the tellurite glass. However, when the pump power exceeded 0.887 mW, a single-mode laser output was observed from the Tm-Ho doped tellurite glass microsphere. The output wavelength of the single-mode laser was observed to be around 2.1 μm. Figure 13b shows the relationship between the output laser power of the tellurite glass microsphere and the absorbed power. It is clear from the inset in the figure that when the pump power is 0.779 mW or lower, there is no laser output emission, and when the power was increased to 0.887 mW, a single-mode laser output was generated as shown in the inset of Figure 11b, with a center wavelength of the output laser at 2092.56 nm.

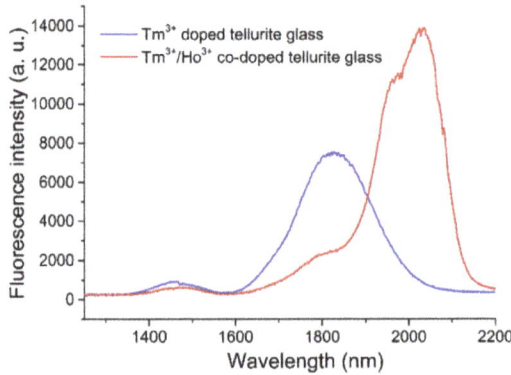

Figure 12. Fluorescence spectrum of the Tm^{3+} doped and Tm-Ho co-doped tellurite glasses. Reprinted/Adapted with permission from [81].

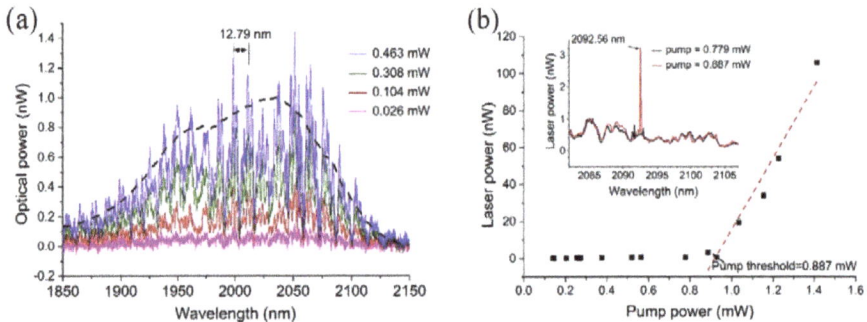

Figure 13. (**a**) Spectrum of WGMs in the tellurite glass microsphere; (**b**) relationship between the output laser power and pump power; inset: spectral output when the pump is below (blue line) and above threshold (red line). Reprinted/Adapted with permission from [81].

4.2. SBS in Tellurite Microsphere

Stimulated Brillion Scattering (SBS) is a non-linear optical phenomenon, caused by the electromagnetic stretching effect generated by high-power incident light which stimulates an ultrasonic wave in the waveguide, and results in the incident light being scattered by the ultrasonic waves.

In the literature [82], the SBS phenomenon has been reported for the first time in a tellurite glass microsphere, and the Q value of the fabricated microsphere exceeded 1.3×10^7. In the experiment, a tellurite glass microsphere was fabricated using melting methods using a CO_2 laser. The diameter of the tapered fiber and microsphere were 2 μm and 116 μm, respectively, and a single-frequency tunable laser with a wavelength of 1550 nm was used to pump the tellurite microsphere. The resulting spectrum is shown in Figure 14a, where the gray line shows the spectrum when the pump was off resonance and the yellow line shows the case when in resonance. It can be clearly seen from the spectrum that the SBS phenomenon is generated and observed at a wavelength that is 65 pm higher than the pump peak wavelength, and this represents the first order Stokes signal. Beyond that, the peak intensity of the yellow line is generally 10 dB larger than the peak intensity of the gray line, and the SBS peak of the yellow line is 15 dB larger than the gray line. These results suggest that most of the power in the strong Rayleigh scattering is feedback, thus inhibiting SBS generation. When the pump power was increased from 0.4 mW to 1.6 mW, the relationship of the coupled pump power and the Brillouin output power was obtained, as shown in Figure 14b, resulting in a threshold for the Brillouin output power of 0.58 mW.

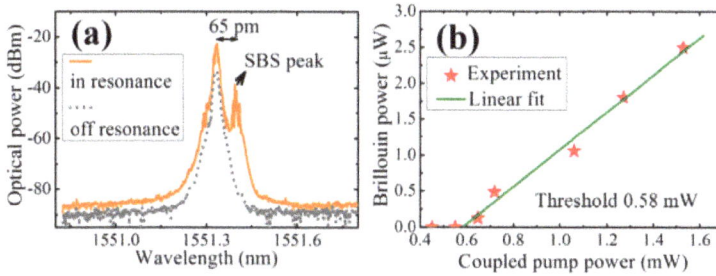

Figure 14. (**a**) Gray line: pump off resonance; yellow line: in resonance; (**b**) relationship between coupled pump power and Brillouin power. Reprinted/Adapted with permission from [82].

4.3. Temperature Sensing Using Chalcogenide Glass Microspheres

The basic properties of chalcogenide glass have been introduced in Section 2.3. Chalcogenide glass has a high photosensitivity, and good transmittance (low attenuation) in the MIR region, making chalcogenide glass microspheres highly sensitive to environmental parameters such as temperature.

In this section, a temperature sensor using chalcogenide glass microspheres in the MIR region is introduced [71]. A Tm^{3+} doped chalcogenide glass microsphere was prepared using the previously mentioned powder floating method. The experimental principle is that the Tm^{3+} transition from the ground state $^3H_6$ to $^3H_4$ occurred, followed by the spontaneous transition of Tm^{3+} from $^3H_4$ to $^3F_4$, and $^3F_4$ to $^3H_6$ generating a fluorescence emission in the 1.5 μm and 1.9 μm wavelength bands.

In the experiment, the diameter of the silica tapered fiber and the chalcogenide glass microsphere were 1.72 μm and 108.52 μm, respectively; pumping was provided from a laser diode with a wavelength of 808 nm. The WGM of the chalcogenide glass microsphere in the wavelength range 1.65 μm to 2 μm was obtained by coupling the pump light into the microsphere, and the spectral output is shown in Figure 15. The blue line is the unmodulated fluorescence spectrum of the chalcogenide glass.

In order to study the sensitivity of the chalcogenide microsphere to temperature, the whole device was placed in an adiabatic environment, and the ambient temperature was increased from 26 °C to 97 °C, resulting in all the resonances in the spectrum shifting to longer wavelengths. For ease of observation, a resonant peak with a wavelength of 1843.91 nm was chosen as a reference. The deviation of the wavelength of the single resonant peak with the change of temperature is shown in Figure 16a, and the relationship between the wavelength shift and temperature in Figure 16b. A linear relationship exists between the observed wavelength shift and temperature, and the resulting temperature sensitivity of the chalcogenide glass was measured to be 26 pm/°C.

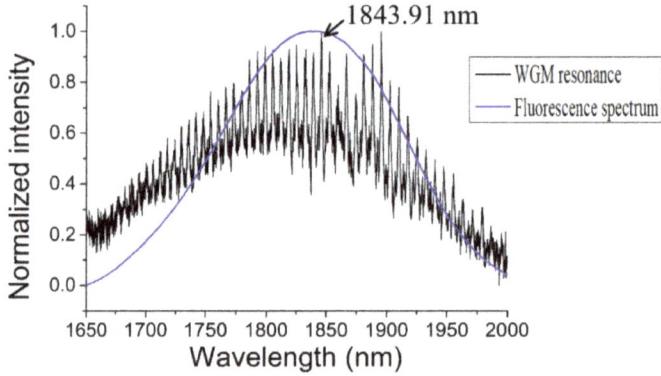

Figure 15. Whispering-gallery mode (WGM) resonance in a chalcogenide glass microsphere (black), compared with the unmodulated fluorescence spectrum of the chalcogenide glass (blue). Reprinted/Adapted with permission from [71].

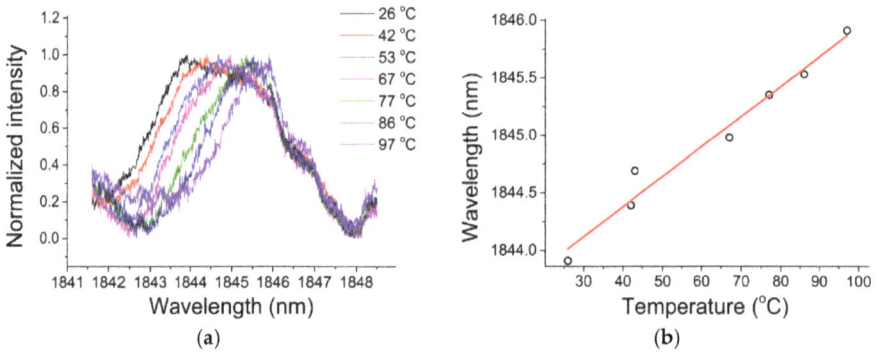

Figure 16. (**a**) Resonance spectra at different temperatures; (**b**) relationship between the wavelength shift and temperature. Reprinted/Adapted with permission from [71].

4.4. Other High-Q Compound Glass Microspheres

Compound glass doped glass microspheres (such as lead silicate and bismuth glasses) possess a high nonlinear coefficient [64,74].

In the experiment conducted to characterize the lead silicate glass microsphere, the diameters of the lead silicate glass microsphere and silica tapered fiber were 109 μm and 2 μm, respectively. A pump source with a wavelength of 1550 nm was coupled into the tapered fiber, and the spectrum of the WGM was achieved in the microsphere by effective coupling of the tapered fiber and the microsphere. Finally, the transmitted light from the fiber was received by a photodetector, and the transmission spectrum is shown in Figure 17. Because the effective refractive index of the tapered fiber and the microsphere are quite different, so many high-order modes exist in the transmission spectrum. A wavelength range of 1554.75–1554.8 nm is selected in the spectrum, and the data points obtained in the experiment are fitted using Lorentz fit; the Q value in the microsphere is calculated to be 0.86×10^7.

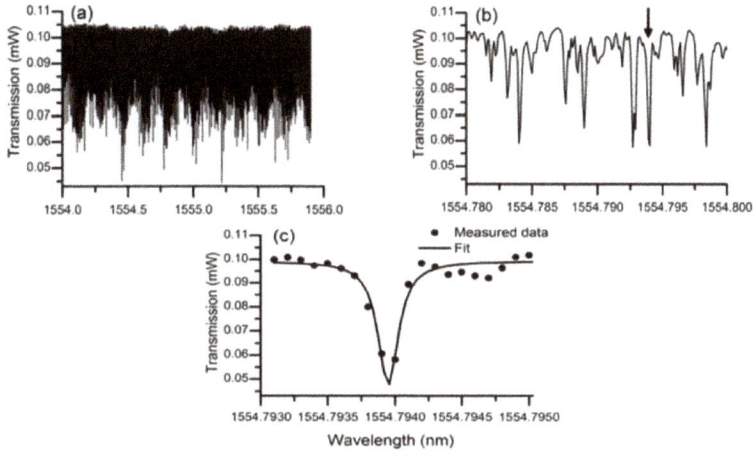

Figure 17. (**a**) Transmission spectra of the lead silicate microsphere in different spectral regions; (**b**) close up spectrum range from 1554.780-1554.800; (**c**) Lorenz fitting curve. Reprinted/Adapted with permission from [64].

The method of characterizing the bismuth glass microsphere is similar to that used for the lead silicate glass microsphere, and is not repeated in detail. The transmission spectrum of bismuth glass microsphere is shown in Figure 18, within the wavelength range from 1548.846 nm to 1548.849 nm, resulting in a Q value of up to 0.6×10^7.

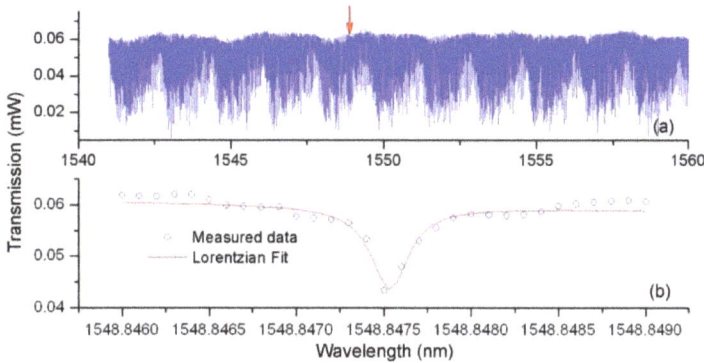

Figure 18. (**a**) Transmission spectra of the bismuth glass microsphere; (**b**) close-up spectrum range from 1548.846–1548.849. Reprinted/Adapted with permission from [74].

5. Conclusions and Outlook

In this paper, a comprehensive review of the properties of compound glasses has been undertaken with a focus on their applications in microsphere devices and sensors. Materials included conventional oxide glasses, heavy metal oxide glasses, and chalcogenide glasses. From their properties, it can be concluded that compound glasses have higher Q, higher rare earth ion doping, and lower phonon energy. The fabrication methods of microspheres using the various compound glass were also reviewed. The traditional method of microsphere fabrication involves melting the glass, which includes fabrication using a CO_2 laser or a ceramic microheater. For the chalcogenide glass fiber with a low melting point, the microsphere can be directly drawn by the "pulling up" method, as well as

Micromachines **2018**, *9*, 356

the more recently discovered powder floating method, the latter method having the advantage that it allows for mass production of higher quality microspheres. The applications of compound glass microspheres were also studied and these include mainly rare earth ion doped microspheres for lasers, high nonlinearity components involving tellurite glass microspheres, and temperature sensing using chalcogenide glass microspheres.

Compound glass microsphere resonators overcome the limitations associated with traditional resonators in terms of glass materials. In the future, it is envisaged that compound glass microsphere resonators will have wide ranging applications in photonics including optical computers and optomechanics due to their high nonlinearity, high Q quality, and high response. Meanwhile, doping rare earth ions in different host materials is expected to achieve higher power output and more efficient lasers accessing different wavelength ranges, most notably in the infrared band. However, most compound glass-based microspheres are coupled using a tapered fiber, whereas prism-based coupling is more conducive to robust packaging. In general, it is fair to state that the prospects for practical application of compound glass microsphere resonators in photonics are very bright and are developing rapidly.

Author Contributions: J.Y. drafted the paper; P.W., G.F. and E.L. provided substantial input to revise and improve the paper.

Acknowledgments: National Natural Science Foundation of China (NSFC) (61575050); The National Key R&D Program (2016YFE0126500); Key Program for Natural Science Foundation of Heilongjiang Province of China (ZD2016012); the Open Fund of the State Key Laboratory on Integrated Optoelectronics (IOSKL2016KF03); the "111" Project (B13015) to the Harbin Engineering University; the Recruitment Program for Young Professionals (The Young Thousand Talents Plan).

Conflicts of Interest: The authors declare no conflict of interest.

References

1. Spillane, S.M.; Kippenberg, T.J.; Vahala, K.J. Ultralow-thresholdRamanlaserusing a spherical dielectric microcavity. *Nature* **2002**, *415*, 621–623. [CrossRef] [PubMed]
2. Jin, X.; Wang, J.; Wang, M.; Dong, Y.; Li, F.; Wang, K. Dispersion engineering of a microsphere via multi-layer coating. *Appl. Opt.* **2017**, *56*, 8023–8028. [CrossRef] [PubMed]
3. Xiong, Q.Q.; Tu, J.P.; Lu, Y.; Chen, J.; Yu, Y.X.; Qiao, Y.Q.; Wang, X.L.; Gu, C.D. Synthesis of hierarchical hollow-structured single-crystalline magnetite (Fe_3O_4) microspheres: The highly powerful storage versus lithium as an anode for lithium ion batteries. *J. Phys. Chem. C* **2012**, *116*, 6495–6502. [CrossRef]
4. Grudinin, I.S.; Maleki, L. Ultralow-threshold Raman lasing with CaF_2 resonators. *Opt. Lett.* **2007**, *32*, 166–168. [CrossRef] [PubMed]
5. Trupke, T.; Green, M.A.; Würfel, P. Improving solar cell efficiencies by down-conversion of high-energy photons. *J. Appl. Phys.* **2002**, *92*, 1668–1674. [CrossRef]
6. Matsko, A.B.; Ilchenko, V.S. Optical resonators with whispering-gallery modes—Part I: Basics. *IEEE J. Sel. Top. Quantum Electron.* **2006**, *12*, 15–32. [CrossRef]
7. Gorodetsky, M.L.; Savchenkov, A.A.; Ilchenko, V.S. Ultimate Q of optical microsphere resonators. *Opt. Lett.* **1996**, *21*, 453–455. [CrossRef] [PubMed]
8. Ward, J.; Benson, O. WGM microresonators: Sensing, lasing and fundamental optics with microspheres. *Laser Photonics Rev.* **2011**, *5*, 553–570. [CrossRef]
9. Ogusu, K.; Yamasaki, J.; Maeda, S.; Kitao, M.; Minakata, M. Linear and nonlinear optical properties of Ag-As-Se chalcogenide glasses for all-optical switching. *Opt. Lett.* **2004**, *29*, 265–267. [CrossRef] [PubMed]
10. Chow, K.K.; Kikuchi, K.; Nagashima, T.; Hasegawa, T.; Ohara, S.; Sugimoto, N. Four-wave mixing based widely tunable wavelength conversion using 1-m dispersion-shifted bismuth-oxide photonic crystal fiber. *Opt. Express* **2007**, *15*, 15418–15423. [CrossRef] [PubMed]
11. Madden, S.J.; Choi, D.-Y.; Bulla, D.A.; Rode, A.V.; Luther-Davies, B.; Ta'eed, V.G.; Pelusi, M.D.; Eggleton, B.J. Long, low loss etched As_2S_3 chalcogenide waveguides for all-optical signal regeneration. *Opt. Express* **2007**, *15*, 14414–14421. [CrossRef] [PubMed]

12. Way, B.; Jain, R.K.; Hossein-Zadeh, M. High-Q microresonators for mid-IR light sources and molecular sensors. *Opt. Lett.* **2012**, *37*, 4389–4391. [CrossRef] [PubMed]

13. Kishi, T.; Kumagai, T.; Shibuya, S.; Prudenzano, F.; Yano, T.; Shibata, S. Quasi-single mode laser output from a terrace structure added on a Nd^{3+}-doped tellurite-glass microsphere prepared using localized laser heating. *Opt. Express* **2015**, *23*, 20629–20635. [CrossRef] [PubMed]

14. Jha, A.; Richards, B.; Jose, G.; Teddy-Fernandez, T.; Joshi, P.; Jiang, X.; Lousteau, J. Rare-earth ion doped TeO_2 and GeO_2 glasses as laser materials. *Prog. Mater. Sci.* **2012**, *57*, 1426–1491. [CrossRef]

15. Messias, D.N.; Vermelho, M.V.D.; de Araujo, M.T.; Gouveia-Neto, A.S.; Aitchison, J.S. Blue energy upconversion emission in thulium-doped-SiO_2-P_2/O_5 channel waveguides excited at 1.064 µm. *IEEE J. Quantum Electron.* **2002**, *38*, 1647–1650. [CrossRef]

16. Rai, V.K.; Menezes, L.D.S.; De Araújo, C.B. Spectroscopy, energy transfer, and frequency upconversion in Tm^{3+}-doped TeO_2-PbO glass. *J. Appl. Phys.* **2007**, *102*, 043505. [CrossRef]

17. Simpson, D.A.; Gibbs, W.E.; Collins, S.F.; Blanc, W.; Dussardier, B.; Monnom, G.; Peterka, P.; Baxter, G.W. Visible and near infra-red up-conversion inTm^{3+}/Yb^{3+} co-doped silica fibers under 980 nm excitation. *Opt. Express* **2008**, *16*, 13781–13799. [CrossRef] [PubMed]

18. Palma, G.; Falconi, M.C.; Starecki, F.; Nazabal, V.; Ari, J.; Bodiou, L.; Charrier, J.; Dumeige, Y.; Baudet, E.; Prudenzano, F. Design of praseodymium-doped chalcogenide micro-disk emitting at 4.7 µm. *Opt. Express* **2017**, *25*, 7014–7030. [CrossRef] [PubMed]

19. Broaddus, D.H.; Foster, M.A.; Agha, I.H.; Robinson, J.T.; Lipson, M.; Gaeta, A.L. Silicon-waveguide-coupled high-Q chalcogenide microspheres. *Opt. Express* **2009**, *17*, 5998–6003. [CrossRef] [PubMed]

20. Eggleton, B.J.; Luther-Davies, B.; Richardson, K. Chalcogenide photonics. *Nat. Photonics* **2011**, *5*, 141–148. [CrossRef]

21. Li, C.R.; Dai, S.X.; Zhang, Q.Y.; Shen, X.; Wang, X.S.; Zhang, P.Q.; Lu, L.W.; Wu, Y.H.; Lv, S.Q. Low threshold fiber taper coupled rare earth ion-doped chalcogenide microsphere laser. *Chin. Phys. B* **2015**, *24*. [CrossRef]

22. Elliott, G.R.; Murugan, G.S.; Wilkinson, J.S.; Zervas, M.N.; Hewak, D.W. Chalcogenide glass microsphere laser. *Opt. Express* **2010**, *18*, 26720–26727. [CrossRef] [PubMed]

23. Yu, Y.; Liu, Z.; Dai, N.; Sheng, Y.; Luan, H.; Peng, J.; Jiang, Z.; Li, H.; Li, J.; Yang, L. Ce-Tb-Mn co-doped white light emitting glasses suitable for long-wavelength UV excitation. *Opt. Express* **2011**, *19*, 19473–19479. [CrossRef] [PubMed]

24. Richards, B.S. Enhancing the performance of silicon solar cells via the application of passive luminescence conversion layers. *Sol. Energy Mater. Sol. Cells* **2006**, *90*, 2329–2337. [CrossRef]

25. Ristić, D.; Berneschi, S.; Camerini, M.; Farnesi, D.; Pelli, S.; Trono, C.; Chiappini, A.; Chiasera, A.; Ferrari, M.; Lukowiak, A.; et al. Photoluminescence and lasing in whispering gallery mode glass microspherical resonators. *J. Lumin.* **2016**, *170*, 755–760. [CrossRef]

26. Sandoghdar, V.; Treussart, F.; Hare, J.; Lefèvre-Seguin, V.; Raimond, J.M.; Haroche, S. Very low threshold whispering-gallery-mode microsphere laser. *Phys. Rev. A At. Mol. Opt. Phys.* **1996**, *54*, R1777–R1780. [CrossRef]

27. Cai, M.; Painter, O.; Vahala, K.J.; Sercel, P.C. Fiber-coupled microsphere laser. *Opt. Lett.* **2000**, *25*, 1430–1432. [CrossRef] [PubMed]

28. Chen, S.Y.; Sun, T.; Grattan, K.T.V.; Annapurna, K.; Sen, R. Characteristics of Er and Er-Yb-Cr doped phosphate microsphere fibre lasers. *Opt. Commun.* **2009**, *282*, 3765–3769. [CrossRef]

29. Dong, C.H.; Yang, Y.; Shen, Y.L.; Zou, C.L.; Sun, F.W.; Ming, H.; Guo, G.C.; Han, Z.F. Observation of microlaser with Er-doped phosphate glass coated microsphere pumped by 780 nm. *Opt. Commun.* **2010**, *283*, 5117–5120. [CrossRef]

30. Wu, T.; Huang, Y.; Huang, J.; Huang, Y.; Zhang, P.; Ma, J. Laser oscillation of Yb^{3+}:Er^{3+} co-doped phosphosilicate microsphere [Invited]. *Appl. Opt.* **2014**, *53*, 4747–4751. [CrossRef] [PubMed]

31. Dong, C.H.; Xiao, Y.F.; Han, Z.F.; Guo, G.C.; Jiang, X.; Tong, L.; Gu, C.; Ming, H. Low-threshold microlaser in Er: Yb phosphate glass coated microsphere. *IEEE Photonics Technol. Lett.* **2008**, *20*, 342–344. [CrossRef]

32. Hu, Y.; Jiang, S.; Luo, T.; Seneschal, K.; Morrell, M.; Smektala, F.; Honkanen, S.; Lucas, J.; Peyghambarian, N. Performance of high-concentration Er^{3+}/-Yb^{3+}/-codoped phosphate fiber amplifiers. *IEEE Photonics Technol. Lett.* **2001**, *13*, 657–659. [CrossRef]

33. Polynkin, P.; Temyanko, V.; Mansuripur, M.; Peyghambarian, N. Efficient and scalable side pumping scheme for short high-power optical fiber lasers and amplifiers. *IEEE Photonics Technol. Lett.* **2004**, *16*, 2024–2026. [CrossRef]

34. Karabulut, M.; Melnik, E.; Stefan, R.; Marasinghe, G.; Ray, C.; Kurkjian, C.; Day, D. Mechanical and structural properties of phosphate glasses. *J. Non Cryst. Solids* **2001**, *288*, 8–17. [CrossRef]

35. Bayya, S.S.; Chin, G.D.; Sanghera, J.S.; Aggarwal, I.D. Germanate glass as a window for high energy laser systems. *Opt. Express* **2006**, *14*, 11687–11693. [CrossRef] [PubMed]

36. Jewell, J.M.; Aggarwal, I.D. Structural influences on the hydroxyl spectra of barium gallogermanate glasses. *J. Non Cryst. Solids* **1995**, *181*, 189–199. [CrossRef]

37. Fang, Z.; Nic Chormaic, S.; Wang, S.; Wang, X.; Yu, J.; Jiang, Y.; Qiu, J.; Wang, P. Bismuth-doped glass microsphere lasers. *Photonics Res.* **2017**, *5*, 740–744. [CrossRef]

38. Stanworth, J.E. Tellurite Glasses. *Nature* **1952**, *169*, 581–582. [CrossRef]

39. Wang, J.S.; Machewirth, D.P.; Wu, F.; Snitzer, E.; Vogel, E.M. Neodymium-doped tellurite single-mode fiber laser. *Opt. Lett.* **1994**, *19*, 1448–1449. [CrossRef] [PubMed]

40. Mori, A.; Ohishi, Y.; Sudo, S. Erbium-doped tellurite glass fibre laser and amplifier. *Electron. Lett. Online* **1997**, *33*, 863–864. [CrossRef]

41. Ohishi, Y.; Mori, A.; Yamada, M.; Ono, H.; Nishida, Y.; Oikawa, K. Gain characteristics of tellurite-based erbium-doped fiber amplifiers for 15-μm broadband amplification. *Opt. Lett.* **1998**, *23*, 274–276. [CrossRef] [PubMed]

42. Lee, J.Y.; Hwang, S.N. A high-gain boost converter using voltage-stacking cell. *Trans. Korean Inst. Electr. Eng.* **2008**, *57*, 982–984. [CrossRef]

43. Pelusi, M.D.; Ta'eed, V.G.; Fu, L.; Mägi, E.; Lamont, M.R.E.; Madden, S.; Choi, D.Y.; Bulla, D.A.P.; Luther-Davies, B.; Eggleton, B.J. Applications of highly-nonlinear chalcogenide glass devices tailored for high-speed all-optical signal processing. *IEEE J. Sel. Top. Quantum Electron.* **2008**, *14*, 529–539. [CrossRef]

44. Stepien, R.; Cimek, J.; Pysz, D.; Kujawa, I.; Klimczak, M.; Buczynski, R. Soft glasses for photonic crystal fibers and microstructured optical components. *Opt. Eng.* **2014**, *53*, 071815. [CrossRef]

45. Tang, G.; Zhu, T.; Liu, W.; Lin, W.; Qiao, T.; Sun, M.; Chen, D.; Qian, Q.; Yang, Z. Tm^{3+} doped lead silicate glass single mode fibers for 20 μm laser applications. *Opt. Mater. Express* **2016**, *6*, 2147–2157. [CrossRef]

46. Frerichs, R. New Optical Glasses with Good Transparency in the Infrared. *J. Opt. Soc. Am.* **1953**, *43*, 1153–1157. [CrossRef]

47. Sanghera, J.S.; Shaw, L.B.; Aggarwal, I.D. Chalcogenide glass-fiber-based mid-IR sources and applications. *IEEE J. Sel. Top. Quantum Electron.* **2009**, *15*, 114–119. [CrossRef]

48. Zhang, X.H.; Guimond, Y.; Bellec, Y. Production of complex chalcogenide glass optics by molding for thermal imaging. *J. Non Cryst. Solids* **2003**, *326–327*, 519–523. [CrossRef]

49. Wang, C.C. Empirical relation between the linear and the third-order nonlinear optical susceptibilities. *Phys. Rev. B* **1970**, *2*, 2045–2048. [CrossRef]

50. Harbold, J.M.; Ilday, F.Ö.; Wise, F.W.; Aitken, B.G. Highly Nonlinear Ge–As–Se and Ge–As–S–Se Glasses for All-Optical Switching. *IEEE Photonics Technol. Lett.* **2002**, *14*, 822–824. [CrossRef]

51. Quemard, C.; Smektala, F.; Couderc, V.; Barthelemy, A.; Lucas, J. Chalcogenide glasses with high non linear optical properties for telecommunications. *J. Phys. Chem. Solids* **2001**, *62*, 1435–1440. [CrossRef]

52. Kosa, T.I.; Wagner, T.; Ewen, P.J.S.; Owen, A.E. Optical properties of silver-doped $As_{33}S_{67}$. *Int. J. Electron.* **1994**, *76*, 845–848. [CrossRef]

53. Shimakawa, K.; Kolobov, A.; Elliott, S.R. Photoinduced effects and metastability in amorphous semiconductors and insulators. *Adv. Phys.* **1995**, *44*, 475–588. [CrossRef]

54. Miyashita, T.; Terunuma, Y. Optical Transmission Loss of As–S Glass Fiber in 1.0–5.5 μm Wavelength Region. *Jpn. J. Appl. Phys.* **1982**, *21*, L75. [CrossRef]

55. Susman, S.; Rowland, S.C.; Volin, K.J. The purification of elemental sulfur.pdf. *J. Mater. Res.* **2018**, *7*, 1526–1533. [CrossRef]

56. King, W.A.; Clare, A.G.; LaCourse, W.C. Laboratory preparation of highly pure As_2Se_3 glass. *J. Non Cryst. Solids* **1995**, *181*, 231–237. [CrossRef]

57. Hilton, A.R.; Hayes, D.J.; Rechtin, M.D. Infrared absorption of some high-purity chalcogenide glasses. *J. Non Cryst. Solids* **1975**, *17*, 319–338. [CrossRef]

58. Adamchik, S.A.; Malyshev, A.Y.; Bulanov, A.D.; Bab'eva, E.N. Fine purification of sulfur from carbon by high temperature oxidation. *Inorg. Mater.* **2001**, *37*, 469–472. [CrossRef]

59. Ležal, D.; Pedlíková, J.; Gurovič, J.; Vogt, R. The preparation of chalcogenide glasses in chlorine reactive atmosphere. *Ceram. Silik.* **1996**, *40*, 55–59.

60. Savage, J.A. Optical properties of chalcogenide glasses. *J. Non Cryst. Solids* **1982**, *47*, 101–115. [CrossRef]

61. Moynihan, C.T.; Macedo, P.B.; Maklad, M.S.; Mohr, R.K.; Howard, R.E. Intrinsic and impurity infrared absorption in As$_2$Se$_3$ glass. *J. Non Cryst. Solids* **1975**, *17*, 369–385. [CrossRef]

62. Bica, L. Formation of glass microspheres with rotating electrical arc. *Mater. Sci. Eng. B Solid-State Mater. Adv. Technol.* **2000**, *77*, 210–212. [CrossRef]

63. Ferreira, M.S.; Santos, J.L.; Frazão, O. New silica microspheres array sensor. *Proc. SPIE* **2014**, *39*, 91571P. [CrossRef]

64. Wang, P.; Murugan, G.S.; Lee, T.; Feng, X.; Semenova, Y.; Wu, Q.; Loh, W.; Brambilla, G.; Wilkinson, J.S.; Farrell, G. Lead silicate glass microsphere resonators with absorption-limited Q. *Appl. Phys. Lett.* **2011**, *98*. [CrossRef]

65. Birks, T.A.; Li, Y.W. The Shape of Fiber Tapers. *J. Lightwave Technol.* **1992**, *10*, 432–438. [CrossRef]

66. Wang, P.F.; Lee, T.; Ding, M.; Dhar, A.; Hawkins, T.; Foy, P.; Semenova, Y.; Wu, Q.; Sahu, J.; Farrell, G.; et al. Germanium microsphere high-Q resonator. *Opt. Lett.* **2012**, *37*, 728–730. [CrossRef] [PubMed]

67. Ballato, J.; Hawkins, T.; Foy, P.; Yazgan-Kokuoz, B.; Stolen, R.; McMillen, C.; Hon, N.K.; Jalali, B.; Rice, R. Glass-clad single-crystal germanium optical fiber. *Opt. Express* **2009**, *17*, 8029–8035. [CrossRef] [PubMed]

68. Grillet, C.; Bian, S.N.; Magi, E.C.; Eggleton, B.J. Fiber taper coupling to chalcogenide microsphere modes. *Appl. Phys. Lett.* **2008**, *92*, 171109. [CrossRef]

69. Elliott, G.R.; Hewak, D.W.; Murugan, G.S.; Wilkinson, J.S. Chalcogenide glass microspheres; their production, characterization and potential. *Opt. Express* **2007**, *15*, 17542–17553. [CrossRef] [PubMed]

70. Wang, P.; Murugan, G.S.; Brambilla, G.; Ding, M.; Semenova, Y.; Wu, Q.; Farrell, G. Chalcogenide microsphere fabricated from fiber tapers using contact with a high-temperature ceramic surface. *IEEE Photonics Technol. Lett.* **2012**, *24*, 1103–1105. [CrossRef]

71. Yang, Z.; Wu, Y.; Zhang, X.; Zhang, W.; Xu, P.; Dai, S. Low temperature fabrication of chalcogenide microsphere resonators for thermal sensing. *IEEE Photonics Technol. Lett.* **2017**, *29*, 66–69. [CrossRef]

72. Lissillour, F.; Ameur, K.A.; Dubreuil, N.; Stéphan, G.; Kérampont, D.; Photoniques, M. Whispering-gallery mode Nd-ZBLAN microlasers at 1:05 p.m. In Proceedings of the SPIE 3416, Infrared Glass Optical Fibers and Their Applications, Quebec, QC, Canada, 28 September 1998; pp. 150–156. [CrossRef]

73. Ward, J.M.; Wu, Y.; Khalfi, K.; Chormaic, S.N. Short vertical tube furnace for the fabrication of doped glass microsphere lasers. *Rev. Sci. Instrum.* **2010**, *81*, 073106. [CrossRef] [PubMed]

74. Wang, P.; Murugan, G.S.; Lee, T.; Ding, M.; Brambilla, G.; Semenova, Y.; Wu, Q.; Koizumi, F.; Farrell, G. High-Q bismuth-silicate nonlinear glass microsphere resonators. *IEEE Photonics J.* **2012**, *4*, 1013–1020. [CrossRef]

75. Baaske, M.D.; Foreman, M.R.; Vollmer, F. Single-molecule nucleic acid interactions monitored on a label-free microcavity biosensor platform. *Nat. Nanotechnol.* **2014**, *9*, 933–939. [CrossRef] [PubMed]

76. Fürst, J.U.; Strekalov, D.V.; Elser, D.; Lassen, M.; Andersen, U.L.; Marquardt, C.; Leuchs, G. Naturally phase-matched second-harmonic generation in a whispering-gallery-mode resonator. *Phys. Rev. Lett.* **2010**, *104*, 153901. [CrossRef] [PubMed]

77. Lin, G.; Fürst, J.U.; Strekalov, D.V.; Yu, N. Wide-range cyclic phase matching and second harmonic generation in whispering gallery resonators. *Appl. Phys. Lett.* **2013**, *103*, 181107. [CrossRef]

78. Weng, W.; Anstie, J.D.; Stace, T.M.; Campbell, G.; Baynes, F.N.; Luiten, A.N. Nano-Kelvin thermometry and temperature control: Beyond the thermal noise limit. *Phys. Rev. Lett.* **2014**, *112*, 160801. [CrossRef] [PubMed]

79. Yang, L.; Vahala, K.J. Gain functionalization of silica microresonators. *Opt. Lett.* **2003**, *28*, 592–594. [CrossRef] [PubMed]

80. Song, F. Er^{3+}-doped tellurite glass microsphere laser: Optical properties, coupling scheme, and lasing characteristics. *Opt. Eng.* **2005**, *44*, 034202. [CrossRef]

81. Yang, Z.; Wu, Y.; Yang, K.; Xu, P.; Zhang, W.; Dai, S.; Xu, T. Fabrication and characterization of Tm^{3+}-Ho^{3+} co-doped tellurite glass microsphere lasers operating at $\sim$2.1 μm. *Opt. Mater. (Amst.)* **2017**, *72*, 524–528. [CrossRef]

82. Guo, C.; Che, K.; Zhang, P.; Wu, J.; Huang, Y.; Xu, H.; Cai, Z. Low-threshold stimulated Brillouin scattering in high-Q whispering gallery mode tellurite microspheres. *Opt. Express* **2015**, *23*, 32261–32266. [CrossRef] [PubMed]

micromachines

MDPI

Article

Long Period Grating-Based Fiber Coupling to WGM Microresonators

Francesco Chiavaioli [1,*,†], Dario Laneve [2,†], Daniele Farnesi [1], Mario Christian Falconi [2], Gualtiero Nunzi Conti [1], Francesco Baldini [1] and Francesco Prudenzano [2]

1 Institute of Applied Physics "Nello Carrara" (IFAC), National Research Council of Italy (CNR), Via Madonna del Piano 10, Sesto Fiorentino, 50019 Firenze, Italy; d.farnesi@ifac.cnr.it (D.F.); g.nunziconti@ifac.cnr.it (G.N.C.); f.baldini@ifac.cnr.it (F.B.)
2 Department of Electrical and Information Engineering, Polytechnic University of Bari, 70125 Bari, Italy; dario.laneve@poliba.it (D.L.); mariochristian.falconi@poliba.it (M.C.F.); francesco.prudenzano@poliba.it (F.P.)
* Correspondence: f.chiavaioli@ifac.cnr.it; Tel.: +39-055-522-6318
† The authors contributed equally to this work.

Received: 29 June 2018; Accepted: 20 July 2018; Published: 23 July 2018

Abstract: A comprehensive model for designing robust all-in-fiber microresonator-based optical sensing setups is illustrated. The investigated all-in-fiber setups allow light to selectively excite high-Q whispering gallery modes (WGMs) into optical microresonators, thanks to a pair of identical long period gratings (LPGs) written in the same optical fiber. Microspheres and microbubbles are used as microresonators and evanescently side-coupled to a thick fiber taper, with a waist diameter of about 18 μm, in between the two LPGs. The model is validated by comparing the simulated results with the experimental data. A good agreement between the simulated and experimental results is obtained. The model is general and by exploiting the refractive index and/or absorption characteristics at suitable wavelengths, the sensing of several substances or pollutants can be predicted.

Keywords: microresonator; whispering gallery mode; long period grating; fiber coupling; distributed sensing; chemical/biological sensing

---

1. Introduction

In recent years, whispering gallery mode (WGM) microresonators, such as microdisks [1], microbubbles [2] and microspheres [3], have gained much interest among researchers thanks to their capability to strongly confine the light in very compact volumes. In fact, during the several revolutions of the light signal in these resonators, the WGM field evanescently couples to the surrounding environment and even a very small change in the microresonator size and/or in the refractive index can induce significant changes in the quality factor (Q-factor) and/or resonance wavelengths of the microresonator. In view of this, a number of sensing applications, involving WGM microresonators, are reported in literature, such as the sensing of local temperature [2–6], refractive index [7], pressure [8], biological [9] and spectroscopic parameters [10]. Their huge potentiality in biosensing by means of label-free detection down to single molecules was also proved [11]. Moreover, by doping the microresonators with rare-earths, integrated light sources with narrow line emission can be obtained [12].

The combination of the peculiarities of optical fibers with WGM microresonators can provide great opportunities in the field of sensing especially. Thanks to the use of a fiber taper, high-Q WGM resonances in different types of microresonators can be efficiently excited. The taper allows for obtaining a proper evanescent electromagnetic field, which can be coupled into the microresonators, and more than 90% of the fiber mode power can be transferred on the microresonators [13]. This paves the way for several scenarios in telecommunications requiring the generation of narrow line emission. One of these is the possibility to develop an all-in-fiber coupling system for quasi-distributed and wavelength selective addressing of different WGM microresonators located along the same optical link. The first fiber-based setup for efficient coupling of light to a high-Q WGM microresonator is illustrated in [14]. It is based on a long period grating (LPG) followed by a thick fiber taper, both of which are derived from the same fiber. The LPG allows wavelength selective excitation of high-order cladding modes; in this way, thicker and more robust tapers (with waist diameters larger than 15 μm, easier to fabricate than the usual 1–2 μm tapers) can be used for coupling the cladding modes to the WGMs. However, the previous configuration does not allow interrogating more than one microresonator if the transmitted light is used to carry on the signal under investigation. The only chance could be the use of the scattered light from each microresonator, resulting in more difficult and time-consuming setup implementation. Therefore, to overcome these limitations, an improvement of the previous coupling system was recently demonstrated [15]: the system is now constituted by a pair of identical LPGs with a tapered fiber in between. The existence of the second LPG allows the light coupling back into the fiber core. Hence, all the information within the core mode is transmitted up to the end of the fiber segment and it can be collected by a single photodetector. It is important to underline that the pair of identical LPGs can operate in different wavelength bands within the range of the detector, thus allowing multiple selective coupling of spatially distributed or quasi-distributed WGM microresonators by means of different wavelengths [15].

The design of the coupling system previously mentioned requires an exhaustive model, especially for developing sensing applications [14,16]. In this work, two different sensing setups, similar to those described in [15], are considered. The two setups consist of either a microsphere or a microbubble coupled to a tapered fiber. An analytical model for simulating the two setups is detailed. The proposed approach is complete and it is well validated with the experimental results reported in literature [15].

2. Overview of the All-in-Fiber Coupling System

Figure 1 shows a sketch of the coupling system used in this work. The system is constituted by an optical WGM microresonator coupled to a tapered fiber. On both sides of the taper, there are the two identical LPGs. The first LPG allows the coupling between the fundamental core mode and a specific cladding mode depending on the grating parameters. Then, the evanescent field of this cladding mode excites the WGMs in the microresonator. Finally, the light is coupled back from the foregoing cladding mode into the fiber fundamental mode via the second LPG.

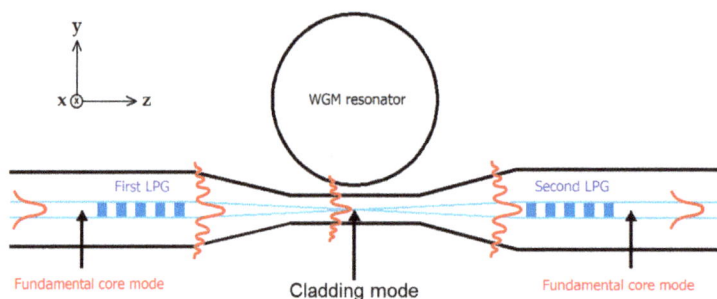

Figure 1. Sketch of the coupling principle of an optical whispering gallery mode (WGM) microresonator by means of a tapered fiber in between two identical long period gratings (LPGs).

In this work, two different types of WGM microresonator are considered: microspheres and microbubbles. The cross-sections of the microresonators, together with a sketch of the tapered fiber waist, are shown in Figure 2. Figure 2a refers to the microsphere-based coupling system, whereas Figure 2b refers to the microbubble-based coupling system. The gap between each microresonator and the taper waist is *g*.

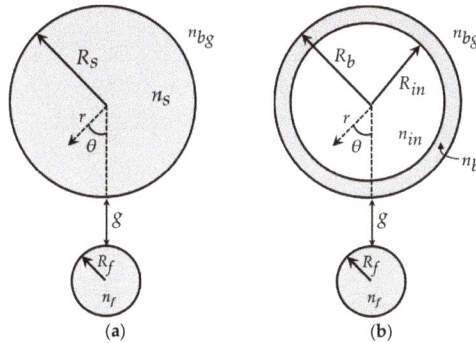

Figure 2. Geometrical configurations of the simulated coupling systems consisting of a tapered fiber coupled to (**a**) a microsphere and (**b**) a microbubble. The cross-section of the tapered fiber waist is shown. The light grey color represents the silica glass.

3. Materials and Methods

3.1. Manufacturing of WGM Microresonators and Optical Fiber LPGs

The model is realistic and refers to two experimental setups described in [15]. Adiabatic tapered fibers were fabricated using heating and pulling of a commercially available boron-germanium co-doped single-mode optical fiber (Fibercore PS1250/1500, Fibercore Ltd., Surrey, UK) [17]. In particular, the fiber core and cladding diameters are 6.9 µm and 124.6 µm, respectively. Tapers with a diameter among 15–18 µm are manufactured to guide/handle the cladding modes of interest. It is worth pointing out that, for an efficient coupling of these cladding modes to WGM microresonators, a partial tapering of the optical fiber is essential to shrink the optical field size and to increase the evanescent field [14]. However, the average diameter of the manufactured tapered fibers is one order of magnitude thicker than that of standard fiber tapers (1–2 µm). This allows for improving the robustness and a decreased fragility of the coupling structure in practical applications.

The coupling system depicted in Figure 1 is completed by manufacturing the pair of identical LPGs on both ends of the tapered fiber. A point-to-point technique employing a KrF excimer laser (Lambda Physic COMPex 110, Lambda Physik AG, Goettingen, Germany) is used to inscribe the gratings [18]. Two different pairs of LPGs are manufactured with a grating period, Λ, of 340 µm and 365 µm, respectively, and a grating length, *L*, of 18.7 mm and 20.1 mm, respectively (55 grating planes).

The model includes two different types of microresonators: silica microspheres and microbubbles, manufactured as in [17,19]. The diameters of these resonator range from 260 µm to 290 µm, for microspheres, and from 380 µm to 500 µm, for microbubbles. In both cases, the microresonator size is large enough, thus the free spectral range (FSR) can be considered significantly smaller than the bandwidth of the LPGs [14].

The experimental setup we used for the monitoring and registration of the transmission spectra consists of two fiber pigtailed tunable external cavity lasers (Anritsu Tunics Plus, linewidth 300 KHz, Anritsu Corporation, Kanagawa Prefecture, Japan), covering the spectral range from 1390 nm to 1640 nm and of an optical spectrum analyzer (OSA–Ando AQ6317B, Yokogawa Test & Measurement Corporation, Tokyo, Japan) for detecting the signals. The coupling mechanism is finally tested by using the same laser sources, which can be finely and continuously swept in the spectral range within few GHz, and a single photodetector connected to a commercially available oscilloscope.

3.2. Theoretical Analysis

In Figure 2, the cross-sections of two different kinds of optical spherical microresonators, coupled to a tapered fiber, are shown. The developed analytical model found the electromagnetic (e.m.) fields of the microresonators and the fiber by solving the Helmholtz equation in spherical and cylindrical coordinates, respectively. In particular, the e.m. fields of the microresonators are described by the well-known WGM theory [20]. Then, the coupled-mode theory allows for modeling the optical interaction between the calculated microresonator modes and fiber modes [21].

3.2.1. Analytical Model of a Dielectric Microsphere

The e.m. fields of a dielectric microsphere is found by solving the following Helmholtz equation in spherical coordinates:

$$\nabla^2 \Psi(r,\theta,\phi) + k^2 n_s \Psi(r,\theta,\phi) = 0 \tag{1}$$

where $k = \omega \sqrt{\mu_0 \varepsilon_0}$ is the wave vector in vacuum, n_s is the refractive index of the microsphere (Figure 2a) and $\Psi(r,\theta,\phi)$ is the electric or magnetic field component. If the polarization of the e.m. fields is supposed to be constant throughout all points in space, the solutions of Equation (1) can be expressed in the following form [20]:

$$\Psi(r,\theta,\phi) = N_s \psi_r(r) \psi_\theta(\theta) \psi_\phi(\phi) \tag{2}$$

where N_s is a normalization factor calculated by assuming equal to 1 the integral of $|\Psi|^2$, over all space, divided by $2\pi R_s$, R_s is the microsphere radius (Figure 2a), and ψ_r, ψ_θ and ψ_ϕ are the radial, polar and azimuthal contributions of the field, respectively [20]. The separation of variables in Equation (2) allows for dividing the microsphere modes in transverse electric (TE) and transverse magnetic (TM) modes. The TE modes are characterized by having the electric field parallel to the microsphere surface, i.e., $\Psi \equiv E_\theta$, $E_r = E_\phi = 0$. The TM modes are characterized by having the electric field perpendicular to the microsphere surface, i.e., $\Psi \equiv H_\theta$, $H_r = H_\phi = 0$ [20]. The other field components (H_r, H_ϕ for TE modes; E_r, E_ϕ for TM modes) are determined by the boundary conditions at the interface between the microsphere and the surrounding background medium (see Figure 2a). Finally, the substitution of Equation (2) in Equation (1) gives the well-known WGM field contributions [20]:

$$\psi_\phi(\phi) = e^{\pm jm\phi} \tag{3}$$

$$\psi_\theta(\theta) = e^{-\frac{m}{2}\theta^2} H_n(\sqrt{m}\theta), \quad m \gg 1 \gg \theta \tag{4}$$

$$\psi_r(r) = \begin{cases} A\, j_l(kn_s r), & r \le R_s \\ B e^{-\alpha_s(r - r_s)}, & r > R_s \end{cases} \tag{5}$$

where H_N is the Hermite polynomial of order $N = l - m$, l and m are the mode numbers, j_l is the spherical Bessel function of the first kind of order l, A and B are two constants evaluated by imposing the boundary conditions at the microsphere surface, $\alpha_s = (\beta - k^2 n_{bg}^2)^{1/2}$ is the exponential decay constant of the evanescent field in the background medium with refractive index $n_{bg} < n_s$ and β_l is the propagation constant of the mode parallel to the microsphere surface [20]. The microsphere modes, $WGM_{l,m,n}$, are uniquely described by the three integers l, m and n. The value m is the number of field maxima along the ϕ-direction; the value $l - m + 1$ is the number of field maxima along the θ-direction; the value n is the number of absolute field maxima along the r-direction.

By matching the tangential field components of TE and TM modes at $r = R_s$, a homogeneous linear equation system of the form $\mathbf{M}x = 0$ is obtained, where $x = [A,B]^T$ and

$$\mathbf{M} = \begin{bmatrix} j_l(kn_s r_s) & -1 \\ kn_s j_l'(kn_s R_s) & \chi\alpha_s \end{bmatrix} \tag{6}$$

where $\chi = 1$, for TE modes, and $\chi = n_s^2/n_{bg}^2$, for TM modes. The first row of the matrix M is obtained by matching either E_θ, for TE modes, or H_θ, for TM modes, at $r = R_s$. The second row is obtained by matching either H_ϕ, for TE modes (assuming $\partial E_\theta/\partial r \gg E_\theta/r$), or E_ϕ, for TM modes (assuming $\partial H_\theta/\partial r \gg H_\theta/r$) [20]. By imposing equal to zero the determinant of the matrix M, the microsphere characteristic equation is obtained:

$$\left(\frac{l}{r_s} + \chi \alpha_s\right) j_l(kn_s R_s) = kn_s j_{l+1}(kn_s R_s) \tag{7}$$

where the recursion formula, $j_l'(x) = lx^{-1}j_l(x) - j_{l+1}(x)$, has been used [20]. Equation (7) relates the resonant wavelengths of the WGMs to the mode numbers l and n.

3.2.2. Analytical Model of a Dielectric Microbubble

A microbubble can be seen as a hollow microsphere with a glass shell in which the WGMs propagate [19]. As illustrated in Figure 2b, the thickness of the glass shell is determined by the inner, R_{in}, and outer, R_b, radii of the microbubble. The refractive index of the glass shell is n_b, whereas the refractive index of the medium inside the microbubble is n_{in}. The e.m. analysis of a microbubble follows the same procedure of the previous Section 3.2.1. In this case, the radial field contribution, $\psi_r(r)$, takes a different form to account for the two separation interfaces:

$$\psi_r(r) = \begin{cases} A\,j_l(kn_{in}r), & r \leq R_{in} \\ Bj_l(kn_b r) + C\,y_l(kn_b r), & R_{in} < r \leq R_b \\ De^{-\alpha_s(r-R_b)}, & r > R_b \end{cases} \tag{8}$$

where y_l is the spherical Bessel function of the second kind of order l, and A, B, C and D are constants to be determined by applying the boundary conditions to the θ- and ϕ-polarized field components at $r = r_{in}$ and $r = R_b$. As described in the previous Section 3.2.1, the boundary conditions lead to the homogeneous linear equation system $\mathbf{M}x = 0$, where $x = [A,\ B,\ C,\ D]^T$ and

$$\mathbf{M} = \begin{bmatrix} j_l(kn_{in}R_{in}) & -j_l(kn_b R_{in}) & -y_l(kn_b R_{in}) & 0 \\ n_{in}j_l'(kn_{in}R_{in}) & -\chi_1 n_b j_l'(kn_b R_{in}) & -\chi_1 n_b y_l'(kn_b R_{in}) & 0 \\ 0 & j_l(kn_b R_b) & y_l(kn_b R_b) & -1 \\ 0 & kn_b j_l'(kn_b R_b) & kn_b y_l'(kn_b R_b) & \chi_2 \alpha_s \end{bmatrix} \tag{9}$$

where $\chi_1 = \chi_2 = 1$, for TE modes, and $\chi_1 = n_{in}^2/n_b^2$, $\chi_2 = n_b^2/n_{bg}^2$, for TM modes. The first two rows of the matrix M are obtained by matching either E_θ, H_ϕ, for TE modes, or H_θ, E_ϕ, for TM modes, at $r = R_{in}$. The third and fourth rows are obtained by matching the same tangential fields at $r = R_b$. The assumptions $\partial E_\theta/\partial r \gg E_\theta/r$ and $\partial H_\theta/\partial r \gg H_\theta/r$ hold in this case too. By imposing equal to zero the determinant of the matrix M in Equation (9), the characteristic equation for the microbubble is obtained.

3.2.3. Coupling Model

The coupled mode theory (CMT) [21] is applied to model the optical coupling between the microresonator WGM$_{l,m,n}$ modes and the tapered fiber LP$_{0,X}$ cladding modes, where the 0 and X subscripts represent the azimuthal and radial orders, respectively, of the linearly-polarized (LP) cladding modes. The analytical model takes into account the coupling of modes both in space and in time formulations [20,21].

The optical interaction between the fiber field, $\mathbf{F}_X$, and the microresonator field, $\mathbf{\Psi}_{l,m,n}$, is obtained and calculated from the following overlap integral [20]:

$$\kappa_{xy}(z) = \frac{k^2}{2\beta_f}(n_{eff}^2 - n_{bg}^2)\iint \mathbf{\Psi}_{l,m,n} \cdot \mathbf{F}_X^* \, dxdy \tag{10}$$

where β_f is the propagation constant of the LP mode and n_{eff} is the effective refractive index of the microresonator, which is related to the propagation constant of the WGM [20]. The integration in Equation (10) is carried out over the transverse xy-plane at a fixed point along the tapered fiber, whose longitudinal axis is directed along the z-axis (see Figure 1). Then, $\kappa_{xy}(z)$ is integrated along the z-axis over the interaction length, L, i.e., $\kappa = \int_L \kappa_{xy}(z)dz$. κ is the power coupling constant, whereas κ^2 is the fraction of the power transferred from the fiber to the microresonator over the interaction region [21]. It should be noted that κ_{xy} is proportional to $e^{-j\Delta\beta z}$, where $\Delta\beta = \beta_f - \beta_m$ is the phase mismatch between the LP mode and the WGM, whose propagation constant is either $\beta_m = m/R_s$, for a microsphere, or $\beta_m = m/R_b$, for a microbubble [20].

The power coupling constants κ is also related to the time evolution of the coupled modes. By considering the microresonator as a lumped oscillator of energy amplitude $a_{WGM}(t)$, the (weak) power coupling with the fiber induces a (slow) time variation of $a_{WGM}(t)$, which can be expressed by means of the following rate equation [21]:

$$\frac{da_{WGM}(t)}{dt} = \left(j\omega_{WGM} - \frac{1}{\tau_0} - \frac{2}{\tau_e}\right)a_{WGM}(t) - j\sqrt{\frac{2}{\tau_e\tau_r}}a_{in}(t) \tag{11}$$

where $\tau_0 = Q_0/\omega_{WGM}$ is the amplitude decay time-constant due to the intrinsic loss phenomena of the microresonator (including surface scattering and absorption and curvature losses), Q_0 is the intrinsic quality factor, ω_{WGM} is the WGM$_{l,m,n}$ resonant frequency, $\tau_e = Q_e/\omega$ is the decay time-constant related to the coupling with the fiber, $Q_e = m\pi/\kappa^2$ is the external quality factor, κ is the foregoing power coupling constant, ω is the input excitation frequency, $\tau_r = 2\pi R/v_g \cong 2\pi R n_{eff}/c$ is the revolution time of either the microsphere ($R = R_s$) or the microbubble ($R = R_b$) and $a_{in}(t)$ is the energy amplitude of the excitation signal at the taper input section [20,21].

The transfer characteristic of the coupling system is found by considering the steady state form of the Equation (11), thus obtaining a_{WGM}, and then applying the following power conservation rule [21]:

$$\frac{|a_{in}|^2}{\tau_r} = \frac{1}{\tau_r}\left|a_{in} - j\sqrt{\frac{2\tau_r}{\tau_e}}a_{WGM}\right|^2 + \frac{2}{\tau_e}|a_{WGM}|^2 \tag{12}$$

where the first term on the right-hand side is the non-resonant power transmitted directly to the fiber output section, while the second term is the resonant power coupled out of the microresonator [21]. Therefore, the transmittance of the system can be expressed as:

$$T = \frac{|a_{out}|^2}{|a_{in}|^2} \tag{13}$$

where a_{out} is the amplitude of the signal at the fiber output section. i.e., $|a_{out}|^2/\tau_r$ is equivalent to the right-hand side of Equation (12).

In the foregoing analytical model, effects of the LPGs in the transmittance calculation can be neglected, since the LPGs simply allow the selective fiber mode excitation. The calculation of the mode excitation strength is not significant and can be avoided since in the transmittance calculation the output light is normalized with respect to the input one (see Equation (13)).

4. Results and Discussion

This section consists of two parts: Section 4.1 presents the numerical results achieved by using the theoretical analysis described in Section 3.2, whereas Section 4.2 shows some experimental results related to possible distributed sensing with WGM microresonators.

4.1. Numerical Results

The e.m. fields and the propagation constants of the microresonator WGMs and the fiber LP modes simulated with the analytical model were successfully validated via a finite element method (FEM) commercial code. Moreover, the results simulated with the overall developed analytical model have been validated with the experiment reported in [15]. In the following, the simulated results are obtained via the analytical model.

The simulations are carried out by employing the experimental parameters detailed in Section 3.1. In particular, an adiabatic fiber taper with a radius $R_f = 9$ μm is considered. The simulated microsphere and microbubble are made of silica glass. The dispersion effect on the refractive index of silica is taken into account via a proper Sellmeier formula [22]. The microsphere radius is $R_s = 145$ μm, whereas the microbubble external and internal radii are $R_b = 200$ μm and $R_{in} = 196.7$ μm, respectively. The simulations are performed by considering air as the surrounding background medium. Moreover, the microbubble is considered empty. According to [15], the simulated wavelength range is centered on $\lambda_c = 1613.3$ nm.

To find the optimal value of the gap g, a number of simulations are performed by considering different gap values, $g = 0, 10, 100, 200, 500, 1000$ nm. For each value of g, the simulated transmittance T is compared with the experimental one. Both setups employing the microsphere and the microbubble are taken into account.

Figure 3a shows the transmittance T of the microsphere-based setup, calculated for three different $WGM_{l,m,n}$, as a function of the radial order X of the $LP_{0,X}$ cladding modes, considering a gap $g = 0$ nm (i.e., taper and microsphere in mechanical contact, as in [15]). The lowest transmittance dip, which corresponds to the highest coupling with the microsphere, can be attained for the $WGM_{774,774,3}$ by exciting with the fiber $LP_{0,5}$ cladding mode through the LPG. The resonance of the $WGM_{774,774,3}$, expressed in terms of the detuning $\Delta\omega$, is shown in Figure 3b. The simulated results are in good agreement with the experiment reported in [15]. In fact, by exciting the $WGM_{774,774,3}$, the simulated transmittance of the microsphere-based coupling system reaches a minimum of $T = 0.52$, while, in the experiment, the measured transmittance is about $T = 0.65$. The small discrepancy can be explained by considering that the actual total losses are higher. However, it is worth nothing that, in Figure 3a, the transmittance simulated for the $WGM_{774,774,3}$ coupled with the $LP_{0,7}$ is $T = 0.66$, practically coincident with the measured transmittance [15], for the same fiber modal order.

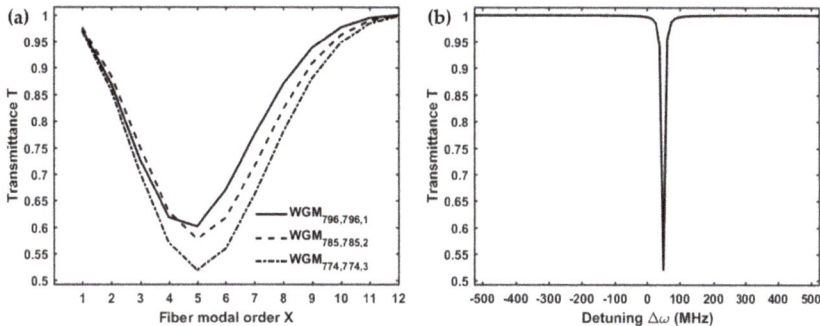

Figure 3. (a) Transmitted power of the microsphere-based setup, calculated for three resonant $WGM_{l,m,n}$, as a function of the modal order X of the $LP_{0,X}$ cladding modes, with a gap $g = 0$ nm; (b) transmission of the $WGM_{774,774,3}$, excited by the $LP_{0,5}$, as a function of the detuning $\Delta\omega$, with a gap $g = 0$ nm; microsphere-based setup.

Figure 4 shows the transmittance T of the microsphere-based setup, calculated for the gap values, $g = 0, 10, 100, 200, 500, 1000$ nm. For each value of g, the transmittance due to the $WGM_{774,774,3}$, which exhibits the lowest dip among the coupled WGMs, is plotted as a function of the radial order X of the fiber modes. In other words, the $WGM_{774,774,3}$ transmittance is predominant with respect to the contribution of the other WGMs. Except for large gap values ($g = 1000$ nm), the simulated minimum transmittance is almost the same in all cases. Instead, the radial order of the fiber modes slightly increases as the gap increases, revealing the influence of g on the phase matching between the WGMs and LP modes.

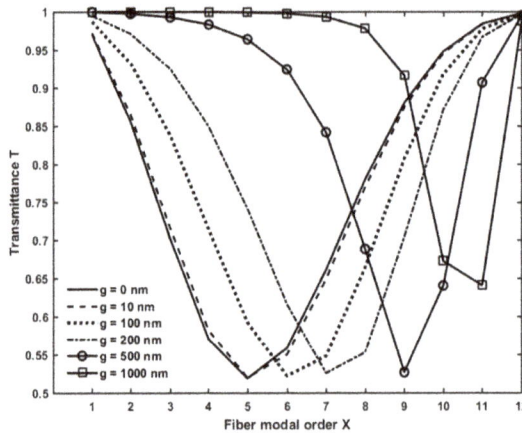

Figure 4. Transmitted power of the microsphere-based setup, calculated for different gap values, ranging from g = 0 nm to g = 1000 nm.

Figure 5a reports on the transmittance T of the microbubble-based setup, calculated for three different $WGM_{l,m,n}$, as a function of the radial order X of the $LP_{0,X}$ cladding modes, considering a gap $g = 0$ nm (i.e., taper and microbubble in mechanical contact, as in [15]). The lowest transmittance dip, which corresponds to the highest coupling with the microbubble, can be attained for the $WGM_{998,998,3}$ by exciting with the fiber $LP_{0,4}$ cladding mode through the LPG. The resonant detuning of the $WGM_{998,998,3}$, corresponding to the wavelength $\lambda = 1618.4$ nm, is shown in Figure 5b. The simulated transmittance of the microbubble-based coupling system reaches a minimum of $T = 0.51$, practically coincident with the measured value [15] even if referring to a slightly lower fiber modal order.

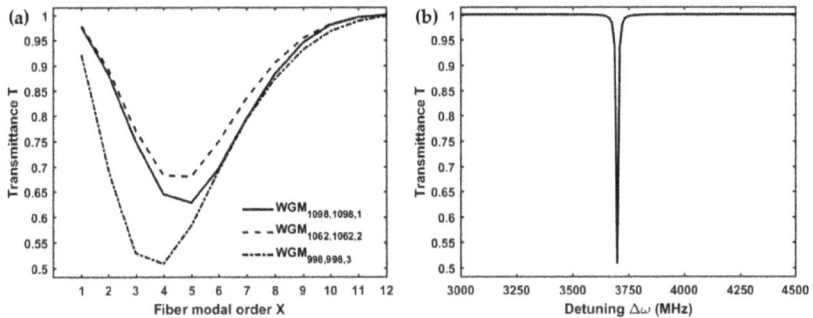

Figure 5. (a) Transmitted power of the microbubble-based setup, calculated for three resonant $WGM_{l,m,n}$, as a function of the modal order X of the $LP_{0,X}$ cladding modes, with a gap g = 0 nm; (b) transmission of the $WGM_{998,998,3}$, excited by the $LP_{0,4}$, as a function of the detuning $\Delta\omega$, with a gap g = 0 nm; microbubble-based setup.

Figure 6 shows the transmittance T of the microbubble-based setup, calculated for the gap values, g = 0, 10, 100, 200, 500, 1000 nm. For each value of g, the transmittance due to the WGM$_{998,998,3}$, which exhibits the lowest dip among the coupled WGMs, is plotted as a function of the radial order X of the fiber modes. In this case, the WGM$_{998,998,3}$ transmittance is predominant with respect to the contribution of the other WGMs. As in the microsphere case, except for large gap values (g = 1000 nm), the simulated minimum transmittance is almost the same in all cases, while the radial order of the fiber modes slightly increases as the gap increases. It is worthwhile noting that, in Figures 4 and 6, the critical coupling condition can be achieved by considering different fiber optic modal order.

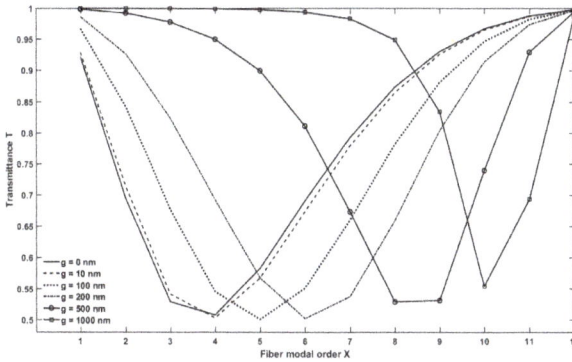

Figure 6. Transmitted power of the microbubble-based setup, calculated for different gap values, ranging from g = 0 nm to g = 1000 nm.

Figure 7 illustrates the normalized electric field of the microbubble WGM$_{998,998,3}$ evanescently coupled to the normalized electric field of the fiber LP$_{0,4}$ cladding mode. The four radial maxima of the LP$_{0,4}$ is evident as well as the third radial order of WGM$_{998,998,3}$.

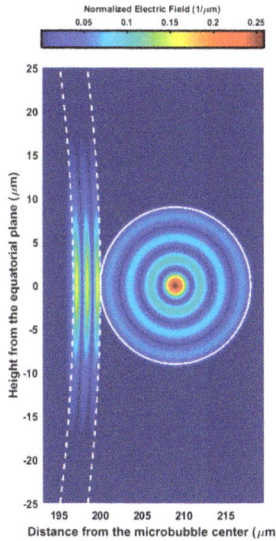

Figure 7. Distribution of the normalized electric fields confined in the microbubble glass layer (on the left) and in the fiber taper (on the right) for the WGM$_{998,998,3}$ coupled with the LP$_{0,4}$, gap g = 0 nm.

4.2. Experimental Results

Towards the development of an all-in-fiber distributed sensing system, the coupling system illustrated in Figure 1 has been tested and high-Q WGM resonances in both microspheres and microbubbles have been proved to be effectively excited [15]. The mechanical contact between the microresonators and the tapered fiber is provided in order to avoid any environmental perturbation. Moreover, the phase-matching conditions between the fiber cladding modes and the WGMs are satisfied due to the azimuthal and radial high-order modes of the spherical microresonators [14]. The transmission dips has been fitted by a Lorentzian function obtaining typical Q-factor values ranging from 10^6 up to 10^8, for both types of microresonators, with a maximum coupling efficiency (or resonance contrast) of about 50%–60%.

Afterwards, the coupling system in Figure 1 is doubled along the same fiber link as a proof-of-concept test. In fact, by a proper design of the LPGs, the coupling system in Figure 1 can be replicated as many times as the effective bandwidth of both source and detector allows. A total bandwidth allocation not less than 40 nm for each pair of LPGs should be taken into account [15]. Figure 8a accounts for the resonances achieved by scanning about 2 GHz around the LPG central wavelengths (1518.9 nm for the first coupling unit and 1613.3 nm for the second one), when the microresonators of the two coupling units are in mechanical contact with their respective fiber tapers (as sketched on the top of Figure 8a). The Q-factor values are comparable to those obtained with the analytical model. In order to prove that the microresonators of the two coupling units can be independently excited without cross-talk, a selectivity test has been performed. By alternatively de-coupling one of the microresonators and by looking at the transmission spectrum of the other one, it is possible to prove the selective excitation, as detailed in Figure 8b (the second microresonator is in contact, while the first is not in contact) and in Figure 8c (the first microresonator in contact, while the second not in contact). As further proof and evidence of this, additional measurements performed by varying the coupling position of the tapered fiber along the azimuthal axis of the microresonators confirm our findings. Therefore, the proposed all-in-fiber coupling system can be effectively used for distributed sensing.

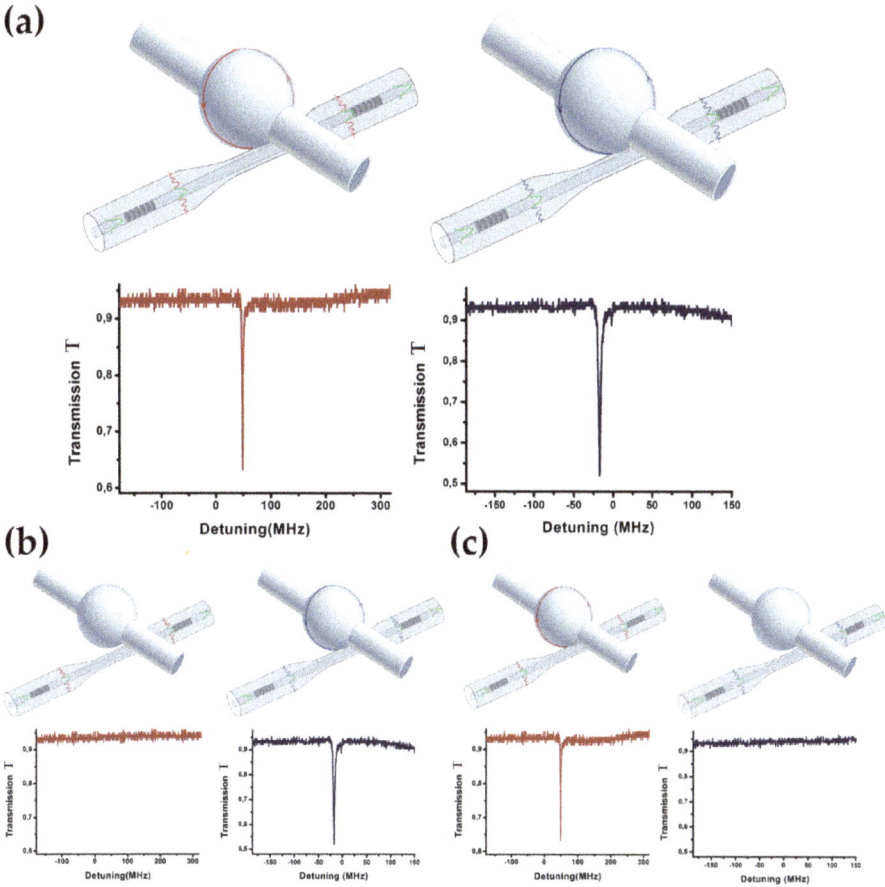

Figure 8. (**a**) Sketch and results of two in-series coupling systems with both the microresonators (microspheres in this case) coupled to each tapered fiber. The corresponding resonances achieved by scanning a laser source around the resonant wavelengths of the LPGs (0 MHz detuning) are also detailed below each sketch. The other two cases are shown to prove the zero cross-talk of the approach proposed; (**b**) the second microsphere is in contact, while the first is not; (**c**) the first microsphere is in contact, while the second is not.

5. Conclusions

A complete model for designing an all-in-fiber coupling system allowing the wavelength selective excitation of spatially distributed or quasi-distributed optical WGM microresonators has been developed. The microresonators are evanescently side-coupled with a fiber taper to increase the coupling efficiency. A pair of identical LPGs, with the microresonator and the tapered fiber in between, can be used to excite the WGMs by means of peculiar cladding modes. The pair of identical LPGs can operate in different wavelength bands allowing multiple selective interrogation of several microresonators along the same optical fiber. The model has been validated with experimental data by considering microsphere- and microbubble-based setups. The simulated results suggest that the coupling system can be effectively used for distributed sensing applications of chemical/biological fluids. The theoretical and experimental analysis and results could open up novel opportunities in the field of chemical/biological sensing [23]. In particular, the all-in-fiber coupling system could bring about a very promising optical platform for multiplexing hollow WGM microstructures, such as microbubble-based resonators, for which an integrated microfluidics is perfectly fitted.

Author Contributions: F.C. and D.L. contributed equally to this work. Conceptualization, F.C., G.N.C. and F.P.; Software and Validation, D.L. and M.C.F.; Formal Analysis, D.L., M.C.F., F.P.; Investigation, F.C. and D.F.; Data Curation, D.F.; Writing—Original Draft Preparation, F.C.; Writing—Review and Editing, all the authors; Supervision, G.N.C, F.B. and F.P.

Funding: This research was partially funded by the Ministero dello Sviluppo Economico (Grant No. F/050425/02/X32), by Horizon 2020 (PON I&C 2014-2020, "ERHA—Enhanced Radiotherapy with HAdron"), by EU COST Action (Grant No. MP1401, "Advanced fibre laser and coherent source as tools for society, manufacturing and life science"), by Ente Cassa di Risparmio di Firenze (Grant No. 2014.0770A2202.8861) and by Centro Fermi (project "MiFo").

Acknowledgments: We thank Simone Berneschi, Franco Cosi, Ambra Giannetti and Cosimo Trono belonging to the Institute of Applied Physics "Nello Carrara" of the National Research Council of Italy (IFAC-CNR) for the insightful discussions and technical assistance.

Conflicts of Interest: The authors declare no conflict of interest. The founding sponsors had no role in the design of the study; in the collection, analyses, or interpretation of data; in the writing of the manuscript, and in the decision to publish the results.

References

1. Palma, G.; Falconi, M.C.; Starecki, F.; Nazabal, V.; Ari, J.; Bodiou, L.; Charrier, J.; Dumeige, Y.; Baudet, E.; Prudenzano, F. Design of praseodymium-doped chalcogenide micro-disk emitting at 4.7 μm. *Opt. Express* **2017**, *25*, 7014–7030. [CrossRef] [PubMed]

2. Ward, J.M.; Yang, Y.; Chormaic, S.N. Highly sensitive temperature measurements with liquid-core micro bubble resonators. *IEEE Photonics Technol. Lett.* **2013**, *25*, 2350–2353. [CrossRef]

3. Ozel, B.; Nett, R.; Weigel, T.; Schweiger, G.; Ostendorf, A. Temperature sensing by using whispering gallery modes with hollow core fibers. *Meas. Sci. Technol.* **2010**, *21*, 094015–094020. [CrossRef]

4. Ma, Q.; Rossmann, T.; Guo, Z. Whispering-gallery mode silica microsensors for cryogenic to room temperature measurement. *Meas. Sci. Technol.* **2010**, *21*, 025310–025317. [CrossRef]

5. Cai, Z.P.; Xiao, L.; Xu, H.Y.; Mortier, M. Point temperature sensor based on green decay in an Er:ZBLALiP microsphere. *J. Lumin.* **2009**, *129*, 1994–1996. [CrossRef]

6. Dong, C.H.; He, L.; Xiao, Y.F.; Gaddam, V.R.; Ozdemir, S.K.; Han, Z.F.; Guo, G.C.; Yang, L. Fabrication of high-Q polydimethylsiloxane optical microspheres for thermal sensing. *Appl. Phys. Lett.* **2009**, *94*, 231119–231121. [CrossRef]

7. Hanumegowda, N.M.; Stica, C.J.; Patel, B.C.; White, I.; Fan, X. Refractometric sensors based on microsphere resonators. *Appl. Phys. Lett.* **2005**, *87*, 201107–201110. [CrossRef]

8. Henze, R.; Seifert, T.; Ward, J.; Benson, O. Tuning whispering gallery modes using internal aerostatic pressure. *Opt. Lett.* **2011**, *36*, 4536–4538. [CrossRef] [PubMed]

9. Soria, S.; Berneschi, S.; Brenci, M.; Cosi, F.; Nunzi Conti, G.; Pelli, S.; Righini, G.C. Optical microspherical resonators for biomedical sensing. *Sensors* **2011**, *11*, 785–805. [CrossRef] [PubMed]

10. Palma, G.; Falconi, C.; Nazabal, V.; Yano, T.; Kishi, T.; Kumagai, T.; Ferrari, M.; Prudenzano, F. Modeling of whispering gallery modes for rare earth spectroscopic characterization. *IEEE Photonics Technol. Lett.* **2015**, *27*, 1861–1863. [CrossRef]

11. Vollmer, F.; Arnold, S. Whispering-gallery-mode biosensing: Label-free detection down to single molecules. *Nat. Methods* **2008**, *5*, 591–596. [CrossRef] [PubMed]

12. Kishi, T.; Kumagai, T.; Shibuya, S.; Prudenzano, F.; Yano, T.; Shibata, S. Quasi-single mode laser output from a terrace structure added on a Nd^{3+}-doped tellurite-glass microsphere prepared using localized laser heating. *Opt. Express* **2015**, *23*, 20629–20635. [CrossRef] [PubMed]

13. Knight, J.C.; Cheung, G.; Jacques, F.; Birks, T.A. Phase-matched excitation of whispering-gallery-mode resonances by a fiber taper. *Opt. Lett.* **1997**, *22*, 1129–1131. [CrossRef] [PubMed]

14. Farnesi, D.; Chiavaioli, F.; Righini, G.C.; Soria, S.; Trono, C.; Jorge, P.; Nunzi Conti, G. Long period grating-based fiber coupler to whispering gallery mode resonators. *Opt. Lett.* **2014**, *39*, 6525–6528. [CrossRef] [PubMed]

15. Farnesi, D.; Chiavaioli, F.; Baldini, F.; Righini, G.C.; Soria, S.; Trono, C.; Nunzi Conti, G. Quasi-distributed and wavelength selective addressing of optical micro-resonators based on long period fiber gratings. *Opt. Express* **2015**, *23*, 21175–21180. [CrossRef] [PubMed]

16. Palma, G.; Falconi, M.C.; Starecki, F.; Nazabal, V.; Yano, T.; Kishi, T.; Kumagai, T.; Prudenzano, F. Novel double step approach for optical sensing via microsphere WGM resonance. *Opt. Express* **2016**, *24*, 26956–26971. [CrossRef] [PubMed]

17. Brenci, M.; Calzolai, R.; Cosi, F.; Nunzi Conti, G.; Pelli, S.; Righini, G.C. Microspherical resonators for biophotonic sensors. In Proceedings of the SPIE Lightmetry and Light and Optics in Biomedicine, Warsaw, Poland, 20–22 October 2004; p. 61580S.

18. Trono, C.; Baldini, F.; Brenci, M.; Chiavaioli, F.; Mugnaini, M. Flow cell for strain- and temperature-compensated refractive index measurements by means of cascaded optical fibre long period and Bragg gratings. *Meas. Sci. Technol.* **2011**, *22*, 1–9. [CrossRef]

19. Berneschi, S.; Farnesi, D.; Cosi, F.; Nunzi Conti, G.; Pelli, S.; Righini, G.C.; Soria, S. High Q silica microbubble resonators fabricated by arc discharge. *Opt. Lett.* **2011**, *36*, 3521–3523. [CrossRef] [PubMed]

20. Little, B.E.; Laine, J.P.; Haus, H.A. Analytic theory of coupling from tapered fibers and half-blocks into microsphere resonators. *J. Lightwave Technol.* **1999**, *17*, 704–715. [CrossRef]

21. Little, B.E.; Chu, S.T.; Haus, H.A.; Foresi, J.; Laine, J.P. Microring resonator channel dropping filters. *J. Lightwave Technol.* **1997**, *15*, 998–1005. [CrossRef]

22. Malitson, I.H. Interspecimen comparison of the refractive index of fused silica. *J. Opt. Soc. Am.* **1965**, *55*, 1205–1208. [CrossRef]

23. Chiavaioli, F.; Gouveia, C.A.J.; Jorge, P.A.S.; Baldini, F. Towards a uniform metrological assessment of grating-based optical fiber sensors: From refractometers to biosensors. *Biosensors* **2017**, *7*, 23. [CrossRef] [PubMed]

micromachines

MDPI

Article

Direct Metal Forming of a Microdome Structure with a Glassy Carbon Mold for Enhanced Boiling Heat Transfer

Jun Kim [1], Dongin Hong [1], Mohsin Ali Badshah [2], Xun Lu [2], Young Kyu Kim [2] and Seok-min Kim [1,2,*]

[1] Department of Mechanical System Engineering, Chung-Ang University, Heukseok-dong, Dongjak-gu, Seoul 06974, Korea; zuhn@cau.ac.kr (J.K.); hdi2305@naver.com (D.H.)

[2] Department of Mechanical Engineering, Chung-Ang University, Heukseok-dong, Dongjak-gu, Seoul 06974, Korea; mohsinali@cau.ac.kr (M.A.B.); luxun@cau.ac.kr (X.L.); kykdes@cau.ac.kr (Y.K.K.)

* Correspondence: smkim@cau.ac.kr; tel.: +82-2-820-5877

Received: 2 July 2018; Accepted: 26 July 2018; Published: 28 July 2018

Abstract: The application of microtechnology to traditional mechanical industries is limited owing to the lack of suitable micropatterning technology for durable materials including metal. In this research, a glassy carbon (GC) micromold was applied for the direct metal forming (DMF) of a microstructure on an aluminum (Al) substrate. The GC mold with microdome cavities was prepared by carbonization of a furan precursor, which was replicated from the thermal reflow photoresist master pattern. A microdome array with a diameter of 8.4 µm, a height of ~0.74 µm, and a pitch of 9.9 µm was successfully fabricated on an Al substrate by using DMF at a forming temperature of 645 °C and an applied pressure of 2 MPa. As a practical application of the proposed DMF process, the enhanced boiling heat transfer characteristics of the DMF microdome Al substrate were analyzed. The DMF microdome Al substrate showed 20.4 ± 2.6% higher critical heat flux and 34.1 ± 5.3% higher heat transfer coefficient than those of a bare Al substrate.

Keywords: direct metal forming; glassy carbon micromold; enhanced boiling heat transfer; metallic microstructure

1. Introduction

Various research studies for enhancing pool boiling heat transfer using micropatterned surfaces have been conducted for the heat dissipation of very large-scale integrated circuits (VLSI) [1–4]. Wei et al. [1] obtained 4.2× enhancement in critical heat flux (CHF) using micro-fin-pin structures. Chu et al. [2] analyzed the effect of the micropatterned surface area on the CHF and concluded that the increase in capillary force in micropatterns promotes the circulation of the working fluid, thus increasing the CHF. The micropattern structures for enhanced boiling heat transfer have been commonly fabricated using a semiconductor fabrication process (e.g., photolithography) on a silicon substrate because it is a well-established fabrication method for microstructures and is compatible with integrated circuits.

Pool boiling heat transfer has also been used in traditional heat exchange systems, such as refrigerators [5], power plants [6], and batteries [7]. For these systems, a method to fabricate a microstructure on a metallic substrate is required for achieving the enhanced pool-boiling phenomenon. Electrochemical etching is a typical method for fabricating a micropattern on a metallic substrate, in which a photolithographed or a laser-patterned barrier mask layer is used for selective etching of the metallic substrate [8–10]. However, the shape of the micropattern obtained by electrochemical etching is limited, owing to its isotropic etching characteristic. To fabricate engineering-designed micropatterns on a metallic substrate, researchers have proposed direct micromachining techniques, such as focused ion beam machining [11], laser machining [12], and electrochemical discharge machining [13] for metallic substrates. However, these techniques are not suitable for large-area micropatterning with a high production rate, which is important for traditional heat exchange systems.

As an alternative method to fabricate a microstructure on a metallic substrate, a direct metal forming (DMF) method is proposed, in which a high hardness mold with microcavities is pressed against the metallic substrate at room or high temperature to transfer the micropattern onto the substrate. Tran et al. [14,15] successfully fabricated a micropattern on an aluminum (Al) alloy substrate by using a DMF process at or near the melting temperature of the substrate with a silicon micromold fabricated by a deep reactive ion etching process. Buzzi et al. [16] and Hirai et al. [17] extended the DMF process to nanoscale patterns on silver and gold substrates using silicon molds. Although silicon molds were used for DMF in previous research studies, silicon is a brittle material and, thus, is not suitable for repeated DMF processing. Nagato et al. [18] fabricated a reverse-pyramid micropattern on an Al substrate using a DMF process at room temperature with an electroformed nickel mold. Although a reverse-pyramid micropattern can be formed by a DMF process at room temperature owing to its geometrical advantage, a hot DMF process is preferred for other complex microstructures and for reducing applied pressure, which might deteriorate the micromold. Since a micromold with high hot hardness is preferred for a hot DMF process, the developments of a mold material and its patterning method are still required.

In this study, we selected glassy carbon (GC) as a mold material for the proposed DMF process. GC (also called vitreous carbon) is a nongraphitizing carbon material [19] and shows high hot hardness and chemical/mechanical resistance [20], which are the required properties for the mold of high-temperature replication processes such as DMF and glass molding. In our previous research, we fabricated a GC micromold using carbonization of a replicated polymer precursor for glass molding application [21–24]. In this research, we applied the GC mold to the DMF of the Al substrate. Figure 1 shows the schematics of the fabrication process of the GC mold and of the DMF process with the GC mold. A microdome structure photoresist (PR) master pattern was fabricated by photolithography and a thermal reflow process, and a polydimethylsiloxane (PDMS) and UV-curable photopolymer intermediate molds were sequentially replicated from the master pattern. The furan precursor with a microdome-cavity structure was replicated from the photopolymer mold by using a thermal curing process, and the GC mold was obtained by carbonization of the furan precursor. In the DMF process, an Al substrate was placed on a GC mold and sufficient heat and pressure were applied under a vacuum environment to transfer the micropattern onto an Al substrate without oxidation. For practical application of the DMF process with a GC mold, we examined the enhanced boiling heat transfer characteristics of the fabricated DMF microdome Al substrates and we compared the CHF and the heat transfer coefficient (HTC) of the microdome structure with those of a bare Al substrate.

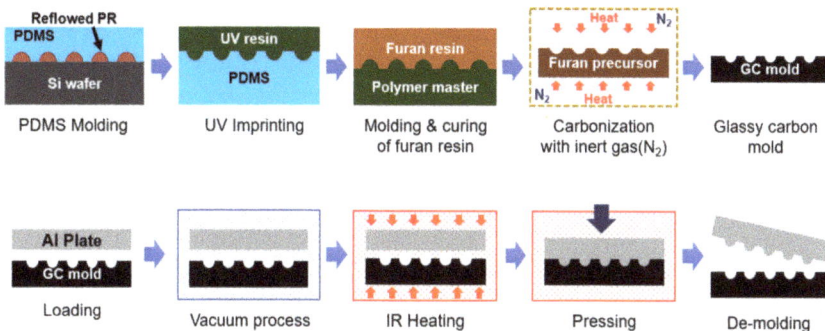

Figure 1. Schematic of a fabrication process for the glassy carbon (GC) mold and the direct metal forming (DMF) process.

2. Fabrication the GC Mold with Macrodome Cavity

A PR master having a microdome array pattern with a pitch of 12.7 µm, a diameter of 10.6 µm, and a sag height of 1.057 µm was fabricated using conventional photolithography and a thermal reflow process [23]. A mixture of a PDMS base material and an initiator (Sylgard 184A and 184B, Dow Corning Korea Ltd., Seoul, Korea) with a mixing ratio of 10:1 was poured onto the PR master pattern and cured at room temperature for 24 h. To obtain an intermediate mold with a positive microdome pattern, we replicated a UV-curable photopolymer mold from the PDMS mold having negative microdome cavities. A UV-curable urethane acrylate photopolymer (UP088, SK Chemicals Co., Ltd., Seongnam, Korea) was poured onto the PDMS mold, and a primer-coated polyethylene terephthalate (PET) film (SH34, SKC Co., Ltd., Seoul, Korea) was used to cover it. Using a rolling process on the PET film, we moved the air bubbles trapped in the photopolymer to the outside of the sample and obtained a uniform photopolymer layer. After the UV irradiation process for 3 min, the cured photopolymer intermediate mold with a size of ~50 × 50 mm² was released from the PDMS mold.

We selected a furan resin (Kangnam Chemical Co. Ltd., Gwangju, Gyeonggi, Korea) as a precursor material for the GC mold owing to its high carbon yield. A furan resin mixture composed of 89.8 wt.% furan resin, 0.2 wt.% p-toluenesulfonic acid ($CH_3C_6H_4SO_3H$ H_2O, PTSA; Kanto Chemical Co. Inc., Tokyo, Japan), and 10 wt.% ethanol was prepared, and a degassing process was conducted for 2 h in a vacuum chamber to remove the entrapped air bubbles in the furan mixture. The degassed furan mixture was poured onto the photopolymer intermediate mold, and a two-step thermal curing process was carried out to avoid warpage of the replicated furan precursor [24]. The furan mixture was first cured under an atmospheric condition for 5 days and then cured again in a convection oven at a maximum temperature of 100 °C. In the second curing process, the rate of temperature increase was set to 0.1 °C/min, and the temperature was maintained for 60 min for every 5 °C temperature increment until the maximum temperature was reached. Finally, the microdome cavity arrayed furan precursor was released from the photopolymer mold and the backside of the furan precursor was polished to obtain the desired thickness and flatness.

To obtain the GC mold, we carried out carbonization of the fabricated furan precursor in a tube furnace under a vacuum environment (modified MIR-TB1001-2; Mirfurnace Co. Ltd., Pocheon, Korea). To allow the slow escape of pyrolysis gases (i.e., H_2, CH_x, CO_2, and CO) from the GC mold and avoid warpage of the GC mold during the carbonization, we gradually increased the temperature of the furnace at a rate of 0.5 °C/min until it reached 600 °C. Beyond 600 °C, the temperature was increased at a rate of 1 °C/min until it reached the maximum temperature of 1000 °C and then maintained at this level for 10 h. After the carbonization, a GC mold with microdome cavities was obtained. Figure 2a shows a scanning electron microscopy (SEM) image of the fabricated microdome cavities on the GC mold. The measured pitch, diameter, and height of the microdome cavity array were 9.9 µm, 8.4 µm, and ~0.76 µm, respectively. The differences in dimensional properties between the PR master and the GC mold were mainly due to the material decomposition in the carbonization process. The shrinkage ratio (reduction ratio of the measured dimensions of the GC mold to the PR master) for the pitch and diameter (inplane) was ~21.4% and that for the height (out-of-plane) was ~28%. Although an anisotropic shrinkage was observed, this shrinkage can be predicted and compensated.

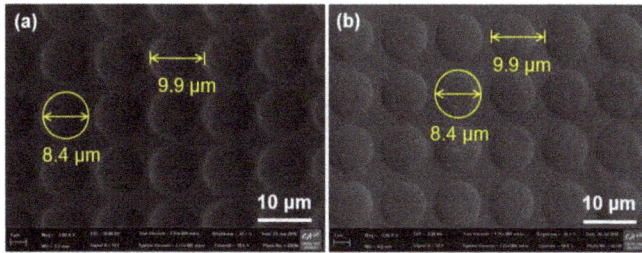

Figure 2. Scanning electron microscope (SEM) images of (**a**) microdome cavities on GC mold and (**b**) microdome array on aluminum (Al) substrate fabricated by DMF with a temperature of 645 °C and a pressure of 2 MPa.

3. Fabrication of a Microdome Patterned Al Substrate by DMF

A DMF system consisting of an infrared (IR) heater for heating up to 1050 °C with a heating rate of 70 °C/min, a motorized pressing unit with a maximum compression force of 150 kgf, and a vacuum chamber was designed and constructed, as shown in Figure 3a. Figure 3b shows the graphite pressing jig unit with an Al substrate (AA1050), a GC mold, and a cover graphite plate. The cover graphite plate was used for applying a preload to prevent any slight movement of the GC mold and Al substrate during the following evacuation process. At the starting point of the DMF process, the pressing jig unit with samples moved up into the vacuum chamber with an IR heater. After the evacuation process, to prevent oxidation, the environmental temperature of the vacuum chamber was increased by the IR heater, as shown in Figure 3c. The DMF process was divided into four stages: heating, holding, pressing, and cooling. Figure 3d illustrates the pressure and temperature history during the DMF process. In the heating stage, the ambient temperature of the GC mold and Al substrate was increased up to the forming temperature of 645 °C. In the holding stage, the temperature was maintained for 30 min to achieve a uniform temperature distribution of the GC mold and Al substrate. In the pressing stage, a compression pressure of 2 MPa was applied to the stack of the GC mold and Al substrate for 10 min. In the cooling stage, the pressure was released and the IR heater was turned off for natural cooling. When the furnace temperature was reduced to room temperature, the pressing jig unit was lowered down from the vacuum chamber and the formed Al substrate was detached from the GC mold.

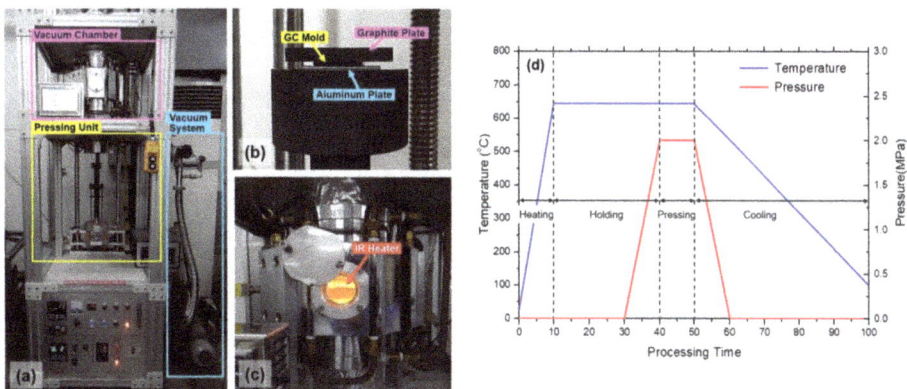

Figure 3. (**a–c**) Photographs of constructed DMF system; (**a**) whole system; (**b**) graphite pressing jig unit with Al substrate, GC mold, and cover graphite plate; and (**c**) vacuum chamber with infrared heater during the heating stage; (**d**) pressure and temperature histories in DMF process.

In the DMF process, the processing temperature and pressure are the important parameters affecting the quality of the metallic microstructure. The ductility of the Al substrate is not enough to lead to a perfect plastic deformation at a low temperature, and the Al substrate can be liquefied at or near the melting temperature. Although higher compression pressure is preferred for improving the replication quality of the metallic microstructure, excessive pressure might deteriorate the GC mold. To determine the appropriate processing condition of the DMF process, we conducted multiple DMF experiments using an Al substrate with a thickness of 1 mm by increasing the processing temperature with a fixed compression pressure of 2 MPa, which was the maximum compression pressure limit to avoid damage to the fabricated GC mold. On the basis of the repeated DMF experiments with various processing temperatures, we selected a processing temperature of 645 °C, which provided sufficient DMF quality. Figure 2b shows the uniformly distributed microdome-structured Al substrate produced by the DMF process at a processing temperature of 645 °C and a compression pressure of 2 MPa. The measured diameter and pitch of the microdome array on the Al substrate obtained from the SEM image were exactly the same as those of the GC mold. Figure 4 shows the three-dimensional surface profiles of (a) the GC mold and (b) the fabricated DMF microdome Al substrate obtained by a laser confocal microscope (OLS-4100, Olympus. Co. Ltd., Tokyo, Japan). The measured height of the microdome on the GC mold was ~0.76 μm and that on the Al substrate was ~0.74 μm. The differences between the GC mold and the DMF microdome Al substrates were negligible, considering the measurement errors in the positive and negative shapes using an optical height measurement system.

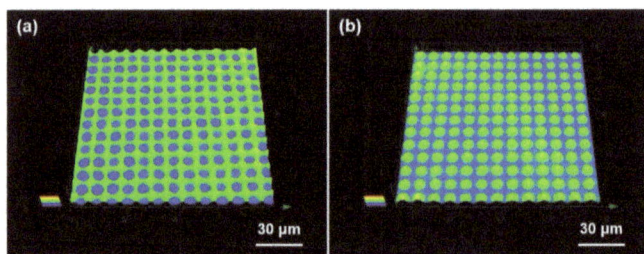

Figure 4. Three-dimensional surface profiles of (**a**) a GC mold and (**b**) a DMF microdome Al substrate obtained by laser confocal microscope.

The thickness of the Al substrate could affect the replication quality of the DMF process. Figure S1 in the supplementary material shows the effect of the Al substrate thickness (0.16, 0.2, 0.3, and 1.0 mm) on the height of the fabricated microdome. It was noted that the effects of Al substrate thickness were negligible when the substrate thickness was thicker than 0.16 mm. It might be because the height of the microdome was much smaller than the thickness of Al substrate.

Since the DMF process was conducted at a high temperature, some chemical reactions including element diffusion could have occurred during the process. To examine the chemical reaction during the DMF process, we analyzed the element compositions of the GC mold and the Al substrate before and after the DMF process using energy-dispersive X-ray spectroscopy (EDX), as shown in Figure S2 (supplementary material). It was noted that there were no significant changes in the element compositions of both the GC mold and the Al substrate during the DMF process owing to the high chemical resistance of the GC mold.

4. Application of the DMF Microdome Al Substrate to Enhanced Boiling Heat Transfer

4.1. Experimental Setup and Measuring Method for Boiling Heat Transfer

As a practical application of the DMF microdome Al substrate, the enhanced boiling heat transfer characteristic of the fabricated sample was examined. Figure 5 shows the schematics of the experimental setup and the analysis system for pool boiling heat transfer. The liquid chamber with a volume of 3.3 L was made of stainless steel (SUS 330L) to prevent corrosion and reaction with other materials. As a heating source, a copper block (thermal conductivity = 401 W/(m·K)) was precisely machined and two cartridge heaters (totaling 600 W maximum) were inserted in the bottom of the block. Moreover, a set of auxiliary heaters was installed inside of the liquid chamber. The power of the cartridge heaters and of the auxiliary heaters was controlled by a proportional-integral-derivative (PID) controller. The sample was attached to the copper block using a thermal grease (MX-4, ARCTIC (HK) Ltd., Hong Kong, China; thermal conductivity = 4.01 W/(m·K)). Finally, the copper block was attached on the bottom of the liquid chamber. Every interface was sealed with a Teflon gasket to prevent any leakage of the boiling liquid. At the top of the liquid chamber, an air-cooled condenser was installed to liquefy the vapor. The outside of the copper block was insulated with silicone (thermal conductivity = 0.2 W/(m·K)) and Teflon (thermal conductivity = 0.25 W/(m·K)) sheets. In this setup, the copper block had a higher thermal conductivity than that of the insulation materials. Therefore, it could be assumed that the heat transfer on the copper block was one dimensional because the heat loss of the side wall was very small and, thus, negligible [25]. Two small holes were drilled in the middle of the copper block, and a couple of thermocouples (J-Type, Omega Engineering Inc., Stamford, CT, USA) were inserted into the holes to calculate the heat flux on the sample surface. The measured data were collected by a multichannel data logger system (Graphtec GL220, DATAQ Instruments Inc., Akron, OH, USA). As a working fluid, the FC-72 coolant (3M Inc., Maplewood, MN, USA) was used. To calculate the heat flux q'', we applied Fourier's law, which is described in Equation (1):

$$q'' = -k\frac{dT}{dz} = -k_{Cu}\frac{T_H - T_L}{d_a} \tag{1}$$

where the temperature gradient (dT/dz) was estimated by measuring the temperature difference $(T_H - T_L)$ and the axial distance d_a between the two thermocouples, as shown in the right-hand side of Equation (1). The temperature of the sample surface T_s can be calculated using the heat flux and the lower temperature measured from the thermocouple T_L from the following equation:

$$T_S = T_L - q''\left(\frac{L_{Cu}}{k_{Cu}} + R_{tc} + \frac{L_{Al}}{k_{Al}}\right) \tag{2}$$

where the contact resistance R_{tc} was calculated as 0.095 cm²·K/W by utilizing the procedure reported by Cooke et al. [26]. Moreover, wall superheat ΔT_S was defined as the temperature difference between the surface temperature Ts and the saturation temperature of the boiling liquid T_{sat} (FC-72, 56 °C):

$$\Delta T_S = T_S - T_{Sat} = T_S - 56\,°C \tag{3}$$

Finally, the HTC h_S was calculated using Newton's law of cooling:

$$h_S = \frac{q''}{T_S - T_{Sat}} = \frac{q''}{\Delta T_S} \tag{4}$$

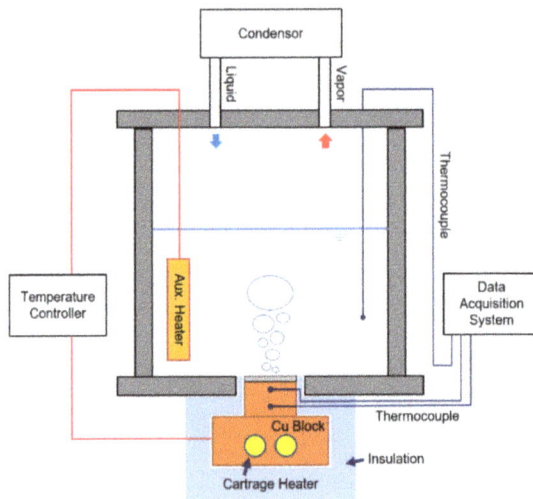

Figure 5. Schematic of experimental setup for boiling heat transfer using fabricated sample.

4.2. Uncertainty Analysis

In the boiling heat transfer experiment, uncertainties existed owing to the errors in the machining of temperature measurement points, errors in the thermocouples for measuring the temperature, and variations in the thermal conductivity of the materials. The positioning accuracy of the holes for temperature measurement was ±0.01 mm, and the measurement error of the J-Type thermocouple was ±0.15 K. The variation in the thermal conductivity with the temperature change was ±2% for copper and ±2.1% for Al, and the uncertainty of the contact resistance R_{tc} was ±2.37%. The overall uncertainty was calculated using the second-power equation proposed by Kline et al. [27]. As a result, the calculated uncertainties of the heat flux and the HTC were 8.72% and 8.74%, respectively. Table 1 summarizes the uncertainty of the pool boiling heat transfer experiment and the calculated values.

Table 1. Uncertainty sources and error values.

Uncertainty Source	Error
Machining error for measuring position	±0.01 mm
J-type thermocouple reading	±0.15 K
Thermal conductivity of Cu	±2%
Thermal conductivity of Al	±2.1%
Thermal contact resistance	±2.37%
Surface temperature reading	±0.62%
Heat flux	±8.72%
Heat transfer coefficient	±8.74%

4.3. Experimental Result and Discussion

A series of boiling experiments was conducted to evaluate the heat transfer performance of the DMF microdome Al substrates. For the comparison, a bare Al substrate was also evaluated using the same experimental procedure. For each set of the boiling experiments, boiling liquid was heated above the saturation temperature (FC-72, 56 °C) for 10 min to eliminate the gases trapped or dissolved in the boiling liquid because they could affect the heat transfer performance in the pool boiling case. After the degasification process, the system was left for an hour at the saturation temperature until it reaches a thermally stable state. Then, the temperature of the copper block was controlled in increments of

1 °C for every minute until the onset of boiling had occurred. Once the boiling had started, the heat flux was increased continuously. From this point, the temperature increment of the copper block was adjusted to 0.5 °C for every minute.

Figure 6 shows the comparison result of the boiling curves for the three DMF microdome Al substrates fabricated at the optimum condition and for the bare Al substrate. In the case of the bare Al substrate, the onset of boiling was started at ~15 °C of the wall superheat temperature. As the wall superheat temperature increased, the heat flux simultaneously increased until the wall superheat temperature reached ~27 °C. As shown in Figure 6, at this point the heat flux was calculated as 171.1 kW/m² as a CHF. Above the CHF point, film boiling occurred and the wall superheat temperature increased rapidly within a short time. Thus, the experiment was terminated when film boiling had occurred. In the case of the DMF microdome Al substrates, the onset of boiling was started at ~14 °C of the wall superheat temperature. It shows that the microdome structure can reduce the overheating temperature leading to the boiling. In the same manner as with the bare Al substrate, boiling experiments were performed until film boiling had occurred. The average value of the CHF for the three DMF microdome Al substrates was calculated as 205.9 kW/m² with ± 4.5 kW/m² of standard deviation, when the wall superheat temperature reached ~25 °C. This value was 20.4 ± 2.6% higher than that for the bare Al substrate. At the wall superheat temperature of 20 °C, the calculated heat flux values of the three DMF microdome Al substrates and the bare Al substrate were ~160.6 ± 2.7 kW/m² and ~97 kW/m², respectively. It shows that the microdome structure enhanced the heat flux by 65.6 ± 2.8% in the moderate boiling regime.

HTC is another important parameter for evaluating the performance of pool boiling heat transfer. The HTC value (h_c) was calculated using Equation (4). The maximum HTC values of the bare Al substrate and the DMF microdome Al substrates were calculated as 5.96 kW/(m²·K) and 8.0 ± 0.3 kW/(m²·K), respectively. Because of the definition of Newton's law of cooling, the HTC is proportional to the heat flux and inversely proportional to the wall superheat temperature. In the case of the bare Al substrate, the overall heat flux was lower and the wall superheat temperature was higher than those of the DMF microdome Al substrates. This explains why the HTC value of the DMF microdome Al substrates was 34.1 ± 5.3% higher than that of the bare Al substrate.

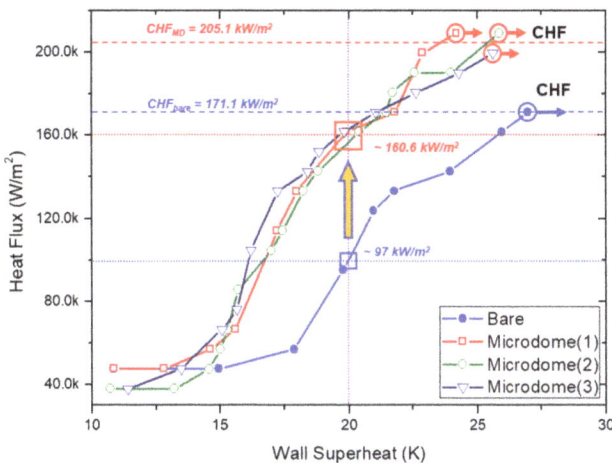

Figure 6. Comparison of the boiling curves of three DMF microdome Al substrates and a bare Al substrate.

5. Conclusions

In this paper, we propose a DMF process using a GC mold. The GC mold was prepared by carbonization of a replicated furan precursor, and a standalone-type DMF system was designed and constructed. To examine the feasibility of the proposed method, we successfully fabricated an array of microdome structure with a pitch of 9.9 μm, a diameter of 8.7 μm, and a height of ~0.74 μm on an Al substrate by using DMF with a processing temperature of 645 °C and a compression pressure of 2 MPa. For a practical application, we examined the enhanced boiling heat transfer characteristics of the DMF microdome Al substrates and compared them with those of the bare Al substrate. The DMF microdome Al substrates showed $20.4 \pm 2.6\%$ higher CHF and $34.1 \pm 5.3\%$ higher HTC than those of the bare Al substrate. Although we successfully demonstrated the fabrication of microdome structure on an Al substrate by DMF with GC mold for enhanced boiling heat transfer, the structural optimization is still required for maximizing CHF and HTC.

The size of the DMF microdome Al substrates (~20×20 mm^2) and the standalone-type DMF system used in this study are not acceptable for a large-area micropatterning process of metallic substrates with a high production rate. However, the proposed DMF with GC mold can be extended to mass production of a large-area micropatterning process of metallic substrate because multiple large-area GC molds can be obtained by the proposed GC mold fabrication process and a batch process concept using multiple molds can be applied to the DMF process. In addition, the proposed method can be extended to the roll-to-roll DMF process, which can provide high patterning speed on a large area because a GC roll mold with microcavities can be obtained by carbonization of the roll-shaped polymer precursor. The optimization of the structural parameter to maximize the boiling heat transfer performance and the development of the roll-to-roll DFM process with a GC roll mold are the subjects of our ongoing research.

Supplementary Materials: The following are available online at http://www.mdpi.com/2072-666X/9/8/376/s1: Figure S1: Effects of the Al substrate thickness (0.16, 0.2, 0.3, and 1.0 mm) on the height of the DMF microdome structure.; Figure S2: EDX analysis results for (a) the GC mold before DMF process, (b) the GC mold after DMF process, (c) a bare Al substrate, and (d) the DMF microdome Al substrate.

Author Contributions: In this paper, S.K. conceived the idea and continuously advised during the experimental and analysis phase. J.K., and S.K. wrote the main manuscript text. J.K., D.H., and Y.K.K. constructed the DFM system and fabricated GC mold and DFM sample. J.K, M.A.B., and X.L. conducted boiling heat transfer experiment and analyzed the data. S.K. proofread and revised the manuscript. All the authors reviewed the manuscript.

Acknowledgments: This research was supported by the Basic Science Research Program through the National Research Foundation of Korea (NRF) funded by the Ministry of Science, ICT and Future Planning (NRF-2017R1A2B4011149) and the Chung-Ang University Research Scholarship Grants in 2017.

Conflicts of Interest: The authors declare no conflict of interest.

References

1. Wei, J.J.; Honda, H. Effects of fin geometry on boiling heat transfer from silicon chips with micro-pin-fins immersed in FC-72. *Int. J. Heat Mass Transfer* **2003**, *46*, 4059–4070. [CrossRef]
2. Chu, K.H.; Enright, R.; Wang, E.N. Structured surfaces for enhanced pool boiling heat transfer. *Appl. Phys. Lett.* **2012**, *100*. [CrossRef]
3. Dong, L.; Quan, X.; Cheng, P. An experimental investigation of enhanced pool boiling heat transfer from surfaces with micro/nano-structures. *Int. J. Heat Mass Transfer* **2014**, *71*, 189–196. [CrossRef]
4. Lu, M.C.; Huang, C.H.; Huang, C.T.; Chen, Y.C. A modified hydrodynamic model for pool boiling CHF considering the effects of heater size and nucleation site density. *Int. J. Therm. Sci.* **2015**, *91*, 133–141. [CrossRef]
5. Javidmand, P.; Hoffmann, K.A. Une modélisation complète d'écoulement diphasique réduit dans des orifices de courts tubes appliquée aux frigorigènes alternatifs HFO-1234yf et HFO-1234ze. *Int. J. Refrig.* **2016**, *69*, 114–135. [CrossRef]
6. Santini, L.; Cioncolini, A.; Butel, M.T.; Ricotti, M.E. Flow boiling heat transfer in a helically coiled steam generator for nuclear power applications. *Int. J. Heat Mass Transfer* **2016**, *92*, 91–99. [CrossRef]

7. Van Gils, R.W.; Danilov, D.; Notten, P.H.L.; Speetjens, M.F.M.; Nijmeijer, H. Battery thermal management by boiling heat-transfer. *Energy Convers. Manag.* **2014**, *79*, 9–17. [CrossRef]

8. Shin, H.S.; Park, M.S.; Chu, C.N. Electrochemical etching using laser masking for multilayered structures on stainless steel. *CIRP Ann. Manuf. Technol.* **2010**, *59*, 585–588. [CrossRef]

9. Zhu, D.; Qu, N.S.; Li, H.S.; Zeng, Y.B.; Li, D.L.; Qian, S.Q. Electrochemical micromachining of microstructures of micro hole and dimple array. *CIRP Ann. Manuf. Technol.* **2009**, *58*, 177–180. [CrossRef]

10. Datta, M.; Landolt, D. Fundamental aspects and applications of electrochemical microfabrication. *Electrochim. Acta* **2000**, *45*, 2535–2558. [CrossRef]

11. Xiao, Y.; Wehrs, J.; Ma, H.; Al-Samman, T.; Korte-Kerzel, S.; Göken, M.; Michler, J.; Spolenak, R.; Wheeler, J.M. Investigation of the deformation behavior of aluminum micropillars produced by focused ion beam machining using Ga and Xe ions. *Scr. Mater.* **2017**, *127*, 191–194. [CrossRef]

12. Gupta, R.K.; Kumar, A.; Nagpure, D.C.; Rai, S.K.; Singh, M.K.; Khooha, A.; Singh, A.K.; Singh, A.; Tiwari, M.K.; Ganesh, P.; et al. Comparison of Stress Corrosion Cracking Susceptibility of Laser Machined and Milled 304 L Stainless Steel. *Lasers Manuf. Mater. Process.* **2016**, *3*, 191–203. [CrossRef]

13. Ramulu, M.; Paul, G.; Patel, J. EDM surface effects on the fatigue strength of a 15 vol% SiCp/Al metal matrix composite material. *Compos. Struct.* **2001**, *54*, 79–86. [CrossRef]

14. Tran, N.K.; Lam, Y.C.; Yue, C.Y.; Tan, M.J. Manufacturing of an aluminum alloy mold for micro-hot embossing of polymeric micro-devices. *J. Micromech. Microeng.* **2010**, *20*, 055020. [CrossRef]

15. Tran, N.; Lam, Y.; Yue, C.; Tan, M. Fabricating protruded micro-features on AA 6061 substrates by hot embossing method. *Int. J. Mech. Aerosp. Ind. Mechatron. Manuf. Eng.* **2012**, *6*, 983–986.

16. Buzzi, S.; Robin, F.; Callegari, V.; Löffler, J.F. Metal direct nanoimprinting for photonics. *Microelectron. Eng.* **2008**, *85*, 419–424. [CrossRef]

17. Hirai, Y.; Ushiro, T.; Kanakugi, T.; Matsuura, T.; Industries, S.E. Fine gold grating fabrication on glass plate by imprint lithography. *Proc. SPIE* **2003**, *5220*, 74–81.

18. Nagato, K.; Miyazaki, S.; Yamada, S.; Nakao, M. Nano/microcomposite surface fabricated by chemical treatment/microembossing for control of bubbles in boiling heat transfer. *CIRP Ann. Manuf. Technol.* **2016**, *65*, 511–514. [CrossRef]

19. Franklin, R.E. Crystallite Growth in Graphitizing and Non-Graphitizing Carbons. *Proc. R. Soc. A Math. Phys. Eng. Sci.* **1951**, *209*, 196–218. [CrossRef]

20. Pierson, H.O. *Handbook of Carbon, Graphite, Diamond and Fullerenes*; Elsevier: New York, NY, USA, 1993; pp. 122–140, ISBN 0-8155-1339-1.

21. Jang, H.; Haq, M.R.; Ju, J.; Kim, Y.; Kim, S.; Lim, J. Fabrication of all glass bifurcation microfluidic chip for blood plasma separation. *Micromachines* **2017**, *8*, 67. [CrossRef]

22. Jang, H.; Haq, M.R.; Kim, Y.; Kim, J.; Oh, P.; Ju, J.; Kim, S.; Lim, J. Fabrication of glass microchannel via glass imprinting using a vitreous carbon stamp for flow focusing droplet generator. *Sensors* **2018**, *18*, 83. [CrossRef] [PubMed]

23. Kim, Y.K.; Ju, J.H.; Kim, S. Replication of a glass microlens array using a vitreous carbon mold. *Opt. Express* **2018**, *26*, 14936–14944. [CrossRef]

24. Ju, J.; Lim, S.; Seok, J.; Kim, S. A method to fabricate Low-Cost and large area vitreous carbon mold for glass molded microstructures. *Int. J. Precis. Eng. Manuf.* **2015**, *16*, 287–291. [CrossRef]

25. Badshah, M.A.; Kim, J.; Jang, H.; Kim, S. Fabrication of Highly Packed Plasmonic Nanolens Array Using Polymer Nanoimprinted Nanodots for an Enhanced Fluorescence Substrate. *Polymers* **2018**, *10*, 649. [CrossRef]

26. Cooke, D.; Kandlikar, S.G. Pool Boiling Heat Transfer and Bubble Dynamics Over Plain and Enhanced Microchannels. *J. Heat Transfer* **2011**, *133*, 052902. [CrossRef]

27. Kline, S.J.; McClintock, F.A. Describing Uncertainties in Single-Sample Experiments. *Mech. Eng.* **1953**, *75*, 3–8.

micromachines

MDPI

Review

Glassy Microspheres for Energy Applications

Giancarlo C. Righini [1,2]

[1]   Enrico Fermi Centre, 00184 Roma, Italy; giancarlo.righini@centrofermi.it
[2]   Nello Carrara Institute of Applied Physics (IFAC CNR), 50019 Sesto Fiorentino, Italy

Received: 27 June 2018; Accepted: 26 July 2018; Published: 30 July 2018

Abstract: Microspheres made of glass, polymer, or crystal material have been largely used in many application areas, extending from paints to lubricants, to cosmetics, biomedicine, optics and photonics, just to mention a few. Here the focus is on the applications of glassy microspheres in the field of energy, namely covering issues related to their use in solar cells, in hydrogen storage, in nuclear fusion, but also as high-temperature insulators or proppants for shale oil and gas recovery. An overview is provided of the fabrication techniques of bulk and hollow microspheres, as well as of the excellent results made possible by the peculiar properties of microspheres. Considerations about their commercial relevance are also added.

Keywords: microspheres; microdevices; glass; polymers; solar energy; nuclear fusion; thermal insulation

1. Introduction

Global energy demand (GED) keeps growing, boosted by a generally strong economic growth. According to the International Energy Agency (IEA), GED grew by 2.1% in 2017, more than twice the 2016 rate; accordingly, global energy-related carbon dioxide emissions increased by 1.4% in 2017, after three years of remaining flat [1]. Still, over 70% of GED growth was met by oil, natural gas and coal; renewable energies, however, exhibited in 2017 the highest growth rate of any energy source. Figure 1 shows the GED average annual growth for the different fuels; the y-axis on the right indicates the net growth rate, while the y-axis on the left reports the energy growth in million tons of oil equivalent (Mtoe). The overall GED in 2017 reached an estimated 14,050 Mtoe.

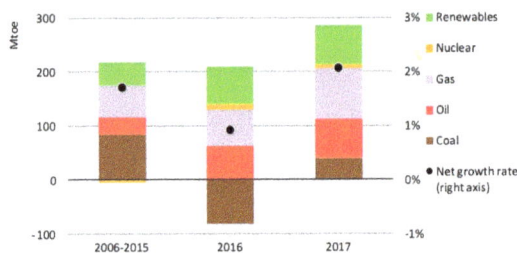

Figure 1. Average annual growth of the global energy demand (GED) by fuel. Reproduced from the IEA report [1].

Continuous advancements in technology are necessary to improve production efficiency, energy security, and—last but not least—environment quality, while maintaining economic competitiveness. A not negligible contribution to some of these goals may be provided by a very simple type of microdevices, namely the microspheres, which can be either solid or hollow, the latter also known as microbubbles. In the following, the term microsphere or microbead will be used when referring to the solid object, and the term hollow microsphere or microbubble for the other type. Thanks to their physical and chemical properties, which include light weight, low thermal conductivity, resistance to compressive stress, and the fact of being almost chemically inert, microspheres and microbubbles (Ms&Mb) have been widely used in pharmaceutical, food, cosmetic, chemical, transportation and construction industrial sectors. Staying more on the research side, Ms&Mb have found advanced applications in optics and photonics; their use as whispering gallery mode (WGM) resonators opened the way to the development of several high-performance lasing and sensing micro devices [2,3]; the search for more compact and robust structures, especially in the biosensing field, is one of the current R&D trends.

Here, the analysis is limited to spherical particles in the micrometer range, which is approximately 0.1 to 1000 μm, or, in other words, from hundreds of nanometers to one millimeter. At the upper end of this range one encounters microbeads which find application in optics as microlenses, e.g., for fiber-to-fiber coupling, while at the lower end one enters into the nanotechnologies, where nanospheres and nanobubbles find several applications in photonics, catalysis, nanoreactors, drug delivery systems. An overview of the fundamentals and the applications of both glass microspheres and glass nanospheres is presented in a forthcoming book [4].

The aim of this article is to provide an overview of the Ms&Mb applications in the energy field, which can be either indirect (solar cells, thermal insulation, low-density drilling fluid for oil and gas extraction, ultra-low-density proppants for shale oil and gas recovery) or direct (in hydrogen storage and in nuclear fusion targets). A short description of the fabrication processes in the laboratory or at an industrial level will be provided, too.

2. Materials and Fabrication Methods

Here only microspheres and microbubbles made in amorphous materials, namely in oxide or chalcogenide glasses and in amorphous polymers, will be considered. For the sake of completeness, however, it should be noted that many other materials, either natural or synthetic, can be used to fabricate Ms&Mb for different applications. A few examples include stainless steel microspheres (for conductive spacers, shock absorption, and micromotor bearings [5]); metallic nickel hollow microspheres (enhanced magnetic properties; Ni/Pt bimetallic microbubbles have potential applications in portable hydrogen generation systems, due to catalytic properties [6]); single-crystal ferrite microspheres (for applications not only as magnetic materials but also in ferrofluid technology and in biomedical fields, e.g., biomolecular separations, cancer diagnosis and treatment, magnetic resonance imaging [7]); single-crystal semiconductor microspheres (for active WGM resonators [8]); ceramic ZrO_2 hollow microspheres (for thermal applications) [9]. Glass, polymer, ceramic, metal solid and hollow microspheres are commercially available; there is a wide choice of quality, sphericity (Sphericity was defined in 1935 by the geologist H. Wadell, with reference to quartz particles (*J. Geology* 1935, *43*, 250) as the ratio of the surface area of a sphere (with the same volume as the given particle) to the surface area of the particle), uniformity, particle size and particle size distribution, to allow the optimal choice for each unique application.

2.1. Oxide Glass Microspheres

Ms&Mb made in pure silica or multi-oxide glass are the most widely used type in research and industrial/commercial applications. At the laboratory level, when a single microsphere is needed, for instance to exploit the characteristics of a discrete WGM resonator [4] or to implement a micro/nano Coordinate Measuring Machine (CMM) probe [10], the fabrication technique may be quite a handcrafted work, requiring the care of a skilled technician. The most common method of fabrication of a high-quality microsphere, in fact, is based on the melting of the tip of a standard optical communication fiber and relies on surface tension's effect to obtain an almost perfect spherical shape. The heating source may be a simple oxygen/butane (or similar) torch, or a high-power laser (especially CO_2 laser), or an electric arc (such as the one produced in a commercial fiber splicer) [4,10–12].

For the fibers drawn from multi-oxide glasses, which have a melting point lower than pure silica, a simple resistive microheater may be sufficient [12]. The use of a commercial or modified optical fiber fusion splicer (e.g., FITEL S182K, Furukawa, Tokyo, Japan) allows a very good control of the process; a cleaved tip of the fiber is inserted in one arm of the splicer and a series of arcs are then produced. The tip partially melts, and the surface tension forces produce the spherical shape. Figure 2 presents a schematic diagram of the experimental apparatus; the heat generated from the electrode discharge can produce a temperature of around 2000 °C, which is sufficient to melt pure silica.

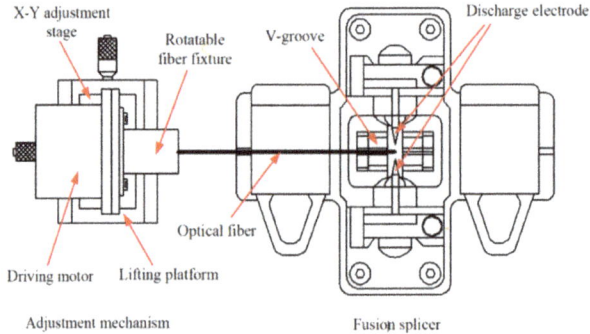

Figure 2. Schematic drawing of the apparatus to produce an integrated optical fiber microsphere in a fusion splicer. Reproduced from [13] under Creative Commons license.

Using this type of apparatus, Yu et al. reported the fabrication of integrated optical fiber microspheres with a diameter smaller than 100 μm, exhibiting 2D roundness error less than 0.70 μm and true sphericity of about 0.5 μm [13]. Such results were achieved by using a fiber tapering technique and a statistical process optimization method (Taguchi method) [14]. The authors here define the true sphericity as a radius difference between a perfect sphere and the 3D fitting surface profile, obtained by a fitting numerical procedure from the photograph of the 2D cross-section of the microsphere under analysis [13]. If the fiber is not tapered in advance, the size of the sphere is larger than 125 μm, namely the cladding diameter of the standard telecom SMF-28 SM silica fiber, and it increases with the number of electric arc shots until approaching saturation at a diameter of about 350 μm, as shown in Figure 3 [15].

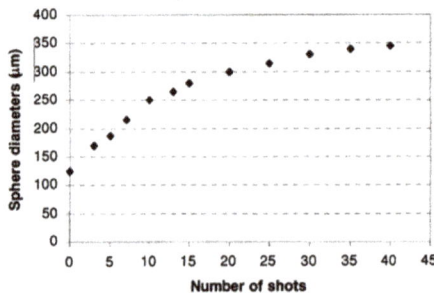

Figure 3. Size of the microspheres produced at the tip of a standard 125 μm telecom fiber, as a function of the arc shots in a fiber fusion splicer. Reproduced from [15].

With the previous methods, only one microsphere can be produced at a time and the size is mostly determined by the fiber size; moreover, the microsphere remains integral with the fiber stem, which is very useful in some applications but may be disadvantageous in others. When a discrete particle is needed, it is convenient to start producing a fine glass powder and then melt it. This approach also permits making microspheres from any oxide glass. As an example, one can crush the glass into particles with sizes ranging from 10 to 100 μm, and single microspheres can be obtained by the localized-laser-heating (LLH) technique, where a cw Ti: sapphire laser with typical power 200 mW at λ = 810 nm is used [16]. Other options include using a microwave plasma torch (the glass grains are dropped through it and the spherical particles are collected at the bottom) [17], or—if starting from the raw components - melting the glass components in a furnace and dropping the viscous glass onto a spinning plate [18]. A disadvantage of these techniques is that one obtains several free spheres with a rather large size distribution; it is, therefore, required to sort the produced spheres by size, while also checking their surface quality.

A similar situation exists for hollow microspheres (aka microbubbles and microballoons). In the laboratory, a usual objective is to fabricate the microbubble integral to a capillary, to form a system of a WGM resonator with integrated microfluidics, which is very convenient for biomedical applications [19,20]. A common technique consists in using a slightly pressurized silica capillary and melting a small volume of it by using a CO_2 laser or an electric arc discharge. Single- and double-pass structures, i.e., spherical shells with one or two openings, can be made [20]. For commercial applications, instead, discrete microballoons are required, and appropriate fabrication methods have been developed over the past several years [21]. Mass production of Ms&Mb with a good control of size dispersion is undoubtedly possible, and the appropriate processes are employed in the industry; basically, solid glass microspheres are produced by direct heating and melting of glass powders, while glass microbubbles are obtained by adding a blowing agent to the glass powder.

The sector of solid and hollow glass microspheres has a relevant commercial interest; according to a market research report, the global glass microspheres market is expected to reach \$1993.36 million by 2019, with an annual growth around 12.4% [22]. As it could be expected, several patents exist, which cover the subject of glass microsphere and microballoons fabrication, in view of various applications. Table 1 gives a representative, and not exhaustive, list of the US patents on this topic.

Table 1. List of United States Patents referring to the fabrication process of glass microspheres. The superscripts s,h indicate solid and hollow microspheres, respectively.

Inventors	US Patent N.	Year	Title
Veatch, F.; Alford, H.E.; Croft, R.D.	2,978,339	1961	Method of producing hollow glass spheres [h]
Beck, W.R.; O'Brien, D.L.	3,365,315	1968	Glass bubbles prepared by reheating solid glass particles
Tung, C.F.; Laird, J.A.	3,946,130	1976	Transparent glass microspheres and products made therefrom [s]
Garnier, P.; Abriou, D.; Coquillon, M.	4,661,137	1987	Process for producing glass microspheres [h]
Block, J.; Lau, J.W.; Rice, R.W.; Colageo, A.J.	5,176,732	1993	Method for making low sodium hollow glass microspheres
Arai, K.; Yamada. K.; Hirano H., Satoh M.	5,849,055	1998	Process for producing inorganic microspheres [s,h]
Henderson, T.M.; Wedding D.K.	6,919,685	2001	Microsphere [h]
Yamada, K.; Hirano, H.; Kusaka, M.; Tanaka, M.	0043996 (Application Publication #)	2001	Hollow aluminosilicate glass microspheres and process for their production [h]
Kirkland, J.J.; Langlois, T.J.; Wang, O.	6,482,324	2002	Porous silica microsphere scavengers [s]
Tanaka, M.; Hirano, H.; Yamada, K.	6,531,222	2003	Fine hollow glass sphere and method for preparing the same [h]
Lipinska-Kalita, K.E.; Hemmers, O.A.	8,663,429	2014	Hollow glass microsphere candidates for reversible hydrogen storage, particularly for vehicular applications [h]

2.2. Chalcogenide Glass Microspheres

Chalcogenide glasses, namely compounds formed predominantly from one or more of the chalcogen elements (Sulfur, Selenium, and Tellurium), are interesting materials in photonics due to their nonlinear properties, photosensitivity, low phonon energy and infrared transparency [23]. Since chalcogenide optical fibers [24] are commercially available, the fabrication method based on the melting of the tip of a fiber can be used in this case as well. A more usual process, however, is to drop the crushed glass through a vertical furnace purged with an inert gas, typically argon [25]. The use of an inert atmosphere is necessary due to the reactive nature of molten chalcogenide glass melting. These solid microspheres find application in biosensing, temperature sensing, lasers and amplifiers [26–28]. It is worth mentioning that binary, ternary and quaternary metal-chalcogenide nanocrystals (e.g., CdSe, PbTe, CuInS$_2$, Cu$_2$ZnSnS$_4$ etcetera) are also of interest in the field of renewable energies, to enhance the efficiency of energy conversion devices [29].

2.3. Polymer Microspheres

Polyethylene (PE), polystyrene (PS) and polymethylmethacrylate (PMMA) microspheres are among the most common types of polymer microspheres; there are, however, many more polymers and synthesis techniques which can be selected depending on the application. As an example, PS microspheres are typically used in biomedical applications due to their ability to facilitate procedures such as cell sorting. PE microspheres are often used as a permanent or temporary filler, but their high sphericity also makes them suitable for various research application (e.g., microscopy techniques, flow visualization, biomedicine). PMMA microspheres (aka acrylic microspheres) have good biocompatibility which allows the particles to be used in many medical and biochemical applications. All these particles are commercially available; for instance, PE microspheres are available from Cospheric (Santa Barbara, CA, USA) in particle size from 1 μm to 1.7 mm [30] and from Polysciences (Warrington, PA, USA) in sizes from 50 nm to 90 μm [31]; PS microspheres are available from MagSphere (Pasadena, CA, USA) in size ranges from 30 nm to 15 μm [32] and from Microspheres-Nanospheres (Cold Spring, NY, USA) in sizes from 50 nm up to 250 μm [33]; PMMA microspheres (Degradex®) can be obtained from Phosphorex (Hopkinton, PA, USA) in the size range 25 nm to 375 μm [34], and from Microbeads (Skedsmokorset, Norway) in standard sizes of 6, 10, 15, 20, 30 and 40 microns (the cross-inking degree of the beads can be adjusted according to application requirements) [35], while Goodfellow (Huntingdon, UK) offers precision PMMA spheres in two diameters of 1.5 mm and 3.18 mm [36]. In many cases, the spheres are available with a coating, but also with opaque, paramagnetic, fluorescent, and phosphorescent properties.

As their glass counterparts, single polymeric spheres as well are of interest as WGM resonators; some authors have reported the fabrication and characterization of polydimethylsiloxane (PDMS) microspheres [37–39]. Single microspheres can be obtained by dipping the tip of an optical fiber into a mixture of PDMS and a curing agent [37]. A larger quantity of PDMS microspheres, with a certain size distribution, can be prepared by mechanical stirring using surfactant solutions [38] or exploiting liquid instabilities [39]. An alternative method, very simple and time-saving, to obtain a high quality polymeric WGM microresonator consists in using a droplet of a commercially available UV-curable adhesive and transferring it onto the tip of an optical (standard or tapered) fiber [40]. On the other side, there exist several polymerization techniques, such as emulsion, dispersion, precipitation and suspension polymerization, which can be used to produce large quantities of polymeric microspheres [41]; many other techniques, such as inkjet printing, electrospraying, and self-assembling processes, may also be adopted for various applications, biology and medicine being one of the most common [42–44].

The importance of polymer microbubbles in biomedicine is also undoubted, being used especially as contrast agents for medical imaging and as therapeutic delivery devices; their fabrication has been the subject of many R&D investigations [45–51]. As an example, Table 2 presents a comparison of different methods for preparing polystyrene microbubbles. It appears that the microencapsulation method is the most suitable for preparing PS hollow microspheres in a quite wide range of sizes [47]. Hollow polymer microspheres with different wall materials, however, may need other, more appropriate, methods.

Table 2. Comparison of different methods for preparing polystyrene hollow microspheres. Reproduced with modifications from [47] under Creative Commons license.

Properties	Liquid Droplet Method	Dried-Gel Droplet Method	Self-Assembly Method	Micro-Encapsulation Method	Emulsion Polymerization Method	Template Method
Equipment cost	High	High	Low	Low	Low	Low
Operation cost	High	High	Low	Low	Low	High
Micromanipulation	Yes	Yes	Yes	No	No	Yes
Batch production	Able	Able	Able	Able	Able	Able
Multiwalled product	No	No	Able	Able	Able	Able
Microsphere diameter, μm	500 ÷ 1500	500 ÷ 1500	≤0.5	50 ÷ 700	≤20	≤5
Sphericity, %	≥97	≥99	≥99	≥99	≥99	≥99
Concentricity, %	≥90	≤90	≥99	≥98	≥98	≥99
Surface roughness, nm	<200	<200	<10	<300	<10	<5

3. Applications in the Field of Energy

As anticipated, solid and hollow microspheres have many applications, which depend on the properties of the constituent material and the size, and involve a wide range of technologies. Their use in several different fields has attracted much interest for many years [52–59]. Nowadays, from an industrial point of view, healthcare and biotechnology are the dominant sector, especially due to the development of drug delivery systems [52,55,57]; together with the sector of cosmetics and personal care, it covers more than 50% of the world market. The construction industry, paints and coatings, and automotive are the other relevant industrial application areas [59]. Depending on the application, sometimes ceramic or crystalline microspheres have better properties; glassy Ms&Mb remain, however, the most used components. This is true also in the case of energy applications, an area that has been becoming increasingly important in recent decades. Here, we can categorize the use of Ms&Mb into three sub-areas: energy saving, energy storage, and energy production. A quantitative feeling of the increased interest in the energy applications of microspheres may be obtained by looking at the number of publications: according to Clarivate Analytics Web of Science, the articles containing the word "microspher*" (i.e., microsphere or microspheres or microspherical) in the title add up to almost 46,000 (~18,000 in the last 5 years, and 4143 in 2017). The articles having the words "microspher*"and "energy" in the title are only 213, but those with "microspher*" in the title and "energy" in the topic are over 3000 (over 1700 in the last 5 years and 443 in 2017); in both cases there has been a continuous growth, as one can see in Figure 4, which summarizes the data in a graphical form. It is interesting to note that, by classifying the 3095 articles of the latter database by country/region, it appears that over 50% of the authors are in Asia, only about 12.5% in America and as many in Europe; this classification, however, is far from being accurate, also because more than 20% of the Clarivate records do not contain data in the relevant field.

Total Publications

3.095

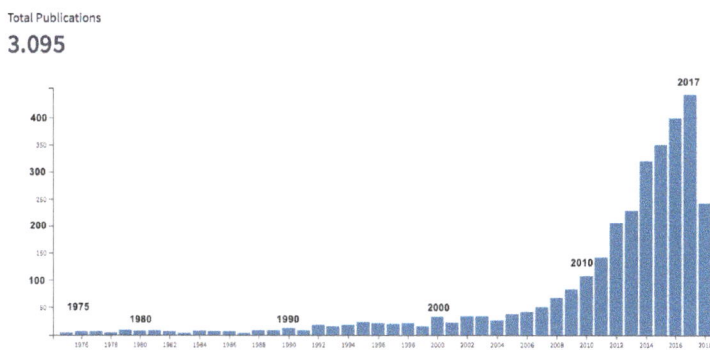

Figure 4. Number of publications with the word "microspher*" in the title and "energy" among the topics. Data from Clarivate Analytics Web of Science; search performed on 21 July 2018.

3.1. Energy Saving

Hollow glass and polymeric microspheres find wide application in the field of thermal insulation, owing to their distinctive properties, such as high compressive strength, low density, low water absorption, low heat conduction, and high chemical resistance. One of the ways hollow glass microspheres (HGM) help us to reduce energy consumption is their use in oil and gas drilling and extraction operations. In fact, HGM have good rolling characteristics, which can significantly improve the drilling performance; moreover, drilling fluids with HGMs exhibit high temperature resistance, high pressure resistance, stability, and durability, also inducing a longer lifetime of the drilling equipment [60,61]. The energy-saving applications of HGM, however, are particularly relevant in the construction sector, since the residential energy consumption is continuously increasing, especially due to the poor insulation of many buildings and to air conditioning, which in some cases can account for over 50% of the total electricity consumption of the building. A cost-effective solution to reduce this energy waste consists in minimizing the solar heat load and the heat dispersion through the roof and walls by using insulator coating materials that have low thermal conductivity and high infrared radiation reflectivity [62,63]; multiple layer thermal insulation coatings may be the most effective solution [63]. The thermal characteristics of HGM have been the subject of several papers, where different aspects were investigated, such as the mechanism of heat transfer in HGM [64,65] or the effect of inclusion of HGM in different materials [66–72]. Figure 5 shows a typical scanning electron microscope (SEM) image of soda-lime silicate glass microbubbles fabricated by Sinosteel Maanshan New Material Technology (Maanshan, China) [64]. In the figure, it can be clearly seen that the bubbles are in perfect spherical shape and with a rather broad size distribution.

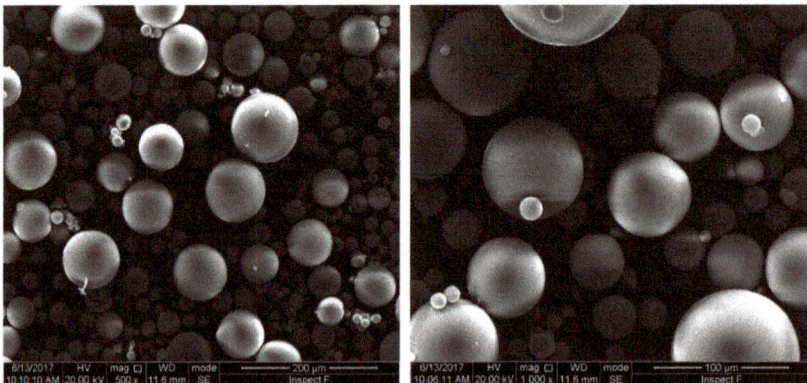

Figure 5. Two SEM (Scanning Electron Microscopy) images of HGM at different magnifications. The size of microbubbles ranges from ~10 μm to ~100 μm, most of them being in the interval 30 to 70 μm. Reproduced from [64] under Creative Commons license.

In the design of insulating structures, it is also important to investigate the long-term durability of the material as a function of different environmental parameters such as water, temperature, and pressure. As an example, Zhang et al. [72] developed a double layer coating composed of an anticorrosive epoxy ester primer and an HGM-containing silicone acrylic topcoat. The HGM size must be properly selected to provide balanced performance on both anticorrosion and heat insulation. An approach to achieve, together with high IR reflection, surface protection from fouling, and therefore longer lifetime, is based on the coating of the HGM by anatase TiO_2 and a superhydrophobic agent (PFOTES-1H,1H,2H,2H-Perfluorooctyltriethoxysilane) [73]. The utility of including HGM to enhance the thermal and mechanical properties of insulating foams has been proved for a long time [74]; in a recent work, a polysiloxane foam was prepared through foaming and crosslinking processes and reinforced with hollow microspheres, which had been modified with vinyl trimethoxysilane (VMS) to improve the compatibility between the filler and the matrix. The thermal stability and the mechanical properties of the reinforced foam were significantly enhanced: the HGM acted offering many nucleation sites, which was favorable in the formation of a uniform cell morphology, with the only disadvantage that they can easily aggregate in the polymer matrix. The foam with 5% VMS–HGM yielded a minimum thermal conductivity of 0.078 W/mK [75]. In the construction sector, HGM may also be used to partially replace Portland cement in a lightweight foamed concrete: depending on the percentage of HGM, one can obtain a higher compressive strength or a lower thermal conductivity, e.g., going from 0.2507 W/mK of the full cement to 0.2029 W/mK of the foam with 6% soda-lime glass HGM [76].

Hollow polymer microspheres (HPM), too, are largely used in insulating materials. As for HGM, surface modification of HPM may be necessary to improve the compatibility of the particles with the matrix; in fact, the low density (25 kg/m^3) of light HPM fillers produces a heterogeneous dispersion in polymer latex, which in turn may result in poor stability of the insulation coatings. Ye et al. [77] developed a simple process to produce nano-TiO_2/HPM core-shell composite particles and applied them in external wall thermal insulation coatings. By adding the novel material to a traditional coating and choosing the optimal volume ratio, the thermal conductivity was reduced about nine-fold, reaching 0.1687 W/mK [77]. Thermally expandable polymer microspheres (EPM) are one of the most widely used foaming additives used today; they consist of core/shell particles in which a blowing agent, typically a low boiling hydrocarbon, is encapsulated by a thermoplastic polymer shell [78–80]. When heated, the hydrocarbon pressure inside the polymer shell increases while the shell itself softens; thus, EPM expand to a target diameter and maintain that diameter after cooling. Fully expanded, each microsphere may increase up to fivefold its original diameter, with over 100 times increase in volume. Two major suppliers of EPM are AzkoNobel (Amsterdam, The Netherlands) [81] and Kureha (Tokyo, Japan) [82]. Sandin et al. [83] proved that EPM reflect solar radiation over a very broad band (UV, Vis, and NIR) much better than dense fillers, not only in traditional white roof coatings but also in tinted coatings. It is the efficient reflection of near-IR radiation which enables tinted cool roof coatings. Figure 6 presents the calculated total solar reflectance (R_{sol}) values for a color-matched blue coating of thickness 600 ± 50 μm which contains 30 vol% of different fillers (namely, $CaCO_3$, glass microspheres, ceramic microspheres, and thermoplastic microspheres). The relatively low reflectance values are obviously due to the absorption in the visible due to the blue color; the performance of the thermoplastic (EPM-containing) coating is clearly superior to the other coatings [83].

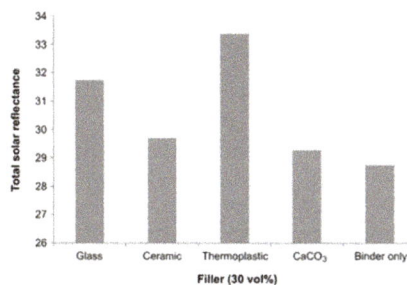

Figure 6. Comparison of the total solar reflectance R_{sol} for similar coatings using different fillers in the same quantity (30 vol%). The value for the binder only is also shown. R_{sol} is calculated by integrating the measured reflectance data in the interval 300 to 2500 nm. Reproduced from [83] under Creative Commons license.

Another interesting application of HGM has been recently proposed by Zhai et al., who demonstrated efficient day- and night-time radiative cooling by using a novel metamaterial which can be manufactured by a high-throughput, economical roll-to-roll process [84]. The concept of radiative cooling is well known: the energy of a hot body is released via the emission of infrared thermal radiation through the atmospheric window, with the heat being dumped directly to the outer space [85–87]. This metamaterial contains SiO_2 microspheres, with size in the range 4 to 8 μm, randomly distributed in a matrix material of polymethylpentene (TPX) that possesses an excellent solar transmittance. Since the encapsulated silica microspheres, too, have negligible absorption in the solar spectrum, the material is not heated by direct solar irradiance; moreover, it exhibits an infrared emissivity greater than 0.93 across the atmospheric window. When backed by a silver coating, 50 μm thick films showed a noontime radiative cooling power of 93 watts per square meter under direct sunshine, thus allowing to cool objects under direct sunlight with zero energy and water consumption [84].

3.2. Energy Storage and Production

In many cases, the technologies for energy production and energy storage are closely interconnected; let us consider here three examples of Ms&Mb use, referring to solar energy, fuel cells, and nuclear energy, respectively.

In the field of solar energy production, glassy microspheres have given only a marginal contribution; an example is represented by a flexible cover glass for solar panels in space applications. The cover glass, patented under the name of Pseudomorphic Glass (PMG) [88], consisted of sphere-like beads typically made of fused silica or ceria-doped borosilicate glasses and diameter of 20–40 μm, embedded in a polymer matrix. The glass can be sprayed onto the solar cells or can be manufactured in the form of microsheets that adhere to the solar cells; it proved to increase the efficiency and the UV transmittance with respect to conventional materials. Further, a multi-layer hybrid PMG cover glass using a thin top layer of ceria-doped borosilicate beads and a bottom layer of fused silica beads guarantees enhanced UV protection and a broadened spectral transmission bandwidth. Another patent was claiming to enhance the properties of an encapsulation adhesive film for a solar cell module; the film was made by mixing transparent microspheres (either polymeric or glassy or ceramic), with an average diameter in the range of 0.1 to 50 μm, together with an adhesive film [89]. The scattering and multiple reflections by the microspheres could improve light harvesting by the solar cell, thus increasing the electrical power generation. An increase in light trapping was also demonstrated in a periodic structure of microspheres deposited using a self-assembly method on the surface of a GaAs solar cell: an increase of about 25% in the conversion power efficiency of the cell was measured when using microspheres with size of 1 μm [90]. It may be interesting to note, even if the device is millimetric and not glassy, that the spherical shape has been adopted to make full solar cells, too: the Japanese company Kyosemi developed Sphelar, a spherical solar-cell technology that captures sunlight in three dimensions [91]. Single spherical cells are produced by dropping molten silicon into a tube, where the silicon droplets become rounded by surface tension during the free fall. Then, a proper process, which includes phosphorus diffusion and deposition of thin electrodes, creates a spherical p–n junction between the inner and outer parts of the crystalline sphere, with diameter of 1–2 mm. These tiny spherical Si solar cells can be incorporated into a variety of transparent materials, creating modules capable of covering a wide range of voltages [91].

In recent years, environment protection has pushed the search for vehicles having less harmful impacts to the environment than internal combustion engine vehicles running on gasoline or diesel. One of the best solutions appears to be that of electric vehicles, which, in turn, are expected to have great advantages from the use of fuel cells [92]; electric vehicles powered by fuel cells can travel for 500 km or more on a tankful of fuel, and—this is the best point—they can be refilled, as with a conventional car, in a matter of minutes rather than hours, unlike battery vehicles. Hydrogen storage in a small volume and light weight, however, is a significant challenge for the development and viability of hydrogen-powered vehicles [93,94]. Here it is where hollow glass microspheres can prove their capabilities: one can exploit the diffusion of hydrogen through the thin wall of an HGM at elevated temperatures and pressures, and then let the gas to be trapped upon cooling to room temperature. HGM with diameter in the range 1 to 100 μm, density between 1.0 and 2.0 gm/cc, and porous-wall structure with wall openings 1 to 100 nm represent a promising material for hydrogen storage, as demonstrated in recent papers and patents that have shown progress in the preparation and use of HGM for this application [95–99]. The storage of hydrogen at pressures up to 100MPa inside an HGM is possible due to the low diffusivity of hydrogen at room temperature; later, to release it, it is necessary to reheat the microspheres. However, a limitation of HGM has been the poor thermal conductivity, which implies unsuitably low release rates of hydrogen gas; to overcome this problem, a proposed solution consisted of doping the glass with transition metals. As an example, cobalt loaded HGM, prepared by mixing cobalt nitrate hexahydrate with the glass powder and using an air-acetylene flame for melting the particles and producing the microspheres, showed an increase of thermal conductivity from 0.072 to 0.198 W/mK when the cobalt loading increased from 0 to 10 wt.% [97]. Hydrogen adsorption capacity, however, has a maximum for cobalt loading at 2 wt.%; beyond 2%, the storage capacity is said to decrease due to the closure of the pores by the uneven deposition of CoO on the surface of the microspheres [97].

Proper strategies must be developed for applications in which rapid storage/release of stored gas is required: one of them is based on the photo-induced outgassing, in which an infrared light lamp is used to accelerate the release rate in comparison with furnace heating. To enhance this outgassing, one must follow the same approach used to increase the thermal conductivity, i.e., doping the glass with "optically active" dopants such as iron, nickel, and cobalt. Rapp and Shelby [100] worked on various borosilicate Corning glasses and various dopants and found a good response by the 0.5 wt.% Fe_3O_4-doped 7070 borosilicate glass; in this case, the amount of hydrogen released was proportional to the lamp intensity. Moreover, the reaction of hydrogen with the iron-doped glass increases the Fe^{2+}/Fe^{3+} ratio, which promotes infrared absorption and thus further enhances the hydrogen yield obtained from photo-induced outgassing. More recently, Shetty et al. [95] developed a facile flame spraying method for producing cobalt-doped HGM using recycled amber glass frit coated by a transition metal salt. They found that doping with 3 wt.% CoO was more effective at photo-induced outgassing than 1.5% wt.% CoO. In addition, a model was developed to estimate the conditions needed to produce HGM with engineered geometric properties, i.e., wall thickness and aspect ratio (diameter divided by the wall thickness).

It may also be worth to mention that crystalline hollow microspheres (e.g., vanadium pentoxide or multishelled TiO_2 and NiO microspheres) may play an important role in electrical energy storage, being used as safe, inexpensive anode materials for lithium ion batteries [101,102] and supercapacitors [103], respectively.

The property of HGM to be permeable to gases and their ability to safely store such high-pressure gases make them attractive for a very important application in the field of nuclear energy. As early as in late 1950s, a group at Lawrence Livermore National Laboratory in the USA was studying how to fill capsules with a mixture of deuterium and tritium (DT) to be compressed until reaching nuclear fusion. In 1972, Nuckolls et al. published a paper in *Nature* showing that the efficient laser thermonuclear burn of small pellets of DT was feasible, opening the way to the development of fusion power reactors [104]. Let us refer to a few recent publications to have an idea of the basic challenges and state of art of the research on inertial confinement fusion (ICF) [105–108]. Since the 1970s, much attention has been focused on the preparation of the nuclear fusion targets filled with hydrogen or its isotopes, since the success of any laser-fusion system depends critically on the low-cost production of suitable fuel capsules that satisfy the overall requirements. Hollow glass or polymer microspheres, with diameters in the range of approximately 50 to 500 μm and wall thickness between 1 and 20 μm appeared to be very good candidates; many papers and patents have been published concerning the fabrication and/or the filling of these microcapsules, and a few of them are cited here [109–120].

According to a recent review on the development of target fabrication for laser-driven ICF at the Research Center of Laser Fusion (RCLF) in China [121], glow discharge polymer (GDP), glass, and polystyrene (PS) hollow microspheres are among the candidates for the ultimate ignition. Let us just refer to the first type: GDP microballoons can be produced by inductive coupled plasma enhanced chemical vapor deposition (ICP-CVD), a method that permits the deposition of high quality dielectric films at low temperature with low damage. Trans-2-butene and H_2 were utilized as the working gases, and the GDP coating was deposited on mandrels made from poly α-methylstyrene (PAMS). A conical quartz tube used as the plasma generator allowed a fast growth rate of ~1.5 m/h; to get homogeneous coating of the mandrels, they were made to roll randomly inside a special designed glass pan. To produce single layered GDP microballoons after the coating deposition, the double layered PAMS/GDP spheres were annealed in vacuum or Ar atmosphere at 300 °C for more than 24 h, so to pyrolyze the mandrel. Typically, 8 μm thick single layered GDP shells with diameter of 450–540 μm were manufactured. Figure 7 summarizes the process [122]. It can be noted that the basic PAMS/GDP process for production of ICF target mandrels had been already tested in 1997 [118].

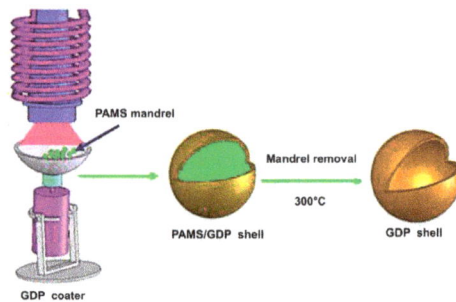

Figure 7. Schematic process for the fabrication of GDP shell. (GDP-glow discharge polymer; PAMS-poly α-methylstyrene). Reproduced with modifications from [122] under Creative Commons license.

The possibility of encapsulating several fuel-filled spheres in a low-density foam was also investigated and patented [123,124]; such a foam was requested to have a cell size smaller than 2 μm, a density of about 0.1×10^3 kg/m^2, and a chemical composition of low average atomic number.

With the increase of the available laser power, the design of the targets has become increasingly complex, and several structured target configurations have been reported, often comprising a multilayer structure [125]. One of the layers usually is a low-atomic-number polymer coating that must ablate as the laser pulse irradiates its surface: the ablation imparts a reaction force to the core material, causing the fuel within to be compressed to high density. The polymeric layers must have a predetermined thickness and a surface finish smoother than 0.1 μm and they must conform perfectly to the glass sphere; the deposition technique is therefore very important [114,126,127]. In recent years, laser-fusion programs seem to have moved to consider larger fuel capsules [121,128]; viable ICF targets are represented by spherical shells with diameter 0.5 to 4 mm, wall thickness 50–100 μm, low density (~250 mg/cm^3), with interconnected voids (each <1 μm diameter), with extreme sphericity (>99.9%, <50 nm roughness variation), and a high degree of concentricity (>99.0%) [129]. Fabricating these pellets with so stringent specifications is a big technical challenge, and even more challenging is the fact that they should be produced at massive scale. In fact, a reliable and economic fuel supply is essential for the viability of future ICF power plants, where the problem nowadays is not the pellet's content, namely DT, but the container itself, namely the spherical capsule. It is estimated that six targets per second, or about 500,000/day, with a cost below 0.25 $/target (orders of magnitude less than current costs), will be required for a power plant with nominal electric output of 1000 MW [130]. The efforts to improve the quality of the targets [131] and to develop the possibility of their large-scale production have made significant progress in recent years. As an example, Li et al. developed a continuous and scalable process for the fabrication of polymer capsules using droplet microfluidics, thus demonstrating that, even with the many remaining limitations, channel-based droplet microfluidics technology has the potential of being applied to ICF target fabrication [129].

4. Conclusions

Scientific and commercial applications of solid and hollow glass and polymer microspheres have been continuously growing in recent decades, in parallel with the advances in their fabrication with high quality and large batches. On the other hand, single microspheres have also gained much attention for their potential as ultra-high quality-factor WGM optical microresonators [2]. One of the advantages of glassy microspheres is that they can be easily doped with chemical elements and compounds to increase their functionality; moreover, they can be made porous or hollow, allowing for encapsulation of other chemical or biomedically relevant components. All these properties open the way to develop microlasers, microsensors, biolabeling or drug-delivery bullets or even to study matter-radiation interactions at the very high power density made possible by the strong light confinement of WGMs [132–136]. Industrial applications are also mature, and it appears that there is a constant or increasing demand from the healthcare and construction (e.g., paints and coatings) sector. Much of the ongoing research relates to the advances in the polymer industry because the possibility of changing the molecular properties (and hence the chemical and physical properties) permits to conceive new applications of polymeric microspheres.

In the field of energy, glassy microspheres (and nanospheres) may offer effective solutions to some of the problems that arise in the advanced technologies of energy generation. At a lower technological level, they can be used for thermal insulation, or as proppants for shale oil and gas recovery, components for solar cells, or anodes for electrical batteries. In all these cases, the synthesis process of organic or inorganic microspheres does not present significant difficulties and the width of size distribution of the produced particles is not a critical factor; current research is aimed at optimizing the choice of the mean size and the usage protocol for the specific application considered. There is, however, a research topic of growing interest where the size of the micrometric and sub-micrometric spheres must be controlled: it is the case of photonic crystal structures and arrays of resonators, whose properties may be exploited to improve light harvesting in solar cells [137–140]. A deeper study of the self-organization processes of colloidal particles and of the process parameters would be important to optimize the design and implementation of photonic crystal structures [137,139].

At a higher technological level, hollow glass and polymer microspheres have a winning potential in two areas, namely hydrogen storage and ICF, which are of great importance for decreasing today's CO_2 and particles' emissions and taking care of the environment. In both applications, the peculiar properties of hollow porous microspheres can be fully exploited. In particular, the fabrication of the shell targets used in ICF experiments is a challenging project that involves different disciplines and advanced material preparation and manufacture process.

Many more challenges and difficulties can be seen ahead, e.g., to develop reliable, very high quality and economic mass-production technologies, but the horizon appears sunny for these microdevices.

Acknowledgments: The support of Museo Storico della Fisica e Centro Studi e Ricerche Enrico Fermi (aka Enrico Fermi Centre) through the Photonic Microcavities (MIFO) project is gratefully acknowledged. This work is part of the activity of the TC20 Committee of ICG (International Commission on Glass).

Conflicts of Interest: The authors declare no conflict of interest.

References

1. IEA. Global Energy & CO_2 Status Report. 2017. Available online: https://www.iea.org/publications/freepublications/publication/GECO2017.pdf (accessed on 10 May 2018).
2. Chiasera, A.; Dumeige, Y.; Féron, P.; Ferrari, M.; Jestin, Y.; Nunzi Conti, G.; Pelli, S.; Soria, S.; Righini, G.C. Spherical whispering-gallery-mode microresonators. *Laser Photonics Rev.* **2010**, *4*, 457–482. [CrossRef]
3. Righini, G.C.; Dumeige, Y.; Féron, P.; Ferrari, M.; Nunzi Conti, G.; Ristic, D.; Soria, S. Whispering gallery mode microresonators: Fundamentals and applications. *Riv. Nuovo Cimento* **2011**, *34*, 435–488.

4. Righini, G.C. *Glass micro- and nanospheres. Physics and applications*; Pan Stanford Publishing: Singapore, 2018.
5. Ghalichechian, N.; Modafe, A.; Beyaz, M.I.; Ghodssi, R. Design, Fabrication, and characterization of a rotary micromotor supported on microball bearings. *J. Microelectromech. Syst.* **2008**, *17*, 632–642. [CrossRef]
6. Yi, R.; Shi, R.; Gao, G.; Zhang, N.; Cui, X.; He, Y.; Liu, X. Hollow metallic microspheres: Fabrication and characterization. *J. Phys. Chem. C* **2009**, *113*, 1222–1226. [CrossRef]
7. Deng, H.; Li, X.; Peng, Q.; Wang, X.; Chen, J.; Li, Y. Monodisperse magnetic single-crystal ferrite microspheres. *Angew. Chem. Int. Ed.* **2005**, *44*, 2782–2785. [CrossRef] [PubMed]
8. Okamoto, S.; Inaba, K.; Iida, T.; Ishihara, H.; Ichikawa, S.; Ashida, M. Fabrication of single-crystalline microspheres with high sphericity from anisotropic materials. *Sci. Rep.* **2014**, *4*, 5186. [CrossRef] [PubMed]
9. Gulyaev, I. Experience in plasma production of hollow ceramic microspheres with required wall thickness. *Ceram. Int.* **2015**, *41*, 101–107. [CrossRef]
10. Fan, K.C.; Cheng, F.; Wang, W.; Chen, Y.; Lin, J.Y. A scanning contact probe for a micro-coordinate measuring machine (CMM). *Meas. Sci. Technol.* **2010**, *21*, 054002. [CrossRef]
11. Ruan, Y.; Boyd, K.; Ji, H.; Francois, A.; Ebendorff-Heidepriem, H.; Munch, J.; Monro, T.M. Tellurite microspheres for nanoparticle sensing and novel light sources. *Opt. Express* **2014**, *22*, 11995–12006. [CrossRef] [PubMed]
12. Wang, P.; Murugan, G.S.; Lee, T.; Ding, M.; Brambilla, G.; Semenova, Y.; Wu, Q.; Koizumi, F.; Farrell, G. High-Q bismuth-silicate nonlinear glass microsphere resonators. *IEEE Photonics J.* **2012**, *4*, 1013–1020. [CrossRef]
13. Yu, H.; Huang, Q.; Zhao, J. Fabrication of an optical fiber micro-sphere with a diameter of several tens of micrometers. *Materials* **2014**, *7*, 4878–4895. [CrossRef] [PubMed]
14. Nair, V.N.; Abraham, B.; MacKay, J.; Box, G.; Kacker, R.N.; Lorenzen, T.J.; Lucas, J.M.; Myers, R.H.; Vining, G.G.; Nelder, J.A.; et al. Taguchi's parameter design: A panel discussion. *Technometrics* **1992**, *34*, 127–161. [CrossRef]
15. Brenci, M.; Calzolai, R.; Cosi, F.; Nunzi Conti, G.; Pelli, S.; Righini, G.C. Microspherical resonators for biophotonic sensors. In Proceedings of the International Society for Optical Engineering (SPIE), Warsaw, Poland, 20–22 October 2004.
16. Yano, T.; Kishi, T.; Kumagai, T. Glass microspheres for optics. *Inter. J. Appl. Glass Sci.* **2015**, *6*, 375–386. [CrossRef]
17. Nunzi Conti, G.; Chiasera, A.; Ghisa, L.; Berneschi, S.; Brenci, M.; Dumeige, Y.; Pelli, S.; Sebastiani, S.; Féron, P.; Ferrari, M.; et al. Spectroscopic and lasing properties of Er^{3+}-doped glass microspheres. *J. Non-Cryst. Solids* **2006**, *352*, 2360–2363. [CrossRef]
18. Peng, X.; Song, F.; Jiang, S.; Peyghambarian, N.; Kuwata-Gonokami, M.; Xu, L. Fiber-taper-coupled L-band Er^{3+}-doped tellurite glass microsphere laser. *Appl. Phys. Lett.* **2003**, *82*, 1497–1499. [CrossRef]
19. Berneschi, S.; Baldini, F.; Barucci, A.; Cosci, A.; Cosi, F.; Farnesi, D.; Nunzi Conti, G.; Righini, G.C.; Soria, S.; Tombelli, S.; et al. Localized biomolecules immobilization in optical microbubble resonators. In Proceedings of the 2016 SPIE, Laer and Applications, San Francisco, CA, USA, 16–18 February 2016.
20. Yang, Y.; Ward, J.; Chormaic, S.N. Quasi-droplet microbubbles for high resolution sensing applications. *Opt. Express* **2014**, *22*, 6881–6898. [CrossRef] [PubMed]
21. Veatch, F.; Alford, H.E.; Croft, R.D. Method of Producing Hollow Glass Spheres. U.S. Patent 2,978,339, 4 April 1961.
22. MicroMarketMonitor. Available online: http://www.micromarketmonitor.com/market-report/glass-microspheres-reports-6570147015.html (accessed on 18 May 2018).
23. Elliott, S.R. Chalcogenide phase-change materials: Past and future. *Int. J. Appl. Glass Sci.* **2015**, *6*, 15–18. [CrossRef]
24. Sanghera, J.; Shaw, L.B.; Aggarwal, I.D. Applications of chalcogenide glass optical fibers. *C. R. Chim.* **2002**, *5*, 873–883. [CrossRef]
25. Elliott, G.R.; Hewak, D.W.; Murugan, G.S.; Wilkinson, J.S. Chalcogenide glass microspheres; their production, characterization and potential. *Opt. Express* **2007**, *15*, 17542–17553. [CrossRef] [PubMed]
26. Ahmad, H.; Aryanfar, I.; Lim, K.S.; Chong, W.Y.; Harun, S.W. Thermal response of chalcogenide microsphere resonators. *Quantum Electron.* **2012**, *42*, 462–464. [CrossRef]

27. Palma, G.; Bia, P.; Mescia, L.; Yano, T.; Nazabal, V.; Taguchi, J.; Moréac, A.; Prudenzano, F. Design of fiber coupled Er^{3+}: Chalcogenide microsphere amplifier via particle swarm optimization algorithm. *Opt. Eng.* **2013**, *53*, 071805. [CrossRef]

28. Palma, G.; Falconi, M.C.; Starecki, F.; Nazabal, V.; Yano, T.; Kishi, T.; Kumagai, T.; Prudenzano, F. Novel double step approach for optical sensing via microsphere WGM resonance. *Opt. Express* **2016**, *24*, 26956–26971. [CrossRef] [PubMed]

29. Aldakov, D.; Lefrançois, A.; Reiss, P. Ternary and quaternary metal chalcogenide nanocrystals: Synthesis, properties and applications. *J. Mater. Chem. C* **2013**, *1*, 3756–3776. [CrossRef]

30. Cospheric. Available online: http://www.cospheric.com/polyethylene_PE_microspheres_beads.htm (accessed on 9 June 2018).

31. Polysciences. Available online: http://www.polysciences.com/default/catalog-products/microspheres-particles/polymer-microspheres/polybead-sup-r-sup-microspheres/polybead-sup-r-sup-non-functionalized-microspheres (accessed on 10 May 2018).

32. Magsphere. Available online: http://www.magsphere.com/Products/Polystyrene-Latex-Particle/polystyrene-latex-particle.html (accessed on 10 May 2018).

33. Microspheres-Nanospheres. Available online: http://www.microspheres-nanospheres.com/Microspheres/Organic/Polystyrene/PS%20Plain.htm (accessed on 10 May 2018).

34. Goodfellow. Available online: http://www.goodfellow.com/pdf/4579_1111010.pdf (accessed on 21 July 2018).

35. Microbeads. Available online: http://www.micro-beads.com/Products.aspx (accessed on 10 May 2018).

36. Degradex. Available online: http://www.degradex.com/pmma-microspheres.html (accessed on 10 May 2018).

37. Dong, C.-H.; He, L.; Xiao, Y.-F.; Gaddam, V.R.; Ozdemir, S.K.; Han, Z.-F.; Guo, G.-C.; Yang, L. Fabrication of high-Q polydimethylsiloxane optical microspheres for thermal sensing. *Appl. Phys. Lett.* **2009**, *94*, 231119. [CrossRef]

38. Grilli, S.; Coppola, S.; Vespini, V.; Merola, F.; Finizio, A.; Ferraro, P. 3D lithography by rapid curing of the liquid instabilities at nanoscale. *Proc. Natl. Acad. Sci. USA* **2011**, *108*, 15106–15111. [CrossRef] [PubMed]

39. Ma, B.; Hansen, J.H.; Hvilsted, S.; Ladegaard Skov, A. Polydimethylsiloxane microspheres with poly(methyl methacrylate) coating: Modelling, preparation, and characterization. *Can. J. Chem. Eng.* **2015**, *93*, 1744–1752. [CrossRef]

40. Gu, G.; Chen, L.; Fu, H.; Che, K.; Cai, Z.; Xu, H. UV-curable adhesive microsphere whispering gallery mode resonators. *Chin. Opt. Lett.* **2013**, *11*, 101401.

41. Omi, S.; Katami, K.; Yamamoto, A.; Iso, M. Synthesis of polymeric microspheres employing SPG emulsification technique. *J. Appl. Polym. Sci.* **1994**, *51*, 1–11. [CrossRef]

42. Nussinovitch, A. *Polymer Macro- and Micro-Gel Beads: Fundamentals and Applications*; Springer Sci. & Business Media: Berlin, German, 2010.

43. Kumacheva, E.; Garstecki, P. *Microfluidic Reactors for Polymer Particles*; John Wiley & Sons: Hoboken, NJ, USA, 2011.

44. Rembaum, A.; Tokes, Z.A. *Microspheres: Medical and Biological Applications (1988)*; CRC Press Revivals: Boca Raton, FL, USA, 2017.

45. Senior, R. Imagify™ (perflubutane polymer microspheres) injectable suspension for the assessment of coronary artery disease. *Expert Rev. Cardiovasc. Ther.* **2007**, *5*, 413–421. [CrossRef] [PubMed]

46. Farook, U.; Edirisinghe, M.J.; Stride, E.; Colombo, P. Novel co-axial electrohydrodynamic in-situ preparation of liquid-filled polymer-shell microspheres for biomedical applications. *J. Microencapsul.* **2008**, *25*, 241–247. [CrossRef] [PubMed]

47. Wei, B.; Wang, S.; Song, H.; Liu, H.; Li, J.; Liu, N. A review of recent progress in preparation of hollow polymer microspheres. *Petrol. Sci.* **2009**, *6*, 306–312. [CrossRef]

48. Cai, P.J.; Tang, Y.J.; Wang, Y.T.; Cao, Y.J. Fabrication of polystyrene hollow spheres in W/O/W multiple emulsions. *Mater. Chem. Phys.* **2010**, *124*, 10–12. [CrossRef]

49. Xiong, X.; Zhao, F.; Shi, M.; Yang, H.; Liu, Y. Polymeric microbubbles for ultrasonic molecular imaging and targeted therapeutics. *J. Biomater. Sci. Polym. Ed.* **2011**, *22*, 417–422. [CrossRef] [PubMed]

50. Skinner, E.K. Sonochemical Production of Hollow Polymer Microspheres for Responsive Delivery. Ph.D. Thesis, University of Bath, Bath, UK, 2013.

51. Jiang, X.; Lin, S.; Rempel, G.L.; Pan, Q. Preparation of Monodisperse Hollow Core Polymer Microspheres via Two-step Dispersion Polymerization. In Proceedings of the 2nd International Conference on Civil, Materials and Environmental Sciences (CMES 2015), London, UK, 13–14 March 2015.

52. Davis, S.S. *Microspheres and Drug Therapy: Pharmaceutical, Immunological, and Medical Aspects*; Elsevier: New York, NY, USA, 1984.

53. Guiot, P.; Couvreur, P. *Polymeric Nanoparticles and Microspheres*; CRC Press: Boca Raton, FL, USA, 1986.

54. Budov, V.V. Hollow glass microspheres. Use, properties, and technology (Review). *Glass Ceram.* **1994**, *51*, 230–235. [CrossRef]

55. Kim, K.K.; Pack, D.W. Microspheres for Drug Delivery. In *BioMEMS and Biomedical Nanotechnology*; Ferrari, M., Lee, A.P., Lee, L.J., Eds.; Springer: Berlin, Germany, 2006.

56. Wilcox, D.L.; Berg, M. Microsphere Fabrication and Applications. An Overview. *MRS Online Proc. Lib. Arch.* **1998**, *372*, 372. [CrossRef]

57. Ma, G. *Microspheres and Microcapsules in Biotechnology-Design, Preparation and Applications*; Pan Stanford Publishing: Singapore, 2013.

58. Ganesan, P.; Johnson, A.J.D.; Sabapathy, L.; Duraikannu, A. Review on Microsphere. *Am. J. Drug Discov. Dev.* **2014**, *4*, 153–179. [CrossRef]

59. Amos, S.E.; Yalcin, B. *Hollow Glass Microspheres for Plastics, Elastomers, and Adhesives Compounds*; Elsevier: New York, NY, USA, 2015.

60. Minhas, A.; Friess, B.; Shirkavand, F.; Hucik, B.; Pena-Bastidas, T.; Ross, B.; Servinski, S.; Angyal, F. Hollow-glass sphere application in drilling fluids: Case study. In Proceedings of the SPE Western Regional Meeting, Garden Grove, CA, USA, 27–30 April 2015.

61. Ari, T.C.; Akin, S. An experimental study on usage of hollow glass spheres (HGS) for reducing mud density in geothermal drilling. In Proceedings of the World Geothermal Congress, Melbourne, Australia, 19–25 April 2015; pp. 1–7.

62. Synnefa, H.; Santamouris, M.; Akbari, H. Estimating the effect of using cool coatings on energy loads and thermal comfort in residential buildings in various climatic conditions. *Energy. Build.* **2007**, *39*, 1167–1174. [CrossRef]

63. Zhang, H.; Wang, F.; Liang, J.; Tang, Q.; Chen, Y. Design of thermal insulation energy-saving coatings for exterior wall. *Chem. Eng. Trans.* **2017**, *61*, 1207–1212.

64. Liu, B.; Wang, H.; Qin, Q.-H. Modelling and characterization of effective thermal conductivity of single hollow glass microsphere and its powder. *Materials* **2018**, *11*, 133. [CrossRef] [PubMed]

65. Li, B.; Yuan, J.; An, Z.G.; Zhang, J.J. Effect of microstructure and physical parameters of hollow glass microsphere on insulation performance. *Mater. Lett.* **2011**, *65*, 1992–1994. [CrossRef]

66. Allen, M.S.; Baumgartner, R.G.; Fesmire, J.E.; Augustynowicz, S.D. Advances in microsphere insulation systems. *AIP Conf. Proc.* **2004**, *710*, 619.

67. Liang, J.Z.; Li, F.H. Measurement of thermal conductivity of hollow glass-bead-filled polypropylene composites. *Polym. Test.* **2006**, *25*, 527–531. [CrossRef]

68. Park, Y.K.; Kim, J.G.; Lee, J.K. Prediction of thermal conductivity of composites with spherical microballoons. *Mater. Trans.* **2008**, *49*, 2781. [CrossRef]

69. Yung, K.C.; Zhu, B.L.; Yue, T.M.; Xie, C.S. Preparation and properties of hollow glass microsphere-filled epoxy-matrix composites. *Compos. Sci. Technol.* **2009**, *69*, 260–264. [CrossRef]

70. Zhu, B.; Ma, J.; Wang, J.; Wu, J.; Peng, D. Thermal, dielectric and compressive properties of hollow glass microsphere filled epoxy-matrix composites. *J Reinf. Plast. Compos.* **2012**, *31*, 1311–1326. [CrossRef]

71. Hsu, C.C.; Chang, K.C.; Huang, T.C.; Yeh, L.C.; Yeh, W.-T.; Ji, W.-F.; Yeh, J.-M.; Tsai, T.-Y. Preparation and studies on properties of porous epoxy composites containing microscale hollow epoxy spheres. *Micropor. Mesopor. Mater.* **2014**, *198*, 15–21. [CrossRef]

72. Zhang, D.; Li, H.; Qian, H.; Wang, L.; Li, X. Double layer water-borne heat insulation coatings containing hollow glass microspheres (HGMs). *Pigment Resin Technol.* **2016**, *45*, 346–353. [CrossRef]

73. Wong, Y.; Zhong, D.; Song, A.; Hu, Y. TiO$_2$-coated hollow glass microspheres with superhydrophobic and high IR-reflective properties synthesized by a soft-chemistry method. *J. Vis. Exp.* **2017**, *122*. [CrossRef]

74. Chukhlanov, V.Yu.; Sysoev, E.P. Use of hollow glass microspheres in organosilicon syntact foam plastics. *Glass Ceram.* **2000**, *57*, 47–48. [CrossRef]

75.	Zhang, C.; Zhang, C.; Huang, R.; Gu, X. Effects of hollow microspheres on the thermal insulation of polysiloxane foam. *J. Appl. Polym. Sci.* **2017**, *134*. [CrossRef]

76.	Shahidan, S.; Aminuddin, E.; Mohd Noor, K.; Ramzi Hannan, N.I.R.; Saiful Bahari, N.A. Potential of hollow glass microsphere as cement replacement for lightweight foam concrete on thermal insulation performance. In Proceedings of the International Symposium on Civil and Environmental Engineering 2016, Wuhan, China, 20–21 December 2016.

77.	Ye, C.; Wen, X.; Lan, J.-L.; Cai, Z.-Q.; Pi, P.-H.; Xu, S.-P.; Qian, Y. Surface modification of light hollow polymer microspheres and its application in external wall thermal insulation coatings. *Pigment Resin Technol.* **2016**, *45*, 45–51. [CrossRef]

78.	Morehouse, D.S., Jr.; Midland, M.; Tetreault, R.J. Expansible Thermoplastic Polymer Particles Containing Volatile Fluid Foaming Agent and Method of Foaming the Same. U.S. Patent 3,615,972, 26 October 1971.

79.	Glorioso, S.Jr.; Burgess, J.H.; Tang, J.; Dimonie, V.L.; Klein, A. Expandable Microspheres for Foam Insulation and Methods. U.S. Patent 8,088,482, 3 January 2012.

80.	Jonsson, M.; Nordin, M.; Malmstrom, M.; Hammer, C. Suspension polymerization of thermally expandable core/shell particles. *Polymer* **2006**, *47*, 3315–3324. [CrossRef]

81.	AzkoNobel. Available online: https://expancel.akzonobel.com (accessed on 5 June 2018).

82.	Kureha. Available online: http://www.kureha.co.jp/en/business/material/microspheres.html (accessed on 5 June 2018).

83.	Sandin, O.; Nordin, J.; Jonsson, M. Reflective properties of hollow microspheres in cool roof coatings. *J. Coat. Technol. Res.* **2017**, *14*, 817–821. [CrossRef]

84.	Zhai, Y.; Ma, Y.; David, S.N.; Zhao, D.; Lou, R.; Tan, G.; Yang, R.; Yin, X. Scalable manufactured randomized glass-polymer hybrid metamaterial for day-time radiative cooling. *Science* **2017**, *355*, 1062–1066. [CrossRef] [PubMed]

85.	Catalanotti, S.; Cuomo, V.; Piro, G.; Ruggi, D.; Silvestrini, V.; Troise, G. The radiative cooling of selective surfaces. *Sol. Energy* **1975**, *17*, 83–89. [CrossRef]

86.	Michell, D.; Biggs, K.L. Radiation cooling of buildings at night. *J. Appl. Energy* **1979**, *5*, 263–275. [CrossRef]

87.	Hossain, M.M.; Gu, M. Radiative cooling: Principles, progress, and potentials. *Adv. Sci.* **2016**, *3*, 1500360. [CrossRef] [PubMed]

88.	Wilt, D.M. Pseudomorphic Glass for Space Solar Cells. U.S. Patent 8,974,899, 10 March 2015.

89.	Lin, J.; Ding, H.; Li, Z. Encapsulation Adhesive Film for Solar Cell Module. U.S. Patent 20,160,329,447, 10 November 2016.

90.	Chang, T.-H.; Wu, P.-H.; Chen, S.-H.; Chan, C.-H.; Lee, C.-C.; Chen, C.-C.; Su, Y.-K. Efficiency enhancement in GaAs solar cells using self-assembled microspheres. *Opt. Express* **2009**, *17*, 6519–6524. [CrossRef] [PubMed]

91.	Taira, K.; Nakata, J. Catching rays. *Nat. Photon.* **2010**, *4*, 602–603. [CrossRef]

92.	O'Hayre, R.; Cha, S.-W.; Prinz, F.B.; Colella, W. *Fuel Cell Fundamentals*; John Wiley & Sons: Hoboken, NJ, USA, 2016.

93.	Lim, K.L.; Kazemian, H.; Yaakob, Z.; Daud, W.R.W. Solid-state materials and methods for hydrogen storage: A critical review. *Chem. Eng. Technol.* **2010**, *33*, 213–226. [CrossRef]

94.	Durbin, D.J.; Malardier-Jugroot, C. Review of hydrogen storage techniques for on board vehicle applications. *Int. J. Hydrogen Energy* **2013**, *38*, 14595–14617. [CrossRef]

95.	Shetty, S.; Hall, M. Facile production of optically active hollow glass microspheres for photo-induced outgassing of stored hydrogen. *Int. J. Hydrogen Energy* **2011**, *36*, 9694–9701. [CrossRef]

96.	Qi, X.; Gao, C.; Zhang, Z.; Chen, S.; Li, B.; Wei, S. Production and characterization of hollow glass microspheres with high diffusivity for hydrogen storage. *Int. J. Hydrogen Energy* **2012**, *37*, 1518–1530. [CrossRef]

97.	Dalai, S.; Savithri, V.; Sharma, P. Investigating the effect of cobalt loading on thermal conductivity and hydrogen storage capacity of hollow glass microspheres (HGMs). *Mat. Today Proc.* **2017**, *4*, 11608–11616. [CrossRef]

98.	Lipinska-Kalita, K.; Hemmers, O. Hollow Glass Microsphere Candidates for Reversible Hydrogen Storage, Particularly for Vehicular Applications. U.S. Patent 8,663,429, 4 March 2014.

99.	Schmid, G.H.S.; Bauer, J.; Eder, A.; Eisenmenger-Sittner, C. A hybrid hydrolytic hydrogen storage system based on catalyst-coated hollow glass microspheres. *Int. J. Energy Res.* **2017**, *41*, 297–314. [CrossRef]

100. Rapp, D.B.; Shelby, J.E. Photo-induced hydrogen outgassing of glass. *J. Non-Cryst. Solids* **2004**, *349*, 254–259. [CrossRef]

101. Tian, P.; Song, Q.; Pang, H.; Ning, G. Hollow microspherical vanadium pentoxide fabricated via non-hydrothermal route for lithium ion batteries. *Mater. Lett.* **2018**, *227*, 13–16. [CrossRef]

102. Ren, H.; Yu, R.; Wang, J.; Jin, Q.; Yang, M.; Mao, D.; Kisailus, D.; Zhao, H.; Wang, D. Multishelled TiO$_2$ hollow microspheres as anodes with superior reversible capacity for lithium Ion batteries. *Nano Lett.* **2014**, *14*, 6679–6684. [CrossRef] [PubMed]

103. Qi, X.; Zheng, W.; Li, X.; He, G. Multishelled NiO hollow microspheres for high-performance supercapacitors with ultrahigh energy density and robust cycle life. *Sci. Rep.* **2016**, *6*, 33241. [CrossRef] [PubMed]

104. Nuckolls, J.; Wood, L.; Thiessen, A.; Zimmerman, G. Laser compression of matter to super-high densities: Thermonuclear (CTR) applications. *Nature* **1972**, *239*, 139–142. [CrossRef]

105. Pfalzner, S. *An Introduction to Inertial Confinement Fusion*; CRC Press: Boca Raton, FL, USA, 2006.

106. Craxton, R.S.; Anderson, K.S.; Boehly, T.R.; Goncharov, V.N.; Harding, D.R.; Knauer, J.P.; McCrory, R.L.; McKenty, P.W.; Meyerhofer, D.D.; Myatt, J.F.; et al. Direct-drive inertial confinement fusion: A review. *Phys. Plasmas* **2015**, *22*, 110501. [CrossRef]

107. Betti, R.; Hurricane, O.A. Inertial-confinement fusion with lasers. *Nat. Phys.* **2016**, *12*, 435–448. [CrossRef]

108. Zohuri, B. *Inertial Confinement Fusion Driven Thermonuclear Energy*; Springer: Berlin, Germany, 2017.

109. Lewkowicz, I. Spherical hydrogen targets for laser-produced fusion. *J. Phys. D Appl. Phys.* **1974**, *7*, L61–L62. [CrossRef]

110. Solomon, D.E.; Henderson, T.M. Laser fusion targets. *J. Phys. D Appl. Phys.* **1975**, *8*, L85–L86. [CrossRef]

111. Hendricks, C.D.; Rosencwaig, A.; Woerner, R.L.; Koo, J.C.; Dressler, J.L.; Sherohman, J.W.; Weinland, S.L.; Jeffries, M. Fabrication of glass sphere laser fusion targets. *J. Nucl. Mater.* **1979**, *85–86*, 107–111. [CrossRef]

112. Koo, J.; Dressler, J.; Hendricks, C. Low pressure gas filling of laser fusion microspheres. *J. Nucl. Mater.* **1979**, *85–86*, 113–115. [CrossRef]

113. Nogami, M.; Moriya, Y.; Hayakawa, J.; Komiyama, T. Fabrication of hollow glass microspheres for laser fusion targets from metal alkoxides. *Rev. Laser Eng.* **1980**, *8*, 793–797. [CrossRef]

114. Peiffre, D.; Corley, T.; Halpern, G.; Brinker, B. Utilization of polymeric materials in laser fusion target fabrication. *Polymer* **1981**, *22*, 450–460. [CrossRef]

115. Deckman, H.W.; Halpern, G.M.; Dunsmuir, J.G. Method for Filling Hollow Shells with Gas for Use as Laser Fusion Targets. U.S. Patent 4,380,855, 26 April 1983.

116. Elsholz, W.E. Fabrication of Glass Microspheres with Conducting Surfaces. U.S. Patent 4,459,145, 10 July 1984.

117. Norimatsu, T.; Kato, Y.; Nakai, S. Target fabrication for laser fusion research in Japan. *J. Vac. Sci. Technol. A* **1989**, *7*, 1165. [CrossRef]

118. McQuillan, B.W.; Nikroo, A.; Steinman, D.A.; Elsner, F.H.; Czechowicz, D.G.; Hoppe, M.L.; Sixtus, M.; Miller, W.J. The PAMS/GDP process for production of ICF target mandrels. *Fusion Technol.* **1997**, *31*, 381–384. [CrossRef]

119. Mishra, K.; Khardekar, R.; Singh, R.; Pant, H.C. Fabrication of polystyrene hollow microspheres as laser fusion targets by optimized density-matched emulsion technique and characterization. *Pramana J. Phys.* **2002**, *59*, 113–131. [CrossRef]

120. Qi, X.; Gao, C.; Zhang, Z.; Chen, S.; Li, B.; Wei, S. Fabrication and characterization of millimeter-sized glass shells for inertial confinement fusion targets. *Chem. Eng. Res. Des.* **2013**, *91*, 2497–2508. [CrossRef]

121. Veselov, A.V.; Drozhin, V.S.; Druzhinin, A.A.; Izgorodin, B.N.; Iiyushechkin, V.M.; Kirillov, G.A.; Komleva, G.V.; Korochkin, A.M.; Medvedev, E.F.; Nikolaev, G.P.; et al. ICF target technology at the Russian federal nuclear center. *Fusion Technol.* **1995**, *28*, 1838–1843. [CrossRef]

122. Wang, T.; Du, K.; He, Z.; He, X. Development of target fabrication for laser-driven inertial confinement fusion at research center of laser fusion. *High Power Laser Sci. Eng.* **2017**, *5*, 25–33. [CrossRef]

123. Rinde, J.A.; Fulton, F.J. Method of Making Foam-Encapsulated Laser Targets. U.S. Patent 4,021,280, 5 March 1977.

124. Hendricks, C.D. Method for Foam Encapsulating Laser Targets. U.S. Patent 4,034,032, 7 May 1977.

125. Yaakobi, B.; Skupsky, S.; McCrory, R.L.; Hooper, C.F.; Deckman, H.; Bourke, P.; Soures, J.M. Symmetric laser compression of argon-filled glass shells to densities of 4–6 g/cm^3. *Phys. Rev. Lett.* **1980**, *44*, 1072–1075. [CrossRef]

126. Liepins, R.; Campbell, M.; Fries, R. Plastic coatings for laser fusion targets. *Progr. Polym. Sci.* **1980**, *6*, 169–186. [CrossRef]

127. Mishra, K.K.; Khardekar, R.K.; Chouhan, R.; Gupta, R.K. A simple and efficient levitation technique for noncontact coating of inertial confinement fusion targets. *Pramana J. Phys.* **2000**, *55*, 919–925. [CrossRef]

128. Cuneo, M.E.; Vesey, R.A.; Bennett, G.R.; Sinars, D.B.; Stygar, W.A.; Waisman, E.M.; Porter, J.L.; Rambo, P.K.; Smith, I.C.; Lebedev, S.V.; et al. Progress in symmetric ICF capsule implosions and wire-array z-pinch source physics for double-pinch-driven hohlraums. *Plasma Phys. Control. Fusion* **2006**, *48*, R1–R35. [CrossRef]

129. Li, J.; Lindley-Start, J.; Porch, A.; Barrow, D. Continuous and scalable polymer capsule processing for inertial fusion energy target shell fabrication using droplet microfluidics. *Sci. Rep.* **2017**, *7*, 6302. [CrossRef] [PubMed]

130. Goodin, D.T.; Alexander, N.B.; Besenbruch, G.E.; Bozek, A.S.; Brown, L.C.; Carlson, L.C.; Flint, G.W.; Goodman, P.; Kilkenny, J.D.; Maksaereekul, W.; et al. Developing a commercial production process for 500 000 targets per day: A key challenge for inertial fusion energy. *Phys. Plasmas* **2006**, *13*, 056305. [CrossRef]

131. Du, K.; Liu, M.; Wang, T.; He, X.; Wang, Z.; Zhang, J. Recent progress in ICF target fabrication at RCLF. *Matter Radiat. Extrem.* **2018**, *3*, 135–144. [CrossRef]

132. Farnesi, D.; Barucci, A.; Righini, G.C.; Nunzi Conti, G.; Soria, S. Generation of hyper-parametric oscillations in silica microbubbles. *Opt. Lett.* **2015**, *40*, 4508–4511. [CrossRef] [PubMed]

133. Ward, J.; Benson, O. WGM microresonators: Sensing, lasing and fundamental optics with microspheres. *Laser Photonics Rev.* **2011**, *5*, 553–570. [CrossRef]

134. He, L.; Özdemir, Ş.K.; Yang, L. Whispering gallery microcavity lasers. *Laser Photonics Rev.* **2013**, *7*, 60–82. [CrossRef]

135. Sheng, W.; Kim, S.; Lee, J.; Kim, S.-W.; Jensen, K.; Bawendi, M.G. In-situ encapsulation of quantum dots into polymer microspheres. *Langmuir* **2006**, *22*, 3782–3790. [CrossRef] [PubMed]

136. Li, T. Fundamental Tests of Physics with Optically Trapped Microspheres. Ph.D. Thesis, University of Texas, Austin, TX, USA, 2013.

137. Atiganyanun, S.; Zhou, M.; Abudayyeh, O.K.; Han, S.M.; Han, S.E. Control of randomness in microsphere-based photonic crystals assembled by Langmuir–Blodgett process. *Langmuir* **2017**, *33*, 13783–13789. [CrossRef] [PubMed]

138. Liu, Z.; Liu, L.; Lu, H.; Zhan, P.; Du, W.; Wan, M.; Wang, Z. Ultra-broadband tunable resonant light trapping in a two-dimensional randomly microstructured plasmonic-photonic absorber. *Sci. Rep.* **2017**, *7*, 43803. [CrossRef] [PubMed]

139. Mikhnev, L.V.; Bondarenko, E.A.; Chapura, O.M.; Skomorokhov, A.A.; Kravtsov, A.A. Influence of annealing temperature on optical properties of the photonic-crystal structures obtained by self-organization of colloidal microspheres of polystyrene and silica. *Opt. Mater.* **2018**, *75*, 453–458. [CrossRef]

140. Kim, D.H.; Dudem, B.; Jung, J.W.; Yu, J.S. Boosting light harvesting in perovskite solar cells by biomimetic inverted hemispherical architectured polymer layer with high haze factor as an antireflective layer. *ACS Appl. Mater. Interfaces* **2018**, *10*, 13113–13123. [CrossRef] [PubMed]

micromachines

MDPI

Article

Ag-Sensitized Yb³⁺ Emission in Glass-Ceramics

Francesco Enrichi [1,2,3,*], **Elti Cattaruzza** [3], **Maurizio Ferrari** [1,4], **Francesco Gonella** [1,3], **Riccardo Ottini** [3], **Pietro Riello** [3], **Giancarlo C. Righini** [1,5], **Trave Enrico** [3], **Alberto Vomiero** [2] and **Lidia Zur** [1,4]

[1] Museo Storico della Fisica e Centro Studi e Ricerche "Enrico Fermi", Roma 00184, Italy; maurizio.ferrari@ifn.cnr.it (M.F.); gonella@unive.it (F.G.); giancarlo.righini@centrofermi.it (G.C.R.); zur@fbk.eu (L.Z.)
[2] Division of Materials Science, Department of Engineering Sciences and Mathematics, Luleå University of Technology, Luleå 97187, Sweden; alberto.vomiero@ltu.se
[3] Dipartimento di Scienze Molecolari e Nanosistemi, Università Ca' Foscari Venezia, Mestre 30172, Venezia, Italy; cattaruz@unive.it (E.C.); ottini-r@live.it (R.O.); riellop@unive.it (P.R.); enrico.trave@unive.it (T.E.)
[4] Istituto di Fotonica e Nanotecnologie del Consiglio Nazionale delle Ricerche (IFN-CNR), Laboratorio CSMFO and Fondazione Bruno Kessler (FBK) Photonics Unit, Povo 38123, Trento, Italy
[5] Istituto di Fisica Applicata Nello Carrara del Consiglio Nazionale delle Ricerche (IFAC-CNR), Sesto Fiorentino 50019, Firenze, Italy
* Correspondence: francesco.enrichi@unive.it

Received: 5 July 2018; Accepted: 26 July 2018; Published: 31 July 2018

Abstract: Rare earth doped materials play a very important role in the development of many photonic devices, such as optical amplifiers and lasers, frequency converters, solar concentrators, up to quantum information storage devices. Among the rare earth ions, ytterbium is certainly one of the most frequently investigated and employed. The absorption and emission properties of Yb³⁺ ions are related to transitions between the two energy levels $^2F_{7/2}$ (ground state) and $^2F_{5/2}$ (excited state), involving photon energies around 1.26 eV (980 nm). Therefore, Yb³⁺ cannot directly absorb UV or visible light, and it is often used in combination with other rare earth ions like Pr³⁺, Tm³⁺, and Tb³⁺, which act as energy transfer centres. Nevertheless, even in those co-doped materials, the absorption bandwidth can be limited, and the cross section is small. In this paper, we report a broadband and efficient energy transfer process between Ag dimers/multimers and Yb³⁺ ions, which results in a strong PL emission around 980 nm under UV light excitation. Silica-zirconia (70% SiO_2-30% ZrO_2) glass-ceramic films doped by 4 mol.% Yb³⁺ ions and an additional 5 mol.% of Na_2O were prepared by sol-gel synthesis followed by a thermal annealing at 1000 °C. Ag introduction was then obtained by ion-exchange in a molten salt bath and the samples were subsequently annealed in air at 430 °C to induce the migration and aggregation of the metal. The structural, compositional, and optical properties were investigated, providing evidence for efficient broadband sensitization of the rare earth ions by energy transfer from Ag dimers/multimers, which could have important applications in different fields, such as PV solar cells and light-emitting near-infrared (NIR) devices.

Keywords: sol-gel; Ag nanoaggregates; Yb³⁺ ions; down-shifting; photonic microdevices

1. Introduction

Rare earth ions (RE³⁺) are widely used in many optical materials and devices, mainly due to their unique spectral properties, which are related to the distribution of their electronic energy levels (spanning from UV to IR), narrow bandwidths and long lifetimes [1]. This makes them excellent candidates for many applications such as lighting [2,3], displays [4], biosensing [5–7], optical amplification [8], anticounterfeiting [9], and photovoltaic (PV) solar cells [10–13]. Among rare earths, Yb³⁺ ions provide light-emitting near-infrared (NIR) absorption and emission features peaked around 980 nm (1.26 eV), related to transitions between the two energy levels $^2F_{7/2}$ (ground state) and $^2F_{5/2}$ (excited state). Therefore, Tb^{3+} ions are often used as co-dopants, allowing the additional possibility of obtaining down-conversion splitting of one 488 nm photon into two 980 nm photons [14–16]. Nevertheless, even in co-doped materials, the limited excitation/absorption bandwidths and the small absorption cross sections of RE^{3+} ions are major limitations for their effective implementation and use in real devices.

In the last two decades, broadband and efficient sensitization of Er^{3+} ions by silicon [17–19] or silver aggregates [20–25] have been reported, showing that multimers and nanoaggregates can act as energy-transfer centres to the RE^{3+} ions. More recently, Ag sensitization was successfully observed in Tb^{3+} [26–28] and Tb^{3+}/Yb^{3+} [29] co-doped materials. In this paper, we report the investigation of the direct interaction between Ag nanoaggregates and Yb^{3+} rare earth ions in sol-gel silica-zirconia glass-ceramic (GC) waveguides. GC films are good candidates for the realization of guided-wave optical planar devices [30]. A GC is constituted by a homogeneous dispersion of ceramic nanocrystals in a glass matrix. According to Extended X-ray Absorption Fine Structure (EXAFS) studies, RE^{3+} ions tend to be incorporated into the ceramic nanocrystals [31], providing a better spectroscopic environment for RE^{3+} ions. Zirconia, for example, has a lower maximum phonon energy than silica [32] and a higher refractive index. The combination of Ag-mediated enhancement with the advantage of the glass-ceramic material is studied, suggesting the possibility to exploit this material for more efficient optical devices.

2. Experimental

Undoped and 4 mol.% Yb doped films of nominal molar composition 70% SiO_2-30% ZrO_2 and additional 5 mol.% Na_2O were prepared by sol-gel technique and deposited by dip-coating. All the reagents were purchased by Sigma Aldrich (Saint Louis, MO, USA). Tetraethyl orthosilicate $Si(OC_2H_5)_4$ (TEOS) and zirconium propoxide $Zr(OC_3H_7)_4$ (ZPO) were used as precursors for silica and zirconia, respectively. TEOS was dissolved in ethanol (EtOH) and hydrolized with H_2O and HCl (TEOS:HCl:H_2O:EtOH = 1:0.01:2:25). For sake of simplicity, we will refer to this solution as Sol:Si, and an analogous label will be used for the other solutions used in the process. For Yb co-doped samples, 4 mol.% ytterbium nitrate $YbNO_3$ was added to the solution (Sol.Si-Yb) and then left stirring for 1 h. In the meanwhile, ZPO was mixed with acetylacetone (Acac) and ethanol (ZPO:Acac:EtOH = 1:0.5:50) and sodium acetate was dissolved in methanol (60 mg/ml), the two solutions will be referred to as Sol.Zr and Sol.Na, respectively. The deposition solution has been obtained by mixing Sol.Si-Yb with Sol.Zr and adding dropwise Sol.Na at room temperature. The solution has been left stirring overnight for about 16 h.

Multi-layer films have been deposited on fused silica substrates by dipping. Each layer was annealed in air at 700 °C for 3 min. A final heat treatment in air at 1000 °C for 1 h was performed after the deposition of the last single layer. The typical thickness of each single layer after heat treatment was about 40 nm, and, by adding 10 layers crack-free films, a total thickness of about 400 nm was achieved. By this thermal treatment, a controlled crystallization occurred and glass-ceramic (GC) film were produced.

Undoped and Yb-doped GC films were identified by GC0 and GC4 labels, respectively. In order to introduce silver in the films, $Ag^+ \leftrightarrow Na^+$ ion-exchange [33] was performed by immersing the samples in a molten salt bath (1 mol.% $AgNO_3$ in $NaNO_3$) at 350 °C for 1 h. After the ion exchange process, the samples were identified by GC0-A and GC4-A. Finally, post-exchange heat treatments for 1 h at 430 °C in air (samples GC0-C and GC4-C) were used to investigate the possibility of migration and aggregation of the metal ions.

Compositional, structural, and optical characterization of the films were obtained by Rutherford Backscattering Spectrometry (RBS), X-Ray Diffraction (XRD, Panalytical, Almelo, the Netherlands), and Photoluminescence Spectroscopy.

RBS was carried out at the National Laboratories of Legaro (Padova, Italy) using a 2.2 MeV ^{4}He$^+$ beam at 160° backscattering angle in IBM geometry. RUMP code was used for the analysis of the experimental spectra [34]. The conversion from areal density (the natural unit of measurement for RBS) to film thickness is based on a molar density of the film equal to a weighted average between silica (2.00 g·cm^{-3}) and zirconia (5.68 g·cm^{-3}), according to the nominal stoichiometric composition of the matrix (70% SiO$_2$-30% ZrO$_2$), confirmed by the RBS analysis.

XRD measurements for crystal phase identification were carried out at room temperature by an X'Pert PRO diffractometer. A Cu anode equipped with a Ni filter was used as a radiation source (Kα radiation, λ = 1.54056 Å). Diffractograms were collected in Bragg-Brentano geometry using a step-by-step scan mode in the 2θ range 10°–100°, with a scanning step of 0.05° and counting time of 30 s/step. Nanocrystals' size was determined by Line Broadening Analysis (LBA) [35], in particular, by using the Warren-Averbach method.

Photoluminescence excitation (PLE) and emission (PL) spectra in the UV-visible range were recorded by an Edinburgh Instruments (Livingston, UK) FLS980 Photoluminescence Spectrometer. A continuous-wave xenon lamp was used as the excitation source for steady-state measurements, coupled to a double-grating monochromator for wavelength selection. In particular, 280 nm excitation was used for the investigation of Ag-related PL emission. The light emitted from the sample was collected by a double-grating monochromator and recorded by a photon counting R928P Hamamatsu photomultiplier tube cooled at −20 °C. The PL emission in the NIR spectral range was instead obtained by exciting the sample with the third harmonic of a pulsed Nd:YAG laser at 355 nm. The emission was analyzed by a single grating monochromator coupled to an InGaAs photodiode and using a standard lock-in technique.

3. Results and Discussion

The elemental composition of the samples before and after Ag introduction and annealing was studied by RBS, confirming the agreement of the film composition with the nominal values for Si, Zr, and Yb. RBS analysis was also used to obtain information about the Ag concentration depth-profile after ion-exchange and annealing, revealing Ag-concentration decreasing from 2 mol.% (at surface) to 1.5 mol.% (inner part of the film) for all the samples. An example is reported in Figure 1, which presents the RBS spectra for undoped silica-zirconia GC samples before and after Ag-exchange (GC0 and GC0-A). The simulated GC0-A spectrum and the specific Ag contribution are also shown.

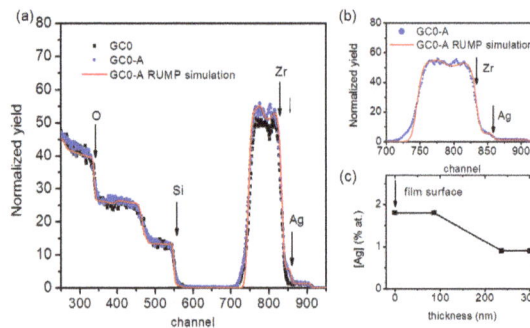

Figure 1. RBS spectra for undoped silica-zirconia GC samples before and after Ag-exchange (GC0 and GC0-A). The simulated GC0-A spectrum and the specific Ag contribution are also shown, resulting in Ag concentrations decreasing from 2 mol.% (at surface) to 1.5 mol.% (inner part of the film).

The XRD analyses of the synthesized GC samples before Ag$^+$↔Na$^+$ ion-exchange are reported in Figure 2. It can be noted that ZrO$_2$ tetragonal-phase nanocrystals were detected in the undoped GC samples (PDF 01-080-0784 50-1089; ICSD 68589), while fluorite-type cubic-phase nanocrystals were observed for Yb doped samples (PDF 01-078-1309; ICSD 62462), attested by the different shape of the characteristic reflection peaks, especially those at $2\theta \approx 35.5°$ and $2\theta \approx 75°$. For GC0 and GC4 samples the zirconia nanocrystals' size determined by Line Broadening Analysis (LBA) [35] was about 14 nm for tetragonal zirconia in GC0 and about 12 nm for cubic zirconia in GC4.

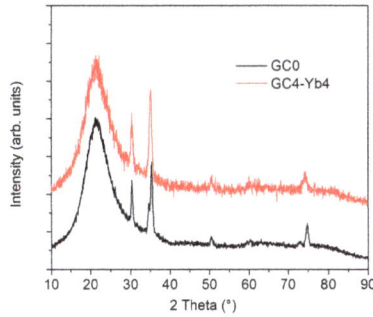

Figure 2. XRD comparison between silica-zirconia-soda GC samples with or without Yb co-doping. Undoped GC samples contain tetragonal-phase zirconia nanocrystals (PDF 01-080-0784; ICSD 68589), while Yb doped samples have a fluorite-type cubic-phase zirconia nanocrystals (PDF 01-078-1309; ICSD 62462), attested by the different shape of the characteristic reflection peaks, especially those at $2\theta \approx 35.5°$ and $2\theta \approx 75°$.

Regarding the optical properties of the synthesized samples, it is well known that silver-doped silicate glasses exhibit broad excitation and emission bands [36–39], in relation to the emitting species. The emission around 330–370 nm is due to isolated emitting Ag^+ ions. The emission around 430–450 nm is due to Ag^+–Ag^+ pairs. The emission around 550–650 nm is due to the formation of $(Ag_3)^{2+}$ trimers or small multimers. These bands are excited by UV and near-UV illumination. The PL emission of undoped GC samples is reported in Figure 3 before an Ag-exchange (GC0), after Ag-exchange (GC0-A), and after Ag-exchange and 1 h annealing at 430 °C (GC0-C). The curves have been obtained by 280 nm excitation. The main contribution, peaked at 425 nm, can be reasonably attributed to Ag^+–Ag^+ pairs. After 1 h post-exchange annealing at 430 °C, the emission shape slightly changes in the following ways: the contribution of Ag^+–Ag^+ pairs decreases, while the emission from trimers and multimers in the red spectral region increases. Noteworthy, and in agreement with a previous work [36], the band around 350 nm of isolated emitting silver ions was not detected, indicating that the probability of isolated ions is low.

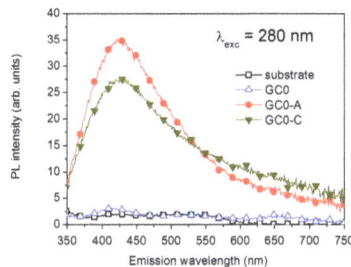

Figure 3. Photoluminescence emission by 280 nm excitation of undoped samples before Ag-exchange (GC0), after Ag-exchange (GC0-A), and after Ag-exchange and annealing (GC0-C). The substrate is also reported as a reference. The main contribution, peaked at 425 nm, is reasonably attributed to Ag^+–Ag^+ pairs. After 1 h annealing at 430 °C, the decreasing of the blue emission in favor of the red emission suggests the formation of Ag trimers and multimers.

The PL emission in the NIR spectral range is reported in Figure 4 for Yb^{3+} doped GC samples before Ag-exchange (GC4), after Ag-exchange (GC4-A), and after Ag-exchange and 1 h annealing at 430 °C (GC4-C). As expected, the direct excitation of Yb^{3+} by 355 nm wavelength light is very weak. However, Ag introduction by ion-exchange results in a strong enhancement of the Yb^{3+} PL emission, possibly due to Ag^+–Ag^+ pairs formed during the ion exchange process. During the annealing, their number decreases to form multimers or bigger aggregates, resulting in a decrease of the PL signal. Furthermore, a lower number of sensitizers means also a higher average distance between them and the Yb^{3+} ions, decreasing the efficiency of the energy-transfer process.

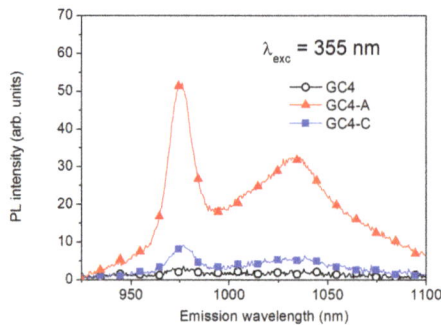

Figure 4. Near-infrared photoluminescence emission of Yb doped samples before Ag-exchange (GC4), after Ag-exchange (GC4-A) and after Ag-exchange and annealing (GC4-C). It can be observed that 355 nm is a very weak excitation wavelength for Yb^{3+} ions, while it results in a strong PL emission in Ag-containing samples.

4. Conclusions

In this paper, we report the synthesis and characterization of efficient NIR emitting glass-ceramic films. In these materials, Yb^{3+} emission around 980 nm was significantly enhanced by $Ag^+ \leftrightarrow Na^+$ ion-exchange, resulting in a strong UV absorption. The significant increase of the PL emission was attributed to energy transfer from Ag^+-Ag^+ pairs and multimers. Ag-sensitized Yb^{3+} doped films could have important applications as spectral downshifters for PV solar cells, converting the UV part of the solar spectrum to NIR photons around 980 nm, which can be efficiently converted to electricity by commercially available crystalline-silicon solar cells. Furthermore, the capability of sol-gel technology to synthesize high quality glass-ceramic waveguides makes these materials suitable for realizing optical devices operating in the NIR spectral range, with significant technological interest in many different fields such as optical amplifiers and lasers, frequency converters, solar concentrators, up to quantum information storage devices.

Author Contributions: Conceptualization—F.E., G.R. and A.V.; Data curation, F.E., R.O., P.R., T.E., A.V. and L.Z.; Formal analysis—R.O.; Funding acquisition—F.E., M.F., G.R. and A.V.; Investigation—F.E., E.C., R.O., T.E., A.V. and L.Z.; Project administration—F.E., G.R. and A.V.; Resources—E.C., M.F., F.G., P.R., G.R. and A.V.; Supervision—F.E. and G.R.; Validation—P.R.; Writing original draft—F.E.; Writing review & editing—E.C., M.F., F.G., G.R., T.E. and L.Z.

Acknowledgments: The research has been partially supported by Centro Fermi through the MiFo (Microcavità Fotoniche) project and the PLESC (Plasmonics for a better efficiency of solar cells) project (Ministero degli Affari Esteri e della Cooperazione Internazionale, MAECI) between South Africa and Italy. Francesco Enrichi acknowledges VINNOVA (Sweden's Innovation Agency) for support, under the Vinnmer Marie Curie Incoming–Mobility for Growth Programme (project "Nano2solar" Ref. N. 2016-02011), Alberto Vomiero acknowledges the Knut & Alice Wallenberg Foundation and the Kempe Foundation for financial support, the European Union's Horizon 2020 research and innovation programme under grant agreement No 654002 and the Laboratori Nazionali di Legnaro of Istituto Nazionale di Fisica Nucleare (INFN-LNL), Padova, Italy for RBS analyses.

Conflicts of Interest: The authors declare no conflict of interest.

References

1. Liu, G.; Jacquier, B. *Spectroscopic Properties of Rare Earths in Optical Materials*; Springer: Berlin, Germany, 2005.
2. Lin, Y.C.; Karlsson, M.; Bettinelli, M. Inorganic phosphor materials for lighting. *Top. Curr. Chem.* **2016**, *2*, 1–47. [CrossRef] [PubMed]
3. Marin, R.; Sponchia, G.; Zucchetta, E.; Riello, P.; Enrichi, F.; De Portu, G.; Benedetti, A. Photoluminescence properties of YAG:Ce^{3+},Pr^{3+} phosphors synthesized via the Pechini method for white LEDs. *J. Nanopart. Res.* **2012**, *14*, 886. [CrossRef]
4. Kim, C.H.; Kwon, I.E.; Park, C.H.; Hwang, Y.J.; Bae, H.S.; Yu, B.Y.; Pyun, C.H.; Hong, G.Y. Phosphors for plasma display panels. *J. Alloys Comp.* **2000**, *311*, 33–39. [CrossRef]
5. Chen, X.; Liu, Y.; Tu, D. *Lanthanide-Doped Luminescent Nanomaterials: From Fundamentals to Bioapplications*; Springer: Berlin, Germany, 2014.
6. Enrichi, F.; Riccò, R.; Meneghello, A.; Pierobon, R.; Cretaio, E.; Marinello, F.; Schiavuta, P.; Parma, A.; Riello, P.; Benedetti, A. Investigation of luminescent dye-doped or rare-earth-doped monodisperse silica nanospheres for DNA microarray labelling. *Opt. Mater.* **2010**, *32*, 1652–1658. [CrossRef]
7. Enrichi, F. Luminescent amino-functionalized or Erbium-doped silica spheres for biological applications. *Ann. N. Y. Acad. Sci.* **2008**, *1130*, 262–266. [CrossRef] [PubMed]
8. Desurvire, E. *Erbium doped fiber amplifiers: principles and applications*; John Wiley and Sons: New York, NY, USA, 1994.
9. Moretti, E.; Pizzol, P.; Fantin, M.; Enrichi, F.; Scopece, P.; Ocaña, M.; Polizzi, S. Luminescent Eu-doped GdVO4 nanocrystals as optical markers for anti-counterfeiting purposes. *Chem. Pap.* **2017**, *71*, 149–159. [CrossRef]
10. Trupke, T.; Green, M.A.; Wurfel, P. Improving solar cell efficiencies by downconversion of high-energy photons. *J. Appl. Phys.* **2002**, *92*, 1668–1674. [CrossRef]
11. Richards, B.S. Enhancing the performance of silicon solar cells via the application of passive luminescence conversion layers. *Sol. Energy Mater. Sol. Cells* **2006**, *90*, 2329–2337. [CrossRef]
12. Strumpel, C.; McCann, M.; Beaucarne, G.; Arkhipov, V.; Slaoui, A.; Cañizo, C.; Tobias, I. Modifying the solar spectrum to enhance silicon solar cell efficiency—An overview of available materials. *Sol. Energy Mater. Sol. Cells* **2007**, *91*, 238–249. [CrossRef]
13. Righini, G.C.; Boulard, B.; Coccetti, F.; Enrichi, F.; Ferrari, M.; Lukowiak, A.; Pelli, S.; Zur, L.; Quandt, A. Light management in solar cells: Recent advances. In Proceedings of the 19th International Conference on Transparent Optical Networks (ICTON), Girona, Spain, 2–6 July 2017; pp. 1–6.
14. Alombert-Goget, G.; Armellini, C.; Berneschi, S.; Chiappini, A.; Chiasera, A.; Ferrari, M.; Guddala, S.; Moser, E.; Pelli, S.; Rao, D.N.; et al. Tb^{3+}/Yb^{3+} co-activated silica-hafnia glass ceramic waveguides. *Opt. Mater.* **2010**, *33*, 227–230. [CrossRef]
15. Bouajaj, A.; Belmokhtar, S.; Britel, M.R.; Armellini, C.; Boulard, B.; Belluomo, F.; Di Stefano, A.; Polizzi, S.; Lukowiak, A.; Ferrari, M.; et al. Tb^{3+}/Yb^{3+} codoped silica-hafnia glass and glass-ceramic waveguides to improve the efficiency of photovoltaic solar cells. *Opt. Mater.* **2016**, *52*, 62–68. [CrossRef]
16. Enrichi, F.; Armellini, C.; Belmokhtar, S.; Bouajaj, A.; Chiappini, A.; Ferrari, M.; Quandt, A.; Righini, G.C.; Vomiero, A.; Zur, L. Visible to NIR downconversion process in Tb^{3+}-Yb^{3+} codoped silica-hafnia glass and glass-ceramic sol-gel waveguides for solar cells. *J. Lumin.* **2018**, *193*, 44–50. [CrossRef]
17. Gourbilleau, F.; Dufour, C.; Levalois, M.; Vicens, J.; Rizk, R.; Sada, C.; Enrichi, F.; Battaglin, G. Room-temperature 1.54 µm photoluminescence from Er-doped Si-rich silica layers obtained by reactive magnetron sputtering. *J. Appl. Phys.* **2003**, *94*, 3869–3874. [CrossRef]
18. Enrichi, F.; Mattei, G.; Sada, C.; Trave, E.; Pacifici, D.; Franzò, G.; Priolo, F.; Iacona, F.; Prassas, M.; Falconieri, M.; et al. Evidence of energy transfer in an aluminosilicate glass codoped with Si nanoaggregates and Er^{3+} ions. *J. Appl. Phys.* **2004**, *96*, 3925–3932. [CrossRef]
19. Enrichi, F.; Mattei, G.; Sada, C.; Trave, E.; Pacifici, D.; Franzò, G.; Priolo, F.; Iacona, F.; Prassas, M.; Falconieri, M.; et al. Study of the energy transfer mechanism in different glasses co-doped with Si nanoaggregates and Er^{3+} ions. *Opt. Mater.* **2005**, *27*, 904. [CrossRef]
20. Strohhöfer, C.; Polman, A. Silver as a sensitizer for Erbium. *Appl. Phys. Lett.* **2002**, *81*, 1414–1416. [CrossRef]

21. Mazzoldi, P.; Padovani, S.; Enrichi, F.; Mattei, G.; Trave, E.; Guglielmi, M.; Martucci, A.; Battaglin, G.; Cattaruzza, E.; Gonella, F.; et al. Sensitizing effects in Ag-Er co-doped glasses for optical amplification. *SPIE Proc.* **2004**, *5451*, 311–326.

22. Martucci, A.; De Nuntis, M.; Ribaudo, A.; Guglielmi, M.; Padovani, S.; Enrichi, F.; Mattei, G.; Mazzoldi, P.; Sada, C.; Trave, E.; et al. Silver sensitized erbium-doped ion exchanged sol-gel waveguides. *Appl. Phys. A* **2005**, *80*, 557–563. [CrossRef]

23. Mattarelli, M.; Montagna, M.; Moser, E.; Vishnubhatla, K.; Armellini, C.; Chiasera, A.; Ferrari, M.; Speranza, G.; Brenci, M.; Conti, G.N.; et al. Silver to erbium energy transfer in phosphate glasses. *J. Non-Cryst. Solids* **2006**, *353*, 498–501. [CrossRef]

24. Mattarelli, M.; Montagna, M.; Vishnubhatla, K.; Chiasera, A.; Ferrari, M.; Righini, G.C. Mechanisms of Ag to Er energy transfer in silicate glasses: a photoluminescence study. *Phys. Rev. B* **2007**, *75*, 125102. [CrossRef]

25. Trave, E.; Back, M.; Cattaruzza, E.; Gonella, F.; Enrichi, F.; Cesca, T.; Kalinic, B.; Scian, C.; Bello, V.; Maurizio, C.; et al. Control of silver clustering for broadband Er^{3+} luminescence sensitization in Er and Ag co-implanted silica. *J. Lumin.* **2018**, *197*, 104–111. [CrossRef]

26. Abbass, A.E.; Swart, H.C.; Kroon, R.E. Effect of silver ions on the energy transfer from host defects to Tb ions in sol–gel silica glass. *J. Lumin.* **2015**, *160*, 22–26. [CrossRef]

27. Li, L.; Yang, Y.; Zhou, D.; Xu, X.; Qiu, J. The influence of Ag species on spectroscopic features of Tb^{3+}-activated sodium–aluminosilicate glasses via Ag^+–Na^+ ion exchange. *J. Non-Cryst. Sol.* **2014**, *385*, 95–99. [CrossRef]

28. Enrichi, F.; Cattaruzza, E.; Ferrari, M.; Gonella Martucci, A.; Ottini, R.; Riello, P.; Righini, G.C.; Trave, E.; Vomiero, A.; Zur, L. Role of Ag multimers as broadband sensitizers in Tb3+/Yb3+ co-doped glass-ceramics. In Proceedings of the SPIE Fiber Lasers and Glass Photonics: Materials through Applications, Strasbourg, France, 22–26 April 2018.

29. Enrichi, F.; Armellini, C.; Battaglin, G.; Belluomo, F.; Belmokhtar, S.; Bouajaj, A.; Cattaruzza, E.; Ferrari, M.; Gonella, F.; Lukowiak, A.; et al. Silver doping of silica-hafnia waveguides containing Tb^{3+}/Yb^{3+} rare earths for downconversion in PV solar cells. *Opt. Mater.* **2016**, *60*, 264–269. [CrossRef]

30. Ferrari, M.; Righini, G.C. Glass-ceramic materials for guided-wave optics. *Int. J. Appl. Glass Sci.* **2015**, *6*, 240–248. [CrossRef]

31. Afify, N.D.; Dalba, G.; Rocca, F. XRD and EXAFS studies on the structure of Er^{3+}-doped SiO_2–HfO_2 glass-ceramic waveguides: Er^{3+}-activated HfO_2 nanocrystals. *J. Phys. D Appl. Phys.* **2009**, *42*, 115416. [CrossRef]

32. Zhao, X.; Vanderbilt, D. Phonons and lattice dielectric properties of zirconia. *Phys. Rev. B* **2002**, *65*, 075105. [CrossRef]

33. Gonella, F. Silver doping of glasses. *Ceram. Inter.* **2015**, *41*, 6693–6701. [CrossRef]

34. Doolittle, L.R. Algorithms for the rapid simulation of Rutherford backscattering spectra. *Nucl. Instrum. Methods B* **1985**, *9*, 344–351. [CrossRef]

35. Enzo, S.; Polizzi, S.; Benedetti, A. Applications of fitting techniques to the Warren-Averbach method for X-ray line broadening analysis. *Z. Kristallogr. Cryst. Mater.* **1985**, *170*, 275–288. [CrossRef]

36. Cattaruzza, E.; Caselli, V.M.; Mardegan, M.; Gonella, F.; Bottaro, G.; Quaranta, A.; Valotto, G.; Enrichi, F. Ag^+-Na^+ ion exchanged silicate glasses for solar cells covering: down-shifting properties. *Ceram. Inter.* **2015**, *41*, 7221–7226. [CrossRef]

37. Cattaruzza, E.; Mardegan, M.; Trave, E.; Battaglin, G.; Calvelli, P.; Enrichi, F.; Gonella, F. Modifications in silver-doped silicate glasses induced by ns laser beams. *Appl. Surf. Sci.* **2011**, *257*, 5434–5438. [CrossRef]

38. Borsella, E.; Battaglin, G.; Garcia, M.A.; Gonella, F.; Mazzoldi, P.; Polloni, R.; Quaranta, A. Structural incorporation of silver in soda-lime glass by the ion exchange process: a photoluminescence spectroscopy study. *Appl. Phys. A* **2000**, *71*, 125–132.

39. Borsella, E.; Gonella, F.; Mazzoldi, P.; Quaranta, A.; Battaglin, G.; Polloni, R. Spectroscopic investigation of silver in soda-lime glass. *Chem. Phys. Lett.* **1998**, *284*, 429–434. [CrossRef]

micromachines

MDPI

Article

Alkali Vapor MEMS Cells Technology toward High-Vacuum Self-Pumping MEMS Cell for Atomic Spectroscopy

Pawel Knapkiewicz

Wroclaw University of Science and Technology, Faculty of Microsystem Electronics and Photonics, Janiszewskiego Str. 11/17, 50-372 Wroclaw, Poland; pawel.knapkiewicz@pwr.edu.pl; Tel.: +48-71-320-48-12

Received: 24 July 2018; Accepted: 12 August 2018; Published: 16 August 2018

Abstract: The high-vacuum self-pumping MEMS cell for atomic spectroscopy presented here is the result of the technological achievements of the author and the research group in which he works. A high-temperature anodic bonding process in vacuum or buffer gas atmosphere and the influence of the process on the inner gas composition inside a MEMS structure were studied. A laser-induced alkali vapor introduction method from solid-state pill-like dispenser is presented as well. The technologies mentioned above are groundbreaking achievements that have allowed the building of the first European miniature atomic clock, and they are the basis for other solutions, including high-vacuum optical MEMS. Following description of the key technologies, high-vacuum self-pumping MEMS cell construction and preliminary measurement results are reported. This unique solution makes it possible to achieve a 10^{-6} Torr vacuum level inside the cell in the presence of saturated rubidium vapor, paving the way to building a new class of optical reference cells for atomic spectroscopy. Because the level of vacuum is high enough, experiments with cold atoms are potentially feasible.

Keywords: alkali cells; MEMS vapor cells; optical cells; atomic spectroscopy; microtechnology; microfabrication; MEMS

---

1. Introduction

The need to develop miniaturized, low power consumption and low-cost instruments/sensors is a current trend. This need is driven by requirements for new applications in which size, weight, and power consumption are key parameters. One example is the development of the miniature, so-called chip-scale atomic clocks (CSAC), applying the Coherent Population Trapping effect (CPT) [1]. The CPT effect is similar to Electromagnetically Induced Transparency (EIT) [2], with the difference being that microwave (electromagnetic) interactions with atoms have been replaced by optical interactions (properly modulated laser light). Using the laser technique, miniaturization of the optical alkali vapor (Cs/Rb) cell and the instrument itself have been made possible. The range of application has expanded as a natural consequence of the availability of small and relatively cheap time and frequency atomic references.

Atomic standards enable very precise control of time and frequency (atomic clocks), as well as high precision of magnetic field measurements. Research on miniature atomic references using the MEMS alkali atom cell is being conducted by several research groups. The common motivation to work on this topic is the growing demand for high-precision and accurate time and frequency references, mostly for telecommunication (terrestrial base stations for telecommunication) and global navigation satellite systems (GNSS), and also for the development of highly sensitive magnetometers. To become competitive solutions for crystal-based time references (TCXO, OCXO), atomic standards must comply with a frequency stability of about $\times 10^{-11} \tau$, low power consumption of ~100 mW, size of a few cubic centimeters (~10–30 cm^3), and be mass producible to give a low price. This is possible through miniaturization and integration using microengineering technology.

The technological development and application extension of atomic time and frequency references has occurred over the last three decades and has resulted in breakthrough discoveries in the field of physics; in particular, optical, laser-based spectroscopy, including the CPT effect mentioned above, and methods for cooling and trapping atoms. The last achievement was awarded the Nobel Prize in 1997 (Steven Chu, Claude Cohen-Tannoudji, William D. P) and was followed by the obtaining of a new state of matter, the so-called Bose-Einstein condensate (Nobel Prize 2001: Eric Cornell, Carl Wieman, Wolfgang Ketterle). Cold atom spectroscopy is most spectacular, due to its future use in the construction of accurate atomic time and frequency standards, and short-term stability (Allan deviation), counted as xE-16, has been impressive. Unfortunately, these solutions exist as laboratory compact setups or laboratory benches only. The key component of such constructions is optical cells, in which 10^{-8} Torr or better vacuum is required inside. Currently, such high vacuum levels can only be obtained by standard pumping systems, because miniature and ready-to-integrate high-vacuum pumping systems do not exist. This is the main reason these solutions have not been miniaturized to date.

Technology of MEMS optical cells for atomic spectroscopy is being developed in several research groups [3–12], including the one represented by the author [13,14]. The author's activity in this field started in 2006. During this period, key technologies of miniature, silicon-glass optical MEMS cells, like non-standard anodic bonding sealing processes in buffer gas atmospheres, as well as the novel cesium vapor introduction during laser-induced dispensing from a solid-state dispenser, have been invented [13–15] and successfully implemented under the MAC-TFC FP7 Project [16], which has resulted in the first European CSAC.

The purpose of this paper is to present important achievements in MEMS vapor cells and high-vacuum MEMS technology toward the development of self-pumping MEMS cells for atomic spectroscopy. The final achievements have no equivalent at the global scale, and are based on the results of work carried out by the team represented by the author. The experience and knowledge gained so far in tandem with the recently developed MEMS ion-sorption pump [17,18] makes it possible to think that achieving the critical condition of high vacuum inside the cell for the development of MEMS optical cell for cold atom spectroscopy is possible.

2. MEMS Alkali Vapor Optical Cell Technology

MEMS alkali vapor cells can be fabricated in several ways [19]. Silicon-glass technology dominates, but low temperature co-fired ceramics (LTCC) ceramic-based solutions have been described in the literature. In this paper, wider analysis of available technologies will not be done. The author will focus on silicon-glass technology, including laser-induced the alkali vapor introduction method, which is crucial for development of the high-vacuum self-pumping MEMS cell.

The MEMS cell technology involves execution of a miniature hermetically sealed silicon-glass structure. The structure consists of deep reactive ion etching (DRIE), or wet etched silicon body, both sides of which are covered with glass wafers (borosilicate glass). The internal structure is composed of an optical chamber, connection channel and chamber for small, pill-like solid-state cesium dispenser (SAES Getters [20]) (Figure 1a,b).

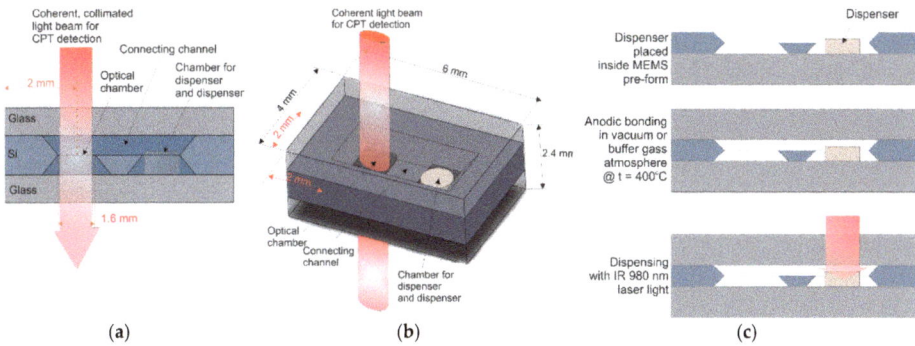

Figure 1. The MEMS cell visualization: (**a**) cross-section view; (**b**) 3D view with dimensions; (**c**) technology path.

Silicon wafer with etched cavities is anodically bonded to the bottom glass (>400 °C, 1 kV) to produce MEMS pre-form (Figure 1c). Next, cesium/rubidium dispenser is placed inside the proper cavity. Subsequently, the MEMS pre-form is anodically bonded at approximately 400 °C in vacuum or buffer gas atmosphere to the top side of the glass plate. The applied voltage is 1.5 kV in vacuum or 0.5–1 kV in the presence of buffer gases. The alkali vapor is introduced with the use of NIR laser light (980 nm wavelength) focused onto a pill-like dispenser, where it is absorbed. The dispenser becomes hot and evaporates the alkali vapors, while the rest of the cell remains cold. The intensity and quantity of the introduction of alkali atoms can be set with laser power and irradiation time.

High-temperature anodic bonding (assembling/sealing process) and laser-induced alkali vapor introduction are key methods in silicon-glass MEMS cell technology. Both methods will be described later.

2.1. High-Temperature Anodic Bonding and Inner Atmosphere Composition

Out-gassing of residual gasses (or particles) from the inner cell walls from the bonded interface might contaminate the inner atmosphere of the MEMS cell and influence the optical properties and long-term stability of the cell. To confirm the quality of anodic bonding, and to obtain information on the atmosphere composition of cells sealed by anodic bonding, several test samples were fabricated, followed by Residual Gas Analysis (RGA). The test samples consist of a silicon substrate with a deep wet-etched cavity forming a large-area Si membrane (5×5 mm^2) and a glass cover (Figure 2).

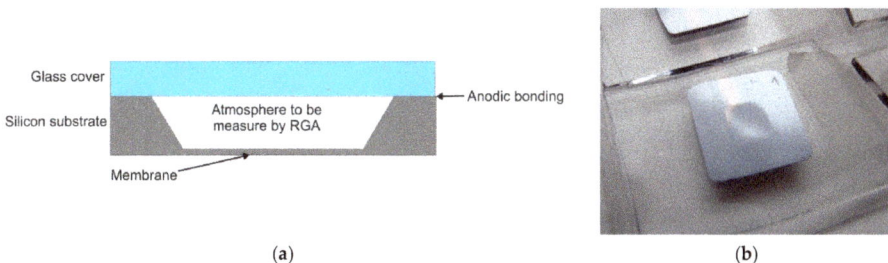

Figure 2. The test structure: (**a**) schematics of the cross-section; (**b**) real view of the structure where the deflected membrane is visible.

Silicon substrate and glass cover were bonded using the anodic bonding process in vacuum with buffer gas (argon) atmosphere as described above. Part of the residual gases—present inside the chamber used for anodic bonding—and the buffer gas were trapped inside the test samples. As is shown in Figure 2b, the deflection of the membrane was visible to the naked eye on the test samples after the anodic bonding process, due to the lower Ar pressure inside the cells compared to the outside atmospheric pressure.

RGA analysis results (Table 1) obtained before or after aging (twin cells, fabricated in the same process, were used; the aging procedure is described later), show that the sealing process—using a modified cleaning and hydrophilization procedure prior to the high-temperature anodic bonding process—gives good results. The composition of the inner atmosphere is stable and shows good repeatability (Figure 3). Low amounts of contaminations (O_2, CO, CO_2, CH_4, C_2H_6) are characteristic for vacuum MEMS structures sealed with the anodic bonding process.

Table 1. Residual Gas Analysis (RGA) of atmosphere composition inside test structures before and after aging: table of partial pressure of different particles.

Gas	Sample No. 1 (mbar)	Sample No. 2 (mbar)	Sample No. 3 (mbar)	Sample No. 4 (mbar)
H_2	1.41×10^{-1}	1.68×10^{-1}	3.75×10^{-2}	4.78×10^{-2}
He	2.30×10^{-3}	1.96×10^{-3}	1.86×10^{-3}	1.68×10^{-3}
CO	3.60×10^{-1}	6.15×10^{-2}	0.00	0.00
N_2	0.00	0.00	3.86×10^{-2}	2.07×10^{-2}
CH_4	2.35×10^{-2}	2.27×10^{-2}	2.18×10^{-2}	9.31×10^{-3}
H_2O	0.00	0.00	0.00	0.00
O_2	4.83×10^{-2}	1.43×10^{-1}	8.95×10^{-1}	1.59×10^{-1}
C_2H_6	4.44×10^{-2}	3.64×10^{-2}	1.07×10^{-2}	1.56×10^{-2}
C_3H_8	0.00	0.00	0.00	0.00
Ar	$1.11 \times 10^{+2}$	$1.16 \times 10^{+2}$	$8.36 \times 10^{+1}$	$8.77 \times 10^{+1}$
CO_2	4.55×10^{-1}	1.75×10^{-1}	$1.87 \times {}^*10^{-1}$	4.35×10^{-1}
Kr	9.80×10^{-4}	1.48×10^{-3}	0.00	0.00
TOT.	$1.12 \times 10^{+2}$	$1.17 \times 10^{+2}$	$8.48 \times 10^{+1}$	$8.84 \times 10^{+1}$

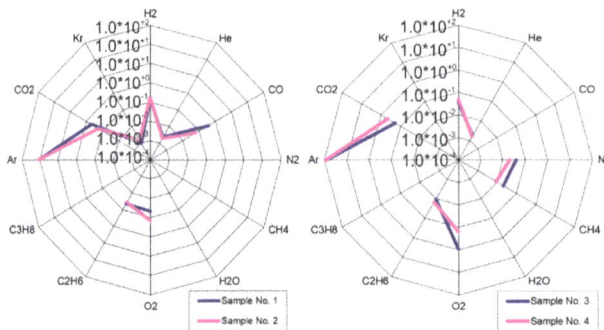

Figure 3. Star-like graphs showing high repeatability of proposed sealing process.

Additionally, analysis of residual gas composition inside the silicon-glass structures after anodic bonding was performed at 2×10^{-3} mbar with and without MEMS getter (SAES Getters, Italy) (Table 2). Results obtained for samples without getter are similar to those presented before. Inner atmosphere inside the test structure with MEMS getter is clean, except for a low (10^{-5} mbar) amount of argon. The internal atmosphere is almost entirely filled with helium (10^{-3} mbar). This vacuum level is the minimum to obtain with the use of MEMS getter.

Table 2. Residual Gas Analysis (RGA) of atmosphere composition inside test structures with and without MEMS getter inside.

Gas	Without Getter		With Getter	
	mbar	%	mbar	%
H_2	9.24×10^{-1}	69.72	0.00	0.00
He	8.55×10^{-4}	0.06	1.06×10^{-3}	96.70
CO	1.89×10^{-1}	14.26	0.00	0.00
N2	2.57×10^{-2}	1.94	0.00	0.00
CH_4	3.23×10^{-2}	2.44	0.00	0.00
H_2O	2.15×10^{-3}	0.16	0.00	0.00
O_2	3.88×10^{-5}	0.00	0.00	0.00
C2H6	0.00	0.00	0.00	0.00
C3H8	0.00	0.00	0.00	0.00
Ar	4.24×10^{-5}	0.00	3.60×10^{-5}	3.30
CO_2	1.51×10^{-1}	11.41	0.00	0.00
Kr	0.00	0.00	0.00	0.00
TOT.	1.32	100.00	1.09×10^{-3}	100.00

The experiment related to the high-temperature anodic bonding and the study of the inner atmosphere revealed the following important facts:

- The inner atmosphere of the MEMS structure is contaminated by products of the anodic bonding process (O_2, CO, CO_2, CH_4, C_2H_6), resulting in a vacuum no better than 10^{-1} mbar,
- Contaminations can be removed from the inner atmosphere with use of MEMS getters, except noble gases, especially helium, whose pressure was maintained at 1×10^{-3} mbar,
- Preprocessing (wafers cleaning, hydrophilization) followed by high-temperature anodic bonding minimizes problems related to out gassing and gas penetration through the bonded silicon-glass interface, which has previously been visible as an unnoticeable effect of the structure's aging.

The obtained results are sufficient for atomic clock technology (CSAC) using the CPT effect. A vacuum level of 1×10^{-3} mbar is far from the value required to cool the atoms, but is a good starting point for a MEMS ion-sorption pump, for which integration with the MEMS alkali vapor optical cell is planned.

2.2. Laser-Induced Alkali Vapor Introduction Method from Solid-State Dispenser

Alkali vapor dispensers are available as resistively heated wires. This solution has been known for years and is still in use. Forming the dispenser into a pill and using it in the MEMS optical cell technology has been invented by author's group and successfully implemented in the MAC-TFC FP6 Project. The purpose of the Project was the development of the first European miniature atomic clock, what was achieved [16]. The laser-induced cesium vapor introduction method from solid-state dispensers is the basis of the technology of MEMS optical cells developed in the author's mother's unit, but also at the Université de Franche-Comté/FEMTO-ST, Besancon, France [21,22].

To evacuate liquid alkali and their vapor effectively, the dispenser must be heated to above 650 °C. The 980 nm and 4 W of maximum power (continues work mode) IR laser-based activation set-up was built. The mechanical part of the set-up ensured precision positioning of a cell. The electronic part ensured precision power control of the IR beam, activation time setting and their automatic timing, and real-time viewing, as well as video recording of the activation process. Effective introduction of alkali vapors depends on the cell design and inner atmosphere conditions. Distance to walls and the presence of buffer gas change heat dissipation, due to the fact that the laser power and irradiation time must be set individually.

2.3. Accelerating Aging Tests of MEMS Cesium Cells

Several MEMS cesium cells of 1.0 mm optical path and 4×6 mm^2 planar dimensions (see Figure 1a,b), filled with Argon as buffer gas with different pressures in the range from 50 mbar to 300 mbar, were successfully fabricated (Figure 4). MEMS getters were intentionally not applied, to check for the influence of residual gases on inner atmosphere quality.

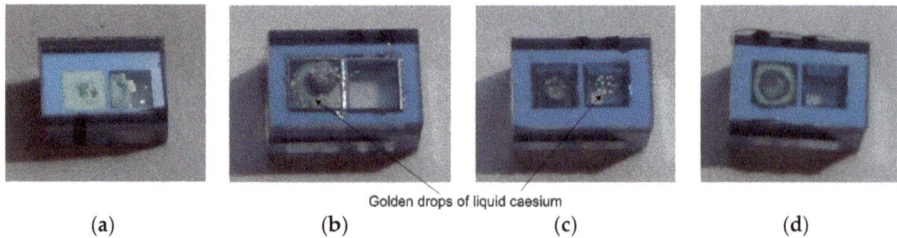

Golden drops of liquid caesium

(a) (b) (c) (d)

Figure 4. Four examples of MEMS cesium cells fabricated at Wroclaw University of Technology: (a) 50 mbar of Ar—large amount of alkali atoms, (b) 100 mbar of Ar—evaporated metal located on dispenser's chamber, (c) 200 mbar of Ar—small amount of cesium in form of golden drops condensate in optical chamber, (d) 300 mbar of Ar—alkali metal visible as golden fog around dispenser.

Alkali vapor dispensing was carried out for fixed values of IR laser power (~0.8 W) and irradiation time (15 s). Different amounts of alkali metal were released depending on the pressure of the buffering gas. Increasing pressure improves heat dissipation; thus, less power is supplied to the dispenser. However, this has positive sides in the form of better control of the amount of released atoms.

All MEMS cells were exposed to accelerated aging tests. There are no existing norms on aging test procedures for miniature MEMS alkali vapor cells. The known aging procedures for electronic devices/components are focused on generation of mechanical tensions as result of temperature changes (temperature cycles). The goal of aging of MEMS cesium cells is to see possible degradation of alkali metal (oxidation—cesium becomes black) caused by contaminations out gassed from bulk materials (mostly glass) and the bonding interface. Out gassing is a function of temperature. We propose a new temperature-accelerated aging procedure for miniature MEMS alkali vapor cells, based on a simplified model for the cell aging. The acceleration factor (AF) is a function of $\Delta T = TX - TN$, where TN is the nominal operating temperature, and TX the aging temperature:

$$AF = 2^{0.1 \times \Delta T} \tag{1}$$

The operating temperature of the MEMS cell should be kept in the range from 40 °C to 85 °C. The normal operating temperature has been set to TN = 75 °C. The applied temperature profile is presented in Figure 5 and consists of cycles with 24 h periods. Each 24 h cycle consists of three steps: heating up the cell from ambient temperature to aging temperature (105 °C or 115 °C) in 1 min, exposition at aging temperature for 23 h and 45 min, cooling down to ambient temperature to make optical observations and pictures—15 min. The cells underwent 13 cycles, i.e., a total of 309 h at TX = 105 °C (corresponding to 2472 h at 75 °C), followed by 16 cycles (380 h) at TX = 115 °C (corresponding to 6080 h at 75 °C). Moreover, after aging at 105 °C and 115 °C, respectively, short temperature shocks were applied where the temperature reaches 250 °C.

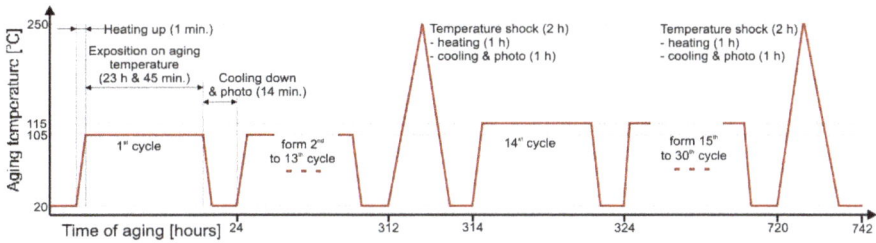

Figure 5. The temperature profile of the accelerated aging test.

During the test, no changes in the amount or color of the cesium drops (as an indication of possible oxidation) were observed. This indicates that the inner atmosphere is clean and stable. The explanation is simple. The inner atmosphere of the cell consists of saturated cesium vapors. Cesium, as a highly reactive atom, reacts with residual gases and neutralizes them. Even if some part of cesium atoms reacts with impurities, the saturation of the cesium vapors is still maintained.

The cesium pill-like dispenser composition contains porous alloys or powder mixtures of Al, Zr, Ti, V or Fe. Those materials are the basis of NEG (Non Evaporable Getters). NEG getters are thermally activated (>200 °C). Therefore, at elevated temperatures (laser-induced dispensing, aging tests) the pill-like dispenser plays a dual role as alkali atom dispenser and NEG getter.

Moreover, the high quality and stability of the applied anodic bonding as a sealing process is confirmed by the resistance of the cells to temperature shocks (250 °C) and possible cesium penetration into the bonded interface. Cesium, as a highly reactive atom, may neutralize impurities, but may also degrade materials and the bonded interface. In the tested MEMS cells, the distance between the cell's inner chamber and the environment was only 1 mm at the narrowest point of the bonded interface. Interface degradation, manifesting as alkali atom oxidation as a result of the atoms' penetration through interface, was not observed. Golden drops/fog of cesium atoms were always visible.

The high-temperature anodic bonding in vacuum or buffer gas atmosphere described here, as well as the investigation of the atmosphere composition inside the MEMS structures and the laser-induced alkali vapor introduction method, are breakthrough technologies for the development of MEMS optical cells for miniature atomic clocks, but they also stand behind the development of future solutions including high-vacuum MEMS. The most important outcomes from the current work are:

- High-temperature anodic bonding provides durable and tight connection of silicon-glass substrates,
- It is possible to obtain a medium vacuum level (10^{-1}–10^{-3}) inside the MEMS structures, which gives a good starting point for the MEMS ion-sorption pump,
- The presence of buffering gases favors better control of laser-induced alkali atom introduction; hence, one should pay particular attention to the process of dispensing in high vacuum.

3. Self-Pumping MEMS Optical Cell for Atomic Spectroscopy

Atomic spectroscopy—cold atoms spectroscopy, for example—requires at least 10^{-8} Torr vacuum and a low concentration of atoms (partial pressure lower than the vapor pressure). This state of the inner atmosphere can only be achieved in a dynamic manner. This means that maintaining high vacuum and dispensing must be done at the same time.

The standard setup consists of an open-sided cell connected to a typical vacuum installation. The inner atmosphere is firstly evacuated with use of turbo-molecular pump (Figure 6). After achieving 10^{-6} Torr, the turbo pump is disconnected and the ion pump starts work. The alkali atoms are delivered continuously from the wire dispenser through thermally activated chemical reaction (current flowing through the dispenser heats it up). To achieve the proper amount of atoms and the required vacuum level, the ion pump must be running in conjunction with dispensing efficiency.

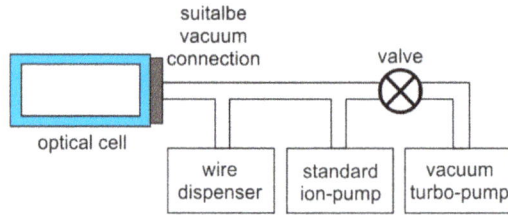

Figure 6. Block diagram of standard setup for atomic spectroscopy at high vacuum and low concentration of atoms.

Development of miniature, integrated MEMS cells for atomic spectroscopy required miniaturization and integration of the pumping system, a proper alkali atoms introduction method, and suitable assembly processes.

The high-temperature anodic bonding described earlier will be used as an assembling method, along with the alkali atoms introduction method from a solid-state pill-like dispenser. Development of integrated, miniaturized pumping system that is suitable for MEMS technology remains a challenge.

The team represented by the author has been conducting research on vacuum microelectronics for many years. This research has resulted in the development of a miniature ion-sorption vacuum pump. The pump makes possible the generation and stabilization of pressure at a 10^{-8} Torr vacuum level, and its construction is fully compatible with MEMS technology. These advantages make the pump usable in new applications [23–25] including self-pumping MEMS optical cells for atomic spectroscopy.

The multilayer silicon-glass spatial structure, consisting of a glass tube with appropriate dimensions connected to a planar structure containing solid-state dispenser and miniaturized ion-sorption pump is proposed (Figure 7). The cell is made of borosilicate (Pyrex-like) glass and silicon. All connections are made with the use of a high-temperature anodic bonding process, which ensures tight and stable leak-proof connections.

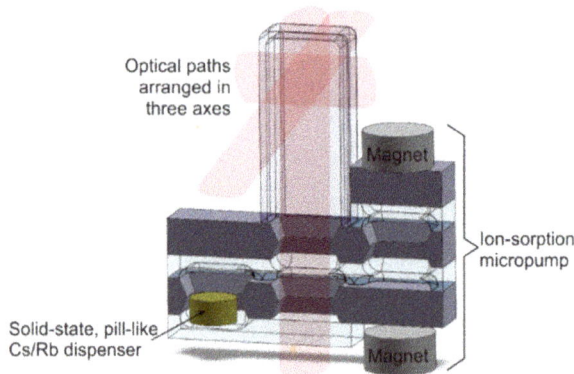

Figure 7. Self-pumping MEMS optical cell for atomic spectroscopy—visualization.

Absorption spectroscopy and cold atom spectroscopy were planned using a self-pumping MEMS cell. Because the author has access to the rubidium atomic spectroscopy laboratory, the dispenser was changed to rubidium. This change has no influence on the cell technology.

A self-pumping MEMS cell filled with rubidium vapor was fabricated. First of all, the open test structure of the self-pumping cell was closed inside the vacuum chamber, and ion current versus pressure was measured. Based on the scaling, the initial vacuum and pumping efficiency was measured (Figure 8).

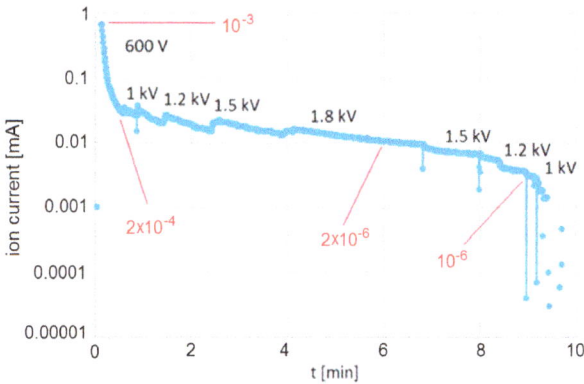

Figure 8. Pumping efficiency in time; vacuum level marked in red.

Initial pressure is equal to 10^{-3} Torr just after the sealing process. After a few minutes of pumping, the vacuum reaches 10^{-6} Torr. Vacuum level was improved by increasing the applied voltage and extending the pumping time. Based on the measured ion current, the vacuum level was at least two orders of magnitude better (blue dots on the graph, Figure 8). Because of the pumping limit of the apparatus used for scaling, precise numbers cannot be given.

After pumping, rubidium atoms were introduced through laser-assisted dispensing from a pill-like solid-state dispenser, reaching saturation of the alkali vapor. A more detailed description of vacuum generation and stabilization can be found in [26].

Doppler-free spectroscopy of rubidium self-pumping MEMS cells was done (D2 line, 780 nm). Characteristics were measured at different test periods (Figure 9). Peak contrast (amplitude) depends on the temperature, and remains comparable to the characteristics of the reference cell. Impurities where partial pressure is higher than 10^{-3} may have an influence on optical spectroscopy. In our experiment, the characteristics are stable: no peak shifts or half-width peaks were observed.

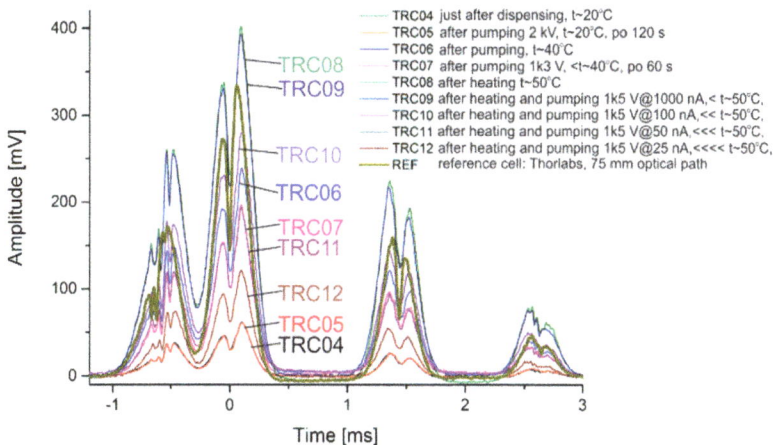

Figure 9. Doppler-free characteristics of the self-pumped rubidium MEMS cell at different test periods.

4. Summary

The solution presented here is the result of work carried out by the team in which the author is the core member developing the technology described in this article. The description of the construction and initial measurement results of the self-pumping high-vacuum cell was preceded by the explanation of key technologies developed earlier for the needs of the European miniature atomic clock. High-temperature anodic bonding and its influence on the atmosphere composition inside MEMS structure were studied, followed by performing accelerated aging tests of MEMS cesium vapor optical cells. It was found that the sealing process ensured high and stable connection. RGA analysis of the inner atmosphere gives important information. After the sealing process, the inner volume of the MEMS structure is depolluted by chemical products of the anodic bonding process (O_2, CO, CO_2, CH_4, C_2H_6), setting the vacuum level at 10^{-1} mbar, even if the bonding process was carried out in a vacuum two orders of magnitude better. MEMS getters improve the situation by eliminating pollution, with the exception of noble gases, whose pressure inside the MEMS structure is set at 10^{-3} mbar, which is the limit using passive vacuum stabilization methods.

The aim of this work is to develop a MEMS optical cell in which it will be possible to generate and stabilize a high vacuum with the simultaneous presence of a small amount of alkaline atoms. Generation of high vacuum inside MEMS structures is possible only with use of active pumping. Integration of the currently developed MEMS ion-sorption pump and MEMS alkali vapor optical cell was proposed, toward a self-pumping, high-vacuum, MEMS alkali vapor optical cell for atomic spectroscopy. The self-pumping high-vacuum cell has no equivalent at a global scale and is probably presented here for the first time ever. High pumping efficiency, making it possible to achieve a 10^{-6} Torr vacuum level, was proved. It must be strongly pointed out here that pumping out of impurities was possible in the presence of saturated rubidium vapor. The presence of saturated alkali atom vapor limited the vacuum level. However, this result makes it possible to think that the generation of a vacuum level of 10^{-8} Torr and the stabilization of partial vapor pressure below saturation is possible.

Doppler-free spectroscopy shows that absorption peaks are comparable to the reference spectrum. Simple absorption tests will not reveal the potential of this new solution. More sophisticated experiments like cold atom spectroscopy, where high vacuum and low concertation of atoms are required, should demonstrate the true possibilities of this solution.

Funding: This work has been partially financed by MAC-TFC FP7 Project (project ID: 224132) and Statutory Grant of Wroclaw University of Science and Technology.

Acknowledgments: Tokens of appreciation go out to all the collaborators who took part in the MAC-TFC FP7 Project. Thanks to John Kitching for the interesting exchange of information that resulted in the launch of new research paths.

Conflicts of Interest: The authors declare no conflicts of interest.

References

1. Dalton, B.J.; McDuff, R. Coherent population trapping: Two unequal phase fluctuating laser fields. *Opt. Acta* **1985**, *32*, 61–70. [CrossRef]
2. Marangos, J.P. Electromagnetically induced transparency. *J. Mod. Opt.* **1998**, *45*, 471–503. [CrossRef]
3. Knappe, S.; Shah, V.; Schwindt, P.D.D.; Hollberg, L.; Kitching, J. A microfabricated atomic clock. *Appl. Phys. Lett.* **2004**, *85*, 1460–1462. [CrossRef]
4. Knappe, S. Chapter 3.18: MEMS Atomic Clocks. In *Comprehensive Microsystems*; Elsevier: New York, NY, USA, 2007; Volume 3, pp. 571–612.
5. Braun, A.M.; Davis, T.J.; Kwakernaak, M.H.; Michalchuk, J.J.; Ulmer, A.; Chan, W.K.; Abeles, J.H.; Shellenbarger, Z.A.; Jau, Y.-Y.; Happer, W.; et al. RF-Integrated and-State Chip-Scale Atomic Clock. In Proceedings of the 39th Annual Precise Time and Time Interval (PTTI) Meeting, Long Beach, CA, USA, 26–29 November 2007; pp. 233–248.

6. Lutwak, R.; Rashed, A.; Varghese, M.; Tepolt, G.; LeBlanc, J.; Mescher, M.; Serkland, D.K.; Geib, K.M.; Peake, G.M.; Römisch, S. The chip-scale atomic clock—Prototype evaluation. In Proceedings of the 39th Annual Precise Time and Time Interval (PTTI) Meeting, Long Beach, CA, USA, 26–29 November 2007; pp. 269–290.

7. Deng, J.; Vlitas, P.; Taylor, D.; Perletz, L.; Lutwak, R. A Commercial CPT Rubidium Clock. In Proceedings of the European Frequency and Time Forum (EFTF), Toulouse, FR, USA, 22–25 April 2008.

8. Wynands, R. The atomic wrist-watch. *Nature* **2004**, *429*, 509–510. [CrossRef] [PubMed]

9. DeNatale, J.F.; Borwick, R.L.; Tsai, C.; Stupar, P.A.; Lin, Y.; Newgard, R.A.; Berquist, R.W.; Zhu, M. Compact, low-power chip-scale atomic clock. In Proceedings of the 2008 IEEE/ION Position, Location and Navigation Symposium, Monterey, CA, USA, 5–8 May 2008; pp. 67–70.

10. Schori, C.; Mileti, G.; Rochat, B.L.P. CPT Atomic Clock based on Rubidium 85. In Proceedings of the 24th European Frequency and Time Forum EFTF, ESA-ESTEC, Noordwijk, The Netherlands, 13–16 April 2010.

11. Violetti, M.; Pellaton, M.; Affolderbach, C.; Merli, F.; Zürcher, J.; Mileti, G.; Skrivervik, A.K. The Microloop-Gap Resonator: A Novel Miniaturized Microwave Cavity for Double-Resonance Rubidium Atomic Clocks. *IEEE Sens. J.* **2014**, *14*, 3193–3200. [CrossRef]

12. Li, S.-L.; Xu, J.; Zhang, Z.-Q.; Zhao, L.-B.; Long, L.; Wu, Y.-M. Integrated physics package of a chip-scale atomic clock. *Chin. Phys. B* **2014**, *23*, 074302. [CrossRef]

13. Knapkiewicz, P.; Dziuban, J.; Walczak, R.; Mauri, L.; Dziuban, P.; Gorecki, C. MEMS caesium vapour cell for European micro-atomic-clock. *Procedia Eng.* **2010**, *5*, 721–724. [CrossRef]

14. Knapkiewicz, P.; Dziuban, J.A.; Gorecki, C.; Dziuban, P.; Walczak, R.; Mauri, L. Komórka cezowa MEMS dla mikrozegara atomowego. *Elektronika* **2010**, *51*, 82–85.

15. Dziuban, J.; Gorecki, C.; Giordano, V.; Nieradko, L.; Maillotte, H.; Moraja, M. Procédé de Fabrication D'une Cellule à Gaz Active Pour L'horloge Atomique à Gaz Ainsi Obtenue. French Patent 06/09089, 17 October 2006.

16. Gorecki, C. Development of first European chip-scale atomic clocks: Technologies, assembling and metrology. *Procedia Eng.* **2012**, *47*, 898–903. [CrossRef]

17. Grzebyk, T.; Górecka-Drzazga, A.; Dziuban, J. Glow-discharge ion-sorption micropump for vacuum MEMS. *Sens. Actuators A* **2014**, *208*, 113–119. [CrossRef]

18. Grzebyk, T.; Knapkiewicz, P.; Szyszka, P.; Górecka-Drzazga, A.; Dziuban, J.A. MEMS ion-sorption high vacuum pump. *J. Phys. Conf. Ser.* **2016**, *773*, 012047. [CrossRef]

19. Knapkiewicz, P. Technological assessment on the MEMS optical alkali vapor cells for atomic references. *Adv. Manuf. Technol.* **2018**, under review.

20. Alkali Metal Dispensers, SAES Getters Electronic Document. Available online: https://www.saesgetters. com/sites/default/files/AMD%20Brochure_0.pdf (accessed on 6 August 2018).

21. Maurice, V. Design, Microfabrication and Characterization of Alkali Vapor Cells for Miniature Atomic Frequency References. Ph.D. Thesis, École Doctorale Sciences Pour I'ingénieur at Microtechniques, Université de Franche-Comté, Besancon, France, 2016.

22. Maurice, V.; Rutkowski, J.; Kroemer, E.; Bargiel, S.; Passilly, N.; Boudot, R.; Gorecki, C.; Mauri, L.; Moraja, M. Microfabricated vapor cells filled with a cesium dispensing paste for miniature atomic clocks. *Appl. Phys. Lett.* **2017**, *110*, 164103. [CrossRef]

23. Grzebyk, T.; Górecka-Drzazga, A. MEMS type ionization vacuum sensor. *Sens. Actuators A* **2016**, *246*, 148–155. [CrossRef]

24. Grzebyk, T.; Górecka-Drzazga, A.; Dziuban, J. MEMS-type self-packaged field-emission electron source. *IEEE Trans. Electron Dev.* **2015**, *62*, 2339–2345. [CrossRef]

25. Grzebyk, T.; Szyszka, P.; Górecka-Drzazga, A.; Dziuban, J. Lateral MEMS-type field-emission electron source. *IEEE Trans. Electron Dev.* **2016**, *63*, 809–813. [CrossRef]

26. Knapkiewicz, P.; Dziuban, J.; Grzebyk, T. Dynamically stabilized high vacuum inside MEMS optical cells for atomic spectroscopy. In Proceedings of the Technical Digest of 31st International Vacuum Nanoelectronics Conference IVNC 2018, Kyoto, Japan, 9–13 July 2018, ISBN 978-1-5386-5715-7.

micromachines

MDPI

Article

Rapid Laser Manufacturing of Microfluidic Devices from Glass Substrates

Krystian L. Wlodarczyk [1,2,*], Richard M. Carter [2], Amir Jahanbakhsh [1], Amiel A. Lopes [2], Mark D. Mackenzie [2], Robert R. J. Maier [2], Duncan P. Hand [2] and M. Mercedes Maroto-Valer [1]

[1] Research Centre for Carbon Solutions (RCCS), Institute of Mechanical, Process and Energy Engineering, School of Engineering and Physical Sciences, Heriot-Watt University, Edinburgh, EH14 4AS, UK; A.Jahanbakhsh@hw.ac.uk (A.J.); M.Maroto-Valer@hw.ac.uk (M.M.M.-V.)
[2] Institute of Photonics and Quantum Sciences, School of Engineering and Physical Sciences, Heriot-Watt University, Edinburgh, EH14 4AS, UK; R.M.Carter@hw.ac.uk (R.M.C.); aal1@hw.ac.uk (A.A.L.); M.Mackenzie@hw.ac.uk (M.D.M.); R.R.J.Maier@hw.ac.uk (R.R.J.M.); D.P.Hand@hw.ac.uk (D.P.H.)
* Correspondence: K.L.Wlodarczyk@hw.ac.uk; Tel.: +44-131-451-3105

Received: 29 June 2018; Accepted: 14 August 2018; Published: 17 August 2018

Abstract: Conventional manufacturing of microfluidic devices from glass substrates is a complex, multi-step process that involves different fabrication techniques and tools. Hence, it is time-consuming and expensive, in particular for the prototyping of microfluidic devices in low quantities. This article describes a laser-based process that enables the rapid manufacturing of enclosed micro-structures by laser micromachining and microwelding of two 1.1-mm-thick borosilicate glass plates. The fabrication process was carried out only with a picosecond laser (Trumpf TruMicro 5×50) that was used for: (a) the generation of microfluidic patterns on glass, (b) the drilling of inlet/outlet ports into the material, and (c) the bonding of two glass plates together in order to enclose the laser-generated microstructures. Using this manufacturing approach, a fully-functional microfluidic device can be fabricated in less than two hours. Initial fluid flow experiments proved that the laser-generated microstructures are completely sealed; thus, they show a potential use in many industrial and scientific areas. This includes geological and petroleum engineering research, where such microfluidic devices can be used to investigate single-phase and multi-phase flow of various fluids (such as brine, oil, and CO_2) in porous media.

Keywords: microfluidic devices; laser materials processing; ultrafast laser micromachining; ultrafast laser welding; enclosed microstructures; glass; porous media; fluid displacement

1. Introduction

Microfluidic devices are used across a wide range of applications in many industrial and research areas, primarily in chemistry, biology, medicine, and pharmacology [1–9]. These devices enable the direct observation and investigation of various physical, chemical, and biological processes occurring at small (even sub-micron) length scales. An operation on such small volumes obviously reduces the amount of testing materials (such as fluids, colloids, cells, etc.) and the time required for analysis, thereby reducing the overall cost of an experiment. Many microfluidic devices also offer temperature control and parallel operation. Simultaneous operations can be executed thanks to the compact size of microfluidic devices. Finally, the hermeticity of microfluidic devices reduces the risk of sample contamination and provides a physical barrier between an operator and an analyzed substance that sometimes can be dangerous (e.g., living cells and bacteria).

Microfluidic devices are also used in geological and petroleum engineering research to investigate various processes governing the macroscopic behavior of subsurface systems at a pore level (i.e., sub-micron scale) [10,11]. These microfluidic devices, often called "micromodels", typically contain structures of interconnected pores whose arrangement and shapes are designed in such a way to represent simplified versions of the geometries typically found in the subsurface systems. In other words, they are constructed to mimic, as close as possible, an internal structure of rocks. Using such microfluidic devices, it is possible to observe and study processes, such as CO_2 injection and trapping [12–16], oil recovery [17–20], dissolution of substances [21], and transport of colloids [22], at a' small (sub-millimeter) scale in the laboratory environment.

Microfluidic devices can be manufactured from a range of materials, such as SU-8 photoresist, polydimethylsiloxane (PDMS), polymethyl methacrylate (PMMA), polylactic acid (PLA), cyclic olefin copolymer (COC), glass, or silicon. Glass, due to its unique combination of high transparency, hardness, thermal stability, electric insulation, surface stability, chemical inertness, and resistance to acids, is often a preferred substrate for the fabrication of microfluidic devices over silicon and polymers. Unfortunately, the conventional manufacturing of microfluidic devices from glass substrates is a complex, multi-step process that involves different fabrication techniques and tools [23–25]. This, in turn, makes the fabrication process time-consuming and expensive, in particular for the prototyping of microfluidic devices in low quantities.

Microfluidic patterns on glass materials are typically generated by either wet (chemical) etching or dry (reactive ion) etching. Wet etching of glass involves the use of strong chemicals, such as hydrofluoric acid (HF), to remove the material [24]. This etching technique enables the manufacturing of deep structures (>500 µm) with an etch rate of several µm/min. The surface roughness of the etched structures can be as low as 10 nm. Unfortunately, the etched structures have a low aspect ratio (close to unity) because this process is isotropic. This means that such structures contain walls with rounded corners and may possess undercuts.

The reactive ion etching (RIE) of glass is an anisotropic process that enables the generation of microfluidic patterns with almost vertical sidewalls (the wall angles up to 88°) [25]. Microchannels generated by RIE can have a very high aspect ratio (up to 40) and low surface roughness (Ra <10 nm). The drawbacks of RIE are a low etching rate (typically <1 µm/min) and a low etch selectivity that forces the use of thick masks. This, in turn, reduces the spatial resolution of etching structures.

Although both etching techniques enable the precision generation of complex microfluidic patterns on glass substrates [24,25], the generation of these patterns is limited to only two dimensions. Moreover, these two fabrication techniques require the preparation of bespoke masks made of either a photoresist or metal. Since such masks are manufactured by photolithography and etching, the entire fabrication process of microfluidic devices is time-consuming and can be expensive, particularly at the prototyping stage.

Lasers enable the generation of microfluidic patterns on various materials, including glass materials [25–32]. Direct (maskless) writing of microstructures on the surface of glass can be obtained using a CO_2 laser, an ultrafast laser, an excimer laser, or an ultraviolet Q-switched solid-state laser. Current laser systems, which often are integrated with a galvo-scanning system and computer-aided design (CAD) software, enable the generation of complex 2.5-dimensional (2.5D) patterns in which each channel may have a different depth and width. The surface elements of such patterns may be generated with different sets of laser parameters, and thus, they may have different widths as well as depths.

Ultrafast (femtosecond) lasers can also be used for the generation of truly three-dimensional (3D) microfluidic patterns inside glass materials, such as fused silica, Borofloat®33, or Pyrex™ [33–38]. This is performed by chemically etching the locally laser-modified regions inside glass. This fabrication technique, often called selective laser-induced etching (SLE), enables the manufacturing of microfluidic devices without the use of a physical mask and additional steps related to the enclosure of the microfluidic patterns. The recent results presented by Gottmann et al. [38] provided strong evidence that the SLE process is an attractive alternative to the conventional etching processes for the fabrication of glass microfluidic devices. Unfortunately, the drawback of this process is a very long etching time (even a few days to complete the development of a microstructure).

In the past, we demonstrated that an ultrafast picosecond laser (Trumpf TruMicro 5×50, Trumpf Ltd., Ditzingen, Germany) can be an effective tool for the direct cutting, drilling, and micromachining of glass plates [39,40], and for joining glass to glass [41,42] or even glass to metal [41] without using any intermediate adhesive layers. In this article, we report on a combination of these processes to manufacture glass microfluidic devices. A picosecond laser was used to generate microfluidic patterns directly on glass (using laser ablation) and to enclose such patterns by microwelding a cover glass plate. Inlet/outlet ports in the cover glass were also generated with the same laser. This process provides a high degree of flexibility in the design of microfluidic devices, which is very important, particularly at the stage of prototyping, and reduces the time and cost associated with their manufacture when a low quantity of the devices is required.

2. Materials and Methods

2.1. Material Used

Borosilicate (Schott Borofloat®33) glass plates with dimensions of 75 mm × 25 mm × 1.1 mm were used for the manufacturing of microfluidic devices. The flatness of the glass plates was ~λ/4. Borofloat®33 contains 81% SiO_2, 13% B_2O_3, 4% Na_2O/K_2O, and ~2% Al_2O_3 [43]. This material has similar optical properties to fused silica, but it is less expensive. This glass is used in many industrial and scientific areas, such as chemistry, optics, photovoltaics, micro-electronics, and biotechnology.

2.2. Laser System

A customized laser processing system based around a 50-W picosecond laser (Trumpf TruMicro 5×50) was used in this work. The laser provides ~6 ps pulses (as measured at full width at half maximum (FWHM)) with a maximum pulse repetition frequency (PRF) of 400 kHz. The model TruMicro 5×50 contains three outputs; each output emits a collimated laser beam of a different wavelength (λ = 1030 nm, 515 nm, or 343 nm). The output laser beams are expanded and delivered to three separate galvo-scanners using appropriate high-reflection (HR)-coated mirrors. Each galvo-scanner is equipped with an approximately 160-mm-focal-length F-theta lens. The laser beams at the focus have different diameters and M^2 values, as listed in Table 1.

Table 1. Laser beam diameters (2 ω_0) and M^2 values measured at the focal points. Measurements were performed using a scanning slit beam profiler (DataRay Beam-Map 2 sensor). The 2 ω_0 values were measured at $1/e^2$ of the peak intensity. Output average power (P), pulse energy (Ep), and peak fluence (F) calculated for each wavelength are also given here.

Wavelength (nm)	P (W)	Ep (µJ)	2 ω_0 (µm)	M^2 (value)	F (J/cm²) [1]
1030	50	125	35 ± 1	1.3 ± 0.1	26.0 ± 1.5
515	30	75	21 ± 1	1.4 ± 0.1	36.3 ± 3.5
343	18	45	20 ± 1	2.1 ± 0.1	28.9 ± 2.9

[1] Peak fluence was calculated as follows: $F = 2 E_P/(\pi\omega_0^2)$, where E_P is pulse energy and ω_0 is the beam radius.

2.3. Laser Micromachining Procedure

Microfluidic patterns and inlet/outlet ports were generated using the 515-nm wavelength. At this wavelength, the peak laser fluence and the machining resolution were the highest. The maximum fluence used for machining the glass plates was 31.1 J/cm². The PRF value, in turn, was limited to 100 kHz because higher values could potentially lead to heat accumulation in the material, and consequently, the fusion of glass particles to the material surface [29].

Prior to the laser treatment, the glass samples were cleaned with isopropanol and wiped off with lens tissues. Following the cleaning process, the glass plates were mounted in a holder that provided a clear aperture underneath the laser machining area, while the holder was fixed to XYZ linear stages

(Aerotech PRO115, Aerotech, Inc., Pittsburgh, PA, USA), as shown in Figure 1. The linear stages provided accurate positioning of the glass samples for machining.

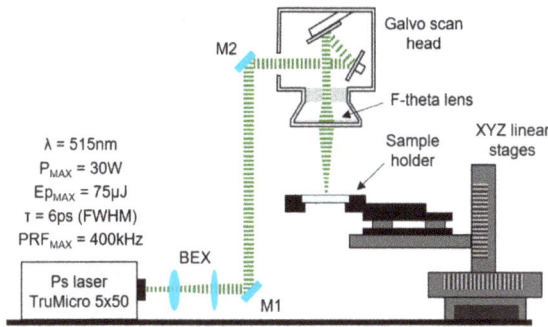

Figure 1. Schematic of the laser system used for machining of the glass plate.

2.4. Laser Microwelding Procedure

The same picosecond laser processing system was also used for the microwelding of glass plates; however, a different processing arrangement was used. In this case the 1030-nm wavelength was used at the maximum PRF of 400 kHz. The processing arrangement was the same as that described in Reference [41]. The only modification was a holder (shown in Figure 2) which provided a large working area (70 mm × 25 mm) and allowed the glass plates to maintain close contact during the welding process. The piston underneath the glass plates, which was under pressure of 1 bar, holds the glass plates in place during the movement of the holder. A 6-mm-thick supporting glass plate prevents the 1.1-mm-thick glass plates from bending.

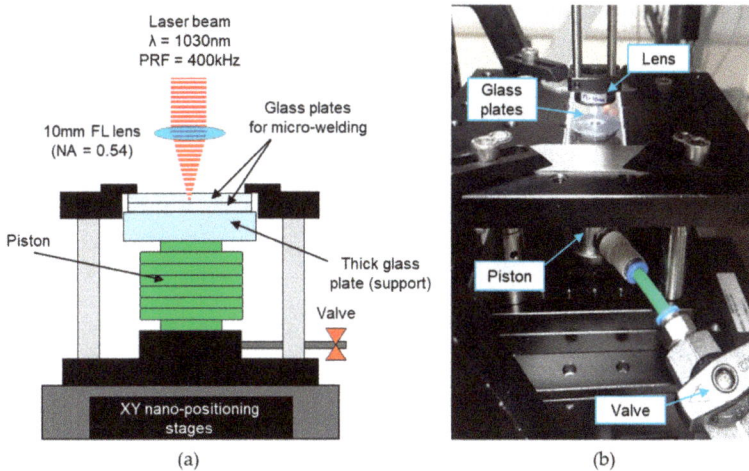

(a)   (b)

Figure 2. Holder used for the microwelding of two glass plates: (**a**) schematic; and (**b**) photograph.

The real challenge in laser microwelding is to bring two glass plates into sufficiently close contact prior to the process. This can be achieved by pressing one plate to the other. Once optical contact is provided, van der Waals forces are capable of holding the two materials together. This, however, requires the glass surfaces to be flat, smooth, and free of debris. Figure 3 shows two glass plates in local contact. Optical fringes (so-called Newtonian rings) visible on this photo indicate a small gap between the two materials, whereas, in the areas where no optical fringes can be seen, the two glass plates can be assumed to be in optical contact.

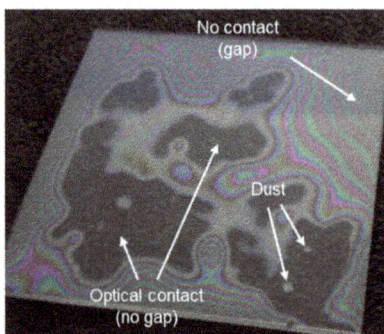

Figure 3. Two glass plates in local contact.

The laser-machined glass plates were contaminated with small amounts of dust and glass particles that would prevent optical contact being achieved. Therefore, it was necessary to clean the plates prior to laser welding to ensure optical contact between the two materials. The cleaning was performed with the use of an ultrasonic bath. The glass plates were individually inserted into a beaker filled with methanol, and the beaker was placed into a water-filled tank of an ultrasonic bath, that was operated at room temperature for ~10 min. After the ultrasonic bath treatment, the samples were dried using a jet of ionized nitrogen. This last cleaning step was performed under an air hood that protected the samples from dust. The cover glass plates were cleaned in the same way.

Recently, more effective cleaning of the laser-machined samples was achieved using hydrofluoric (HF) acid. In this method, the glass plates are placed into a beaker filled with a 5% HF solution, and are kept in this solution at room temperature for a couple of minutes. This removes any debris resulting from the laser machining process, even glass particles that fused to the glass surface.

Following the cleaning process, the glass cover and the laser-machined glass plate are placed into contact, and a force is applied to bring them into optical contact. Such prepared samples are transferred to the holder, while the holder is fixed to XY nano-positioning stages (Aerotech ANT95-XY, Aerotech, Inc., Pittsburgh, PA, USA), as shown in Figure 2b. The laser spot used for microwelding had a diameter of ~3 μm in air. Such a small spot was obtained by focusing the laser beam through a 10-mm-focal-length aspheric lens. Since the laser beam was stationary, the glass samples had to move during the welding process. This was achieved by means of the XY nano-positioning stages which moved the glass plates through the fixed focus of the laser. The incident laser radiation was focused ~80 μm below the glass–glass interface in order to generate a weld seam across the interface. The procedure for the generation of welds in a specific location inside a transparent material was described in Reference [41].

2.5. Testing of the Laser-Manufactured Microfluidic Devices

The laser-manufactured microfluidic devices were tested using the set-up shown in Figure 4. The aim of this test was to determine whether the welds are capable of limiting the flow of fluids to the enclosed microfluidic patterns.

Initially, the microfluidic channels were filled with air. In the test, deionized (DI) water was injected into a microfluidic device through one of the inlet ports. After filling, the device was inspected for any water leakage. Following the visual inspection of the device, nitrogen was injected in order to remove the water. The maximum injection pressure was 1.5 bar. During these experiments, the flow of fluids was recorded using a digital camera (Canon IXUS 60, Canon, Inc., Tokyo, Japan).

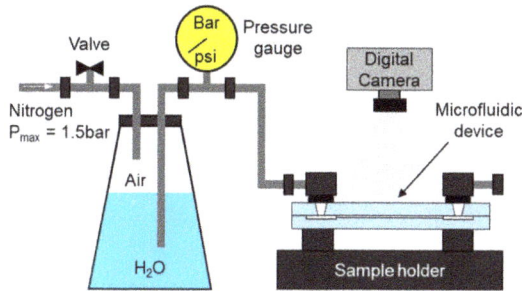

Figure 4. Set-up used for testing the laser-manufactured microfluidic devices.

3. Results and Discussion

3.1. Calibration of the Laser Micro-Machining Process

Figure 5a shows that the picosecond laser (Trumpf TruMicro 5×50) can machine grooves (channels) with a roughly Gaussian cross-section. A peak laser fluence >10 J/cm^2 (ablation threshold) is required, together with a PRF < 400 kHz, and a pulse overlap in the range of 85–95%. If the overlap is too small, the result is partial machining and the generation of damage on the back surface of glass, whereas, if the overlap and PRF are too high, the resultant heat accumulated in the material can be sufficient to cause cracking.

The laser-generated channels were measured and characterized using a 3D surface profilometer (Alicona InfiniteFocus®, Alicona Ltd., Raaba, Austria). This instrument can measure many surface parameters, including depth, width, and surface roughness, with a vertical resolution down to 10 nm. Figure 5a shows a groove that was generated at a peak laser fluence (F) of 19.9 J/cm^2. The PRF value and laser beam scan velocity (v) were 20 kHz and 40 mm/s, respectively. At these laser parameters, the distance between the centers of subsequent laser pulses (Δx) was 2 µm, corresponding to a 90.4% pulse overlap (calculated as O = ((2 w_0 − Δx)/2 w_0) × 100%, where 2 w_0 is the laser beam diameter).

Channels with different depths can be generated by altering peak fluence and the number of laser passes, while maintaining the same velocity and PRF. As can be seen in Figure 5b, the laser-generated channels can have a minimum depth of 6 µm and a minimum width of 10 µm, as measured at FWHM. Surface roughness (Ra) along the bottom of the channels was measured to be approximately 1 µm. This value was calculated for a 0.2-mm length.

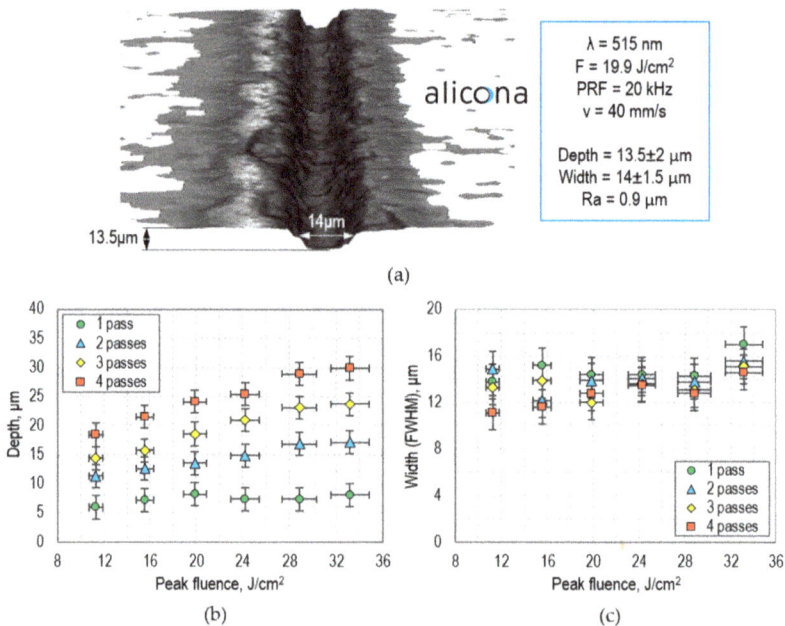

Figure 5. (**a**) 13.5-μm deep and 14-μm wide groove (channel) generated by the picosecond laser using a 21-μm-diameter spot; (**b**) Depth and (**c**) width (measured at full width at half maximum (FWHM)) of the channels generated with a pulse repetition frequency (PRF) of 20 kHz and a laser beam scan velocity (v) of 40 mm/s. Channels were measured using a three-dimensional (3D) surface profilometer (Alicona InfiniteFocus®).

Larger areas (including channels wider than 14 μm) can also be generated by the picosecond laser, typically by raster scanning the beam. The depth can be controlled by selecting an appropriate combination of laser fluence (F), pulse overlap (O), and hatch distance (i.e., a distance between the scanning lines). Figure 6a shows a 1 mm × 1 mm area that was generated using F = 11.3 μJ/cm^2, O = 92.8% (PRF = 100 kHz and v = 150 mm/s), and hatch = 1.8 μm. The depth of this area is 73 μm, and the surface roughness (Sa) is 2.2 μm (as calculated for the central 0.8 mm × 0.8 mm region).

As can be seen in Figure 6b, the depth of the laser-machined areas is well controlled. If areas deeper than 160 μm are required, then the laser scan must be repeated. Using this approach, it is possible to generate very deep micro-wells, reservoirs, or even through holes (that can be used as inlet/outlet ports in glass covers for microfluidic devices). The surface roughness of the micro-wells and reservoirs was measured to be between 1.6 and 2.2 μm, as shown in Figure 6c. Finally, as can be noted from the example shown in Figure 6a, the laser-machined areas have inclined walls. The slope angle of these walls (α) was measured to be ~82°, which limits the aspect ratio of the laser-generated areas to 7 (calculated as tan(α)).

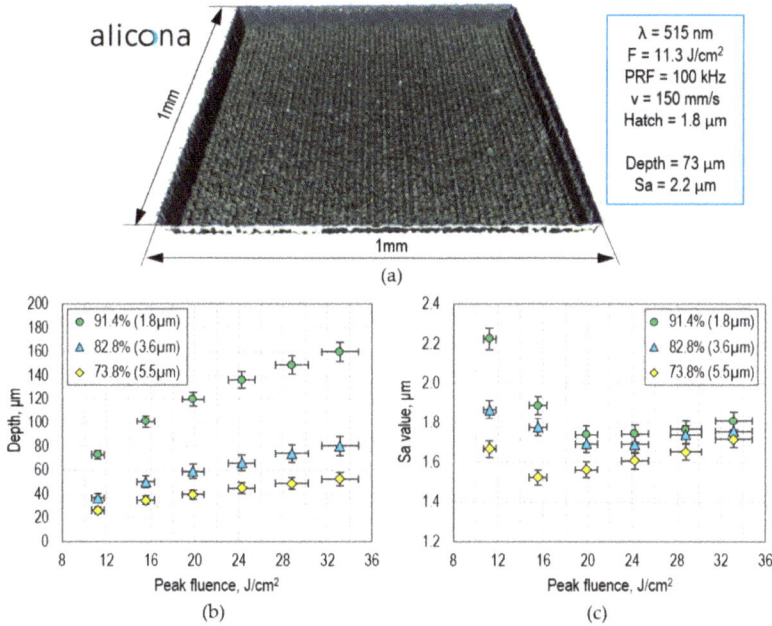

Figure 6. (a) Example of a 1 mm × 1 mm area generated by a moving 21-µm-diameter laser spot; (b) Ablation depth and (c) average surface roughness (Sa) of the 1 mm × 1 mm areas generated using a different combination of laser fluence and hatch distance. The laser machining was performed with PRF = 100 kHz and a scan velocity of 150 mm/s. Surfaces were measured using a 3D surface profilometer (Alicona InfinityFocus®).

3.2. Calibration of the Laser Microwelding Process

The laser microwelding system enables the generation of weld seams at the interface of two glass plates, as demonstrated in References [41,42]. Previously, however, the welds were generated in relatively small areas (typically less than 5 mm × 5 mm). The challenge in this project was to generate welds along the glass–glass interface over a large area (75 mm × 25 mm). To ensure accurate positioning of weld seams during the laser microwelding process, it was necessary to take into account any tilt of the glass plates located in the holder. This was performed by recording the height position of the top glass plate at three different (X,Y) locations at which Fresnel reflection could be observed (see Reference [41] for more details). In this way, it was possible to compensate for the tilt and maintain the laser focus always at the same level (approximately 80 µm) below the glass–glass interface.

Figure 7a shows a photograph that was taken during the laser microwelding of two 1.1-mm-thick Borofloat®33 glass plates. The glass plates are blank, i.e., they do not contain any laser-generated patterns. The glowing lines seen in Figure 7a contain weld seams that were generated at the interface of the two glass plates. The cross-section of two weld seams is shown in Figure 7b, showing that they are teardrop shaped, and are 75 µm wide and 120 µm long. The welds were generated using an average laser power (P) of 2 W, moving the glass plates with a velocity of 2 mm/s. At this velocity, the pulse overlap was very high (nearly 99.98%).

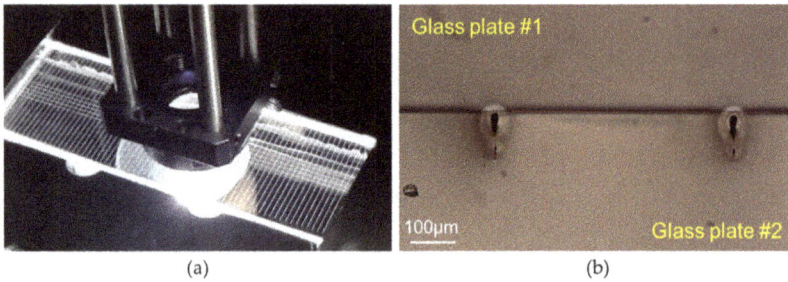

(a)　　　　　　　　　　　　　　　(b)

Figure 7. (**a**) Photograph taken during the laser welding of two 1.1-mm-thick glass plates; (**b**) cross-section of the weld seams generated at P = 2 W and v = 2 mm/s.

The generation of weld seams occurs via multi-photon absorption and subsequent plasma generation inside the glass plates, as described in Reference [41]. By focusing the laser beam below the glass–glass interface, this plasma is generated in the top part of the lower glass (glass plate #2). This plasma expands and locally melts a small volume of surrounding glass that crosses into the bottom part of the cover glass (glass plate #1), and, when this solidifies on cooling, it forms a weld. Our previous work [42] showed that gaps smaller than 3 µm can be closed during the laser welding process.

3.3. Manufacturing of Microfluidic Devices

Figure 8 shows a glass microfluidic device that was manufactured using the process described above. It contains a grid of microchannels that are 36 µm deep and 14 µm wide (FWHM). The distance between the microchannels is only 50 µm. The device also contains two 5-mm-long feed channels (also 36 µm deep and 14 µm wide) and two square inlet/outlet ports (1.5 mm × 1.5 mm).

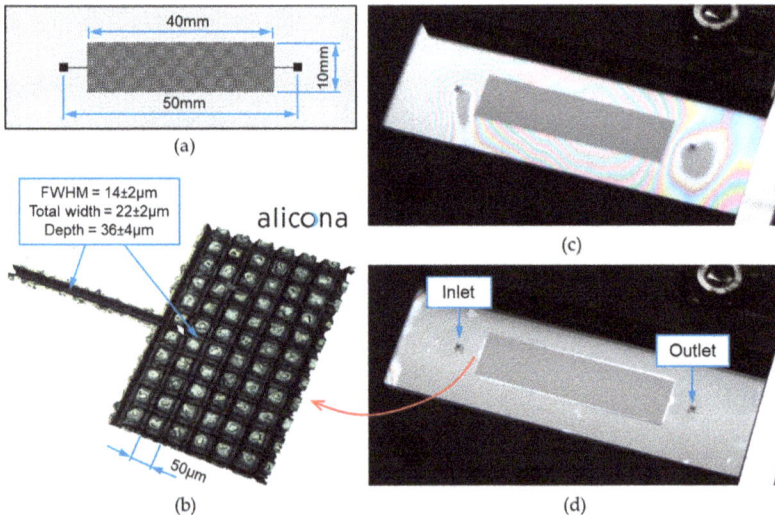

Figure 8. Example of laser-manufactured microfluidic device: (**a**) design; (**b**) 3D surface profile of its internal structure; (**c,d**) microfluidic device before and after laser microwelding, respectively.

The total time required for the generation of the microfluidic pattern and for the drilling of the inlet/outlet ports was approximately 15 min. Laser machining was carried out using three passes of the laser beam with PRF = 100 kHz and v = 150 mm/s. In this case, the laser fluence used had a constant value (F = 26.4 J/cm^2) since a single depth was required. In order to drill two inlet/outlet ports (through holes), a total of nine passes were used. To maintain the laser beam focus during the drilling process, the sample was moved by a 0.1-mm distance toward the galvo-scanner after each laser scan.

The laser-machined glass samples were cleaned in a bath containing 5% HF solution for two minutes in order to remove debris and redeposited material, as can be seen in Figure 9.

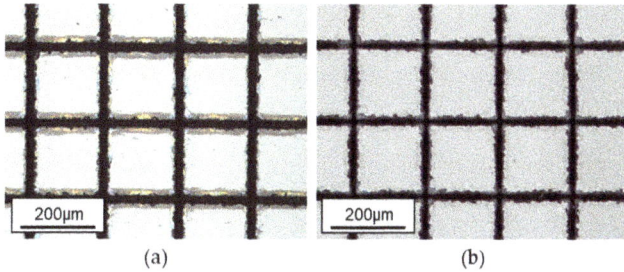

(a)   (b)

Figure 9. Optical microscope image of the laser-generated structure: (**a**) before and (**b**) after the cleaning in 5% hydrogen fluoride (HF) solution. The glass sample was etched for 2 min.

Figure 8c,d show the same microfluidic device before and after the laser microwelding process. Optical fringes visible in Figure 8c indicate a large gap between the two glass plates. Two adjacent bright or dark fringes represents a change in the gap length by $\lambda/2$.

Laser microwelding was started at the location where the gap was the smallest, i.e., near the inlet/outlet port on the right-hand side. In this way, it was possible to generate a plasma and initiate the process. Weld seams were generated around the microfluidic pattern, using the same laser parameters as those provided in Section 3.2. To ensure that the microfluidic device was properly sealed, weld seams were generated along several lines (separated by a 0.5-mm distance from each other) on all sides of the pattern. The location of the welds is shown in Figure 8a—see white dashes. This strategy allowed all existing gaps to be closed during the microwelding process, as can be seen in Figure 8d.

3.4. Fluid Flow Test

The microfluidic channels were filled with DI water during its injection. Water did not leak from the devices, demonstrating that the laser-generated welds provided good sealing. To subsequently displace the water, nitrogen was injected into the devices under different pressures (up to 1.5 bar). These pressures did not cause physical damage to the devices. During the gas injection, nitrogen created preferential paths (so called "fingers") to escape the device through the outlet port, as can be seen in Figure 10a. This phenomenon is called "viscous fingering" and can be observed in porous media (e.g., hydrocarbon-bearing rocks) when a less viscous fluid displaces a more viscous fluid [11,16,44].

Not all water could be removed from the microfluidic devices following the nitrogen injection. In some places, water was trapped in the microstructure, as shown in Figure 10b. This photo was taken following the fluid flow test when the inlet and outlet pressures were equal to the ambient pressure. The image shows a good contrast between water and air.

Figure 10. (**a**) Photograph taken during the testing of a laser-manufactured microfluidic device; (**b**) zoomed image of the microstructure partially filled in with water. This image was obtained using a Leica optical microscope.

During the water injection at a pressure of ~1.5 bar, when the microfluidic device still contained air, it was observed that the pressure inside the device caused a little deformation of the glass plates, giving rise to a couple of optical fringes as a result of gap formation within the laser-machined area. Although the gap was very small (approximately 1 μm) and did not have any visible impact on the behavior of the flow of fluids, the authors consider this may be a problem in particular at high injection pressures, where injected fluids may start bypassing the microchannels by flowing through the gaps. One of the solutions to overcome this problem would be to use thicker (hence, stiffer) glass substrates for manufacturing the microfluidic devices. Another solution would be to place the microfluidic device inside a hermetic vessel and apply an external pressure onto the microfluidic device in order to compensate for the internal pressure. This solution requires a special vessel with a transparent window to provide optical access to the microfluidic device; however, it is feasible, as already shown in several publications [12,16–19]. Alternatively, it is of course possible to create additional weld regions in the interstices between microchannels, at the expense of increasing the manufacturing time.

4. Conclusions

This paper describes a laser-based process suitable for the rapid manufacturing of glass microfluidic devices. Using this process, it is possible to generate almost arbitrary enclosed microfluidic structures. The lateral resolution of the patterns, however, is limited to the laser spot size used for micromachining. The fluid flow tests performed for the microfluidic devices proved that a good sealing of the laser-generated microstructures can be obtained using the picosecond laser microwelding process. Weld seams generated by the laser not only eliminate any existing gaps between two glass plates, bringing the materials into close contact, but they also confine the flow of fluids to the designated areas. Hence, the laser microwelding process seems to be an attractive alternative to processes such as anodic bonding and thermal bonding.

Continuing work will focus on the investigation of the flow of different fluids through a range of pore network patterns generated by the laser. Currently, we are building a workstation dedicated to such experiments. The workstation will be equipped with a high-resolution camera with a high-zoom objective lens in order to observe the flow of fluids in microchannels, while fluids will be injected to the microfluidic devices at controlled flow rates using syringe pumps. In this way, it will be possible to investigate various fluid transport processes and to determine conditions at which injected fluids follow the microfluidic patterns, as well as identifying the limit beyond which fluids start flowing through small gaps between the glass plates, bypassing the microchannels.

Finally, it should be highlighted that the laser microwelding process has some limitations. One of the limitations is the lateral dimension of weld seams that determines the minimum size of the areas suitable for welding. In our case, two parallel microchannels cannot be isolated from each other by producing weld seams between them if the clearance between the microchannels is less than 100 μm. Weld seams of smaller dimensions, however, should be possible to generate using a different combination of laser welding parameters, i.e., using a different scan speed, pulse energy, repetition rate, and pulse duration. Unfortunately, the last two parameters cannot be changed in our laser system. In addition, equipping the laser microwelding system with a visualization system would simplify the necessary precise positioning of the welding samples. Using such a system, weld seams could be readily generated in specific locations, even in small areas between individual microchannels. Another limitation of our laser microwelding system is a welding speed (currently 2 mm/s). Higher welding speeds, however, can be achieved using ultrafast lasers operating at higher PRFs. For instance, using a laser with a PRF of 2 MHz, it should be possible to increase the welding speed to 10 mm/s, reducing the total welding time of microfluidic devices by a factor of five.

Supplementary Materials: The following are available online at http://www.mdpi.com/2072-666X/9/8/409/s1, Table S1: Generation of narrow grooves on Borofloat®33 glass using PRF = 20 kHz. This dataset was used to plot graphs in Figure 5. Table S2: Generation of narrow grooves on Borofloat®33 glass using PRF = 10 kHz. Data not presented in the article. Table S3: Generation of 1 mm × 1 mm areas on Borofloat®33 glass. This dataset was used to plot graphs in Figure 6.

Author Contributions: K.L.W. performed all experiments and wrote this article; R.M.C designed the laser welding system and provided technical support during the laser welding experiments; A.J. assisted during the tests of the microfluidic devices; A.A.L. developed an efficient method for the micromachining of glass; M.D.M. performed HF etching; R.R.J.M. provided scientific and technical advice; D.P.H. provided scientific and technical advice, together with access to the laser facilities; M.M.M.-V. supervised the entire work.

Funding: This project received funding from the European Research Council (ERC) under the European Union's Horizon 2020 Research and Innovation program (MILEPOST, Grant agreement No.: 695070). The received fund also covers the publication costs in open access. The paper reflects only the authors' view and ERC is not responsible for any use that may be made of the information it contains. The authors also thank the EPSRC Centre for Innovative Manufacturing in Laser-based Production Processes (EP/K030884/1) for providing access to the laboratory space and laser facilities.

Conflicts of Interest: The authors declare no conflict of interest. The funding sponsors (ERC and EPSRC) had no role in the design of the study; in the collection, analyses, or interpretation of data; in the writing of the manuscript, and in the decision to publish the results.

References

1. Whitesides, G.M. The origins and the future of microfluidics. *Nature* **2006**, *442*, 368–373. [CrossRef] [PubMed]
2. Sackmann, E.K.; Fulton, A.L.; Beebe, D.J. The present and future role of microfluidics in biomedical research. *Nature* **2014**, *507*, 181–189. [CrossRef] [PubMed]
3. Watanabe, T.; Sassa, F.; Yoshizumi, Y.; Suzuki, H. Review of microfluidic devices for on-chip chemical sensing. *Electron. Commun. Jpn.* **2017**, *100*, 25–32. [CrossRef]
4. Weibel, D.B.; Whitesides, G.M. Applications of microfluidics in chemical biology. *Curr. Opin. Chem. Biol.* **2006**, *10*, 584–591. [CrossRef] [PubMed]
5. Riahi, R.; Tamayol, A.; Shaegh, S.A.M.; Ghaemmaghami, A.M.; Dokmeci, M.R.; Khademhosseini, A. Microfluidics for advanced drug delivery systems. *Curr. Opin. Chem. Eng.* **2015**, *7*, 101–112. [CrossRef]
6. Nan, L.; Jiang, Z.; Wei, X. Emerging microfluidic devices for cell lysis: A review. *Lab Chip* **2014**, *14*, 1060–1073. [CrossRef] [PubMed]
7. Faustino, V.; Catarino, S.O.; Lima, R.; Minas, G. Biomedical microfluidic devices by using low-cost fabrication techniques: A review. *J. Biomech.* **2016**, *49*, 2280–2292. [CrossRef] [PubMed]
8. Bruijns, B.; van Asten, A.; Tiggelaar, R.; Gardeniers, H. Microfluidic devices for forensic DNA analysis: A review. *Biosensors* **2016**, *6*, 41. [CrossRef] [PubMed]
9. Sajeesh, P.; Sen, A.K. Particle separation and sorting in microfluidic devices: A review. *Microfluid. Nanofluid.* **2014**, *17*, 1–52. [CrossRef]
10. Karadimitriou, N.K.; Hassanizadeh, S.M. A review of micromodels and their use in two-phase flow studies. *Vadose Zone J.* **2012**, *11*. [CrossRef]

11. Tsakiroglou, C.; Vizika-Kavvadias, O.; Lenormand, R. Use of Micromodels to Study Multiphase Flow in Porous Media. Available online: http://www.jgmaas.com/SCA/2013/SCA2013-038.pdf (accessed on 15 August 2018).

12. Riazi, M.; Sohrabi, M.; Bernstone, C.; Jamiolahmady, M.; Ireland, S. Visualisation of mechanisms involved in CO2 injection and storage in hydrocarbon reservoirsand water-bearing aquifers. *Chem. Eng. Res. Des.* **2011**, *89*, 1827–1840. [CrossRef]

13. Hu, R.; Wan, J.; Kim, Y.; Tokunaga, T.K. Wettability effects on supercritical CO2–brine immiscible displacement during drainage: Pore-scale observation and 3D simulation. *Int. J. Greenh. Gas Contorl* **2017**, *60*, 129–139. [CrossRef]

14. Kim, Y.; Wan, J.; Kneafsey, T.J.; Tokunaga, T.K. Dewetting of silica surfaces upon reactions with supercritical CO_2 and brine: Pore-scale studies in micromodels. *Environ. Sci. Technol.* **2012**, *46*, 4228–4235. [CrossRef] [PubMed]

15. Bahralolom, I.M.; Bretz, R.E.; Orr, F.M. Experimental investigation of the interaction of phase behavior with microscopic heterogeneity in a CO2 flood. *SPE Reserv. Eng.* **1988**, *3*, 662–672. [CrossRef]

16. Wang, Y.; Zhang, C.; Wei, N.; Oostrom, M.; Wietsma, T.W.; Li, X.; Bonneville, A. Experimental study of crossover from capillary to viscous fingering for supercritical CO_2–water displacement in a homogeneous pore network. *Environ. Sci. Technol.* **2013**, *47*, 212–218. [CrossRef] [PubMed]

17. Campbell, B.T.; Orr, F.M. Flow visualization for CO_2/crude-oil displacements. *Soc. Pet. Eng. J.* **1985**, *25*, 665–678. [CrossRef]

18. Van Dijke, M.I.J.; Sorbie, K.S.; Sohrabi, M.; Danesh, A. Simulation of WAG floods in an oil-wet micromodel using a 2-D pore-scale network model. *J. Pet. Sci. Eng.* **2006**, *52*, 71–86. [CrossRef]

19. Sohrabi, M.; Danesh, A.; Tehrani, D.H.; Jamiolahmady, M. Microscopic mechanisms of oil recovery by near-miscible gas injection. *Transp. Porous Media* **2008**, *72*, 351–367. [CrossRef]

20. Sohrabi, M.; Danesh, A.; Jamiolahmady, M. Visualisation of residual oil recovery by near-miscible gas and SWAG injection using high-pressure micromodels. *Transp. Porous Media* **2008**, *74*, 239–257. [CrossRef]

21. Oostrom, M.; Mehmani, Y.; Romero-Gomez, P.; Tang, Y.; Liu, H.; Yoon, H.; Kang, Q.; Joekar-Niasar, V.; Balhoff, M.T.; Dewers, T.; et al. Pore-scale and continuum simulations of solute transport micromodel benchmark experiments. *Comput. Geosci.* **2016**, *20*, 857–879. [CrossRef]

22. Goldenberg, L.C.; Hutcheon, I.; Wardlaw, N. Experiments on transport of hydrophobic particles and gas bubbles in porous media. *Transp. Porous Media* **1989**, *4*, 129–145. [CrossRef]

23. Leester-Schädel, M.; Lorenz, T.; Jürgens, F.; Richter, C. Fabrication of microfluidic devices. In *Microsystems for Pharmatechnology*; Dietzel, A., Ed.; Springer: New York, NY, USA, 2016; pp. 23–57.

24. Iliescu, C.; Taylor, H.; Avram, M.; Miao, J.; Franssila, S. A practical guide for the fabrication of microfluidic devices using glass and silicon. *Biomicrofluidics* **2012**, *6*, 016505. [CrossRef] [PubMed]

25. Queste, S.; Salut, R.; Clatot, S.; Rauch, J.-Y.; Khan Malek, C.G. Manufacture of microfluidic glass chips by deep plasma etching, femtosecond laser ablation, and anodic bonding. *Microsyst. Technol.* **2010**, *16*, 1485–1493. [CrossRef]

26. Khan Malek, C.G. Laser processing for bio-microfluidics applications (part I). *Anal. Bioanal. Chem.* **2006**, *385*, 1351–1361. [CrossRef] [PubMed]

27. Khan Malek, C.G. Laser processing for bio-microfluidics applications (part II). *Anal. Bioanal. Chem.* **2006**, *385*, 1362–1369. [CrossRef] [PubMed]

28. Yen, M.-H.; Cheng, J.-Y.; Wei, C.-W.; Chuang, Y.-C.; Young, T.-H. Rapid cell-patterning and microfluidic chip fabrication by crack-free CO2 laser ablation on glass. *J. Micromech. Microeng.* **2006**, *16*, 1143–1153. [CrossRef]

29. Nikumb, S.; Chen, Q.; Li, C.; Reshef, H.; Zheng, H.Y.; Qiu, H.; Low, D. Precision glass machining, drilling and profile cutting by short pulse lasers. *Thin Solid Films* **2005**, *477*, 216–221. [CrossRef]

30. Darvishi, S.; Cubaud, T.; Longtin, J.P. Ultrafast laser machining of tapered microchannels in glass and PDMS. *Opt. Laser. Eng.* **2012**, *50*, 210–214. [CrossRef]

31. Fu, L.-M.; Ju, W.-J.; Yang, R.-J.; Wang, Y.-N. Rapid prototyping of glass-based microfluidic chips utilizing two-pass defocused CO_2 laser beam method. *Microfluid. Nanofluid.* **2013**, *14*, 479–487. [CrossRef]

32. Gomez, D.; Goenaga, I.; Lizuain, I.; Ozaita, M. Femtosecond laser ablation for microfluidics. *Opt. Eng.* **2005**, *44*, 051105. [CrossRef]

33. Sugioka, K.; Cheng, Y. Fabrication of 3D microfluidic structures inside glass by femtosecond laser micromachining. *Appl. Phys. A* **2014**, *114*, 215–221. [CrossRef]

34. Serhatlioglu, M.; Ortaç, B.; Elbuken, C.; Biyikli, N.; Solmaz, M.E. CO_2 laser polishing of microfluidic channels fabricated by femtosecond laser assisted carving. *J. Micromech. Microeng.* **2016**, *26*, 115011. [CrossRef]

35. Bellouard, Y.; Said, A.; Dugan, M.; Bado, P. Fabrication of high-aspect ratio, micro-fluidic channels and tunnels using femtosecond laser pulses and chemical etching. *Opt. Express* **2004**, *12*, 2120–2129. [CrossRef] [PubMed]

36. Matsuo, S.; Sumi, H.; Kiyama, S.; Tomita, T.; Hashimoto, S. Femtosecond laser-assisted etching of Pyrex glass with aqueous solution of KOH. *Appl. Surf. Sci.* **2009**, *255*, 9758–9760. [CrossRef]

37. Gottmann, J.; Hermans, M.; Ortmann, J. Digital photonic production of micro structures in glass by in-volume selective laser-induced etching using a high speed micro scanner. *Phys. Procedia* **2012**, *39*, 534–541. [CrossRef]

38. Gottmann, J.; Hermans, M.; Repiev, N.; Ortmann, J. Selective laser-induced etching of 3D precision quartz glass components for microfluidic applications—Up-scaling of complexity and speed. *Micromachines* **2017**, *8*. [CrossRef]

39. Wlodarczyk, K.L.; MacPherson, W.M.; Hand, D.P. Laser Processing of Borofloat®33 Glass. Available online: https://researchportal.hw.ac.uk/en/publications/laser-processing-of-borofloat33-glass (accessed on 15 August 2018).

40. Wlodarczyk, K.L.; Brunton, A.; Rumsby, P.; Hand, D.P. Picosecond laser cutting and drilling of thin flex glass. *Opt. Lasers Eng.* **2016**, *78*, 64–74. [CrossRef]

41. Carter, R.M.; Chen, J.; Shephard, J.D.; Thomson, R.R.; Hand, D.P. Picosecond laser welding of similar and dissimilar materials. *Appl. Opt.* **2014**, *53*, 4233–4238. [CrossRef] [PubMed]

42. Chen, J.; Carter, R.M.; Thomson, R.R.; Hand, D.P. Avoiding the requirement for pre-existing optical contact during picosecond laser glass-to-glass welding. *Opt. Express* **2015**, *23*, 18645–18657. [CrossRef] [PubMed]

43. Borofloat®33—Borosilicate Glass. Available online: https://www.schott.com/borofloat/english/ (accessed on 11 Jun 2018).

44. Rabbani, H.S.; Or, D.; Liu, Y.; Lai, C.-Y.; Lu, N.B.; Datta, S.S.; Stone, H.A.; Shokri, N. Suppressing viscous fingering in structured porous media. *Proc. Natl. Acad. Sci. USA* **2018**, *115*, 4833–4838. [CrossRef] [PubMed]

micromachines

MDPI

Review

Spray Pyrolysis Technique; High-*K* Dielectric Films and Luminescent Materials: A Review

Ciro Falcony [1,*]**, Miguel Angel Aguilar-Frutis** [2] **and Manuel García-Hipólito** [3]

[1] Departamento de Física, CINVESTAV, Apdo. Postal 14-470, Delegación Gustavo A. Madero, Mexico City C.P. 07000, Mexico
[2] Instituto Politécnico Nacional, Centro de Investigación en Ciencia Aplicada y Tecnología Avanzada, Legaría 694 Colonia Irrigación, Mexico City C.P. 11500, Mexico; mafrutis@yahoo.es
[3] Instituto de Investigaciones en Materiales, UNAM, Apdo. Postal 70-360, Delegación Coyoacán, Mexico City C.P. 04150, Mexico; maga@unam.mx
*   Correspondence: cfalcony@fis.cinvestav.mx; Tel.: +52-55-5747-3703

Received: 7 July 2018; Accepted: 23 July 2018; Published: 19 August 2018

Abstract: The spray pyrolysis technique has been extensively used to synthesize materials for a wide variety of applications such as micro and sub-micrometer dimension MOSFET´s for integrated circuits technology, light emitting devices for displays, and solid-state lighting, planar waveguides and other multilayer structure devices for photonics. This technique is an atmospheric pressure chemical synthesis of materials, in which a precursor solution of chemical compounds in the proper solvent is sprayed and converted into powders or films through a pyrolysis process. The most common ways to generate the aerosol for the spraying process are by pneumatic and ultrasonic systems. The synthesis parameters are usually optimized for the materials optical, structural, electric and mechanical characteristics required. There are several reviews of the research efforts in which spray pyrolysis and the processes involved have been described in detail. This review is intended to focus on research work developed with this technique in relation to high-*K* dielectric and luminescent materials in the form of coatings and powders as well as multiple layered structures.

Keywords: spray pyrolysis technique; dielectric materials; luminescent materials

---

1. Introduction

The spray pyrolysis technique is a low-cost, non-vacuum required, way to synthesize materials in the form of powders and films. In the case of films, they are usually deposited over a wide variety of substrates that can be easily adapted for large area deposition and industrial production processes [1–7]. A large amount of the work reported using this technique is concerned with semiconductors, metal and transparent conductive oxides (TCO's) related to their electrical conductivity characteristics. In particular, in the case of TCO's and their relevance for photovoltaic applications, a considerable amount of effort was set to optimize their optical transparency in the visible and electrical conductivity characteristics. This was the case for indium-tin oxides (ITO), indium-Zinc Oxide (IZO), fluorinated-tin oxide (FTO) and many others [1,3]. It was until a few decades ago that different metal oxide and compounds, mixed or in a multiple layer form, incorporating a large variety of dopants were synthesized by this technique for other application purposes [8–10]. Thus, coatings were developed to modify the optical absorption/transmittance, and emissivity of flat glass for the automotive, as well as construction industries. Furthermore, they were also developed for multiple layered structures, such as planar waveguides and resonant optical cavities for photonics [11–13], as well as semiconducting and metal oxide layers. These were doped with a variety of atomic and molecular centers, synthesized by this technique, for the development of light emission devices [14,15]. The dielectric characteristics of many metal oxides were also evaluated for high dielectric constant coatings for dielectric gate layers that might find applications in MOSFET technology, as well [16–20].

This technique is an atmospheric pressure chemical synthesis of materials, in which a precursor solution of chemical compounds in the proper solvent is sprayed through a furnace. In the case of powders, or on a hot substrate in the case of films, where a pyrolysis reaction is achieved, metal oxides is the preferred compound to be obtained by this technique. Nonetheless, metal and semiconductor materials have also been synthesized by a proper deposition ambient and carrier gas choice [1,15]. In this paper, a revision of the work involving the spray pyrolysis technique (published in the later period of time) will be presented. This focuses on the high-*K* dielectric and luminescent properties of coatings and powders as well as multiple layered structures. This review will begin with a brief general description of the basic physical and chemical principles utilized by this technique, and the different experimental arrangements and deposition regimes that are involved in this process. The main characteristics of high-*K* dielectric materials deposited on different type of substrates will then be discussed, as well as the luminescent characteristics of both powders and coatings of materials obtained by the incorporation of dopants in a suitable matrix.

2. Spray Pyrolysis as Materials Synthesis Technique

The spray pyrolysis technique involves three major process stages: Precursor solution composition, aerosol generation and transport, and synthesis process. Every one of these stages is tuned according to of the final chemical and physical characteristics of the material targeted; these adjustments and the choice of materials/processes at each stage will affect the rest of the stages, to some extent. Thus, at the first stage, the chemical composition of the precursor solution will have to involve a compound(s) that will render after the pyrolysis stage the chemical composition required. The selection of the solvent will limit the maximum concentration of the precursor compound in the solution and will determine the best choice for the aerosol generation/transport process and the temperature and rate of synthesis. At the second stage, the aerosol droplet size distribution, determined by the aerosol generation mechanism, will set the morphological characteristics of the final material produced, as well as the proper range of synthesis temperatures. The carrier gas nature and flux rate will propitiate or reduce the probability of a reactive interaction with the precursor compound. At the last stage, the decision whether the final chemical reaction takes place on a gas phase or on a hot substrate will determine if the material synthesized is a powder or a film coating. In general, given an experimental setup, the synthesis parameters that are more relevant are the concentration molarity of the precursor solution, the carrier gas flux rate, and the synthesis temperature.

The solvent in the precursor solution is chosen attending to the solubility of the precursor compound and on its physical properties such as density and viscosity as well as on the final byproducts that will generate and how neutral for their disposal they will be. The preferred choice is water or a mixture of water and an alcohol, which will dissolve many inorganic salts (such as chlorides, some nitrites and fluorides). Organic salts will require organic solvents that, when properly selected, could render excellent precursor solutions, especially for thin films deposition processes [4].

The aerosol generation mechanism could be as simple as a pneumatic system or a more complex but more tunable ultrasonic system. Figure 1 illustrates both systems. In the most common setup for a pneumatic system (Figure 1a), a Venturi nozzle is used in which the precursor solution is fed through a fine (capillary like) inlet into a pressurized carrier gas jet flow. An equation to estimate the average drop diameter has been developed for this type of nozzle [21]:

$$d = 0.64D\left[1 + 0.011\left(\frac{G_l}{G_g}\right)^2\right]\left[\frac{2\gamma}{\rho v^2 D}\right]^{0.45} \tag{1}$$

where; G_l and G_g represent the mass flow rate of liquid and gas, respectively, γ the liquid surface tension, ρ the density of the gas, D the diameter of the spraying solution inlet orifice, and v the velocity of gas. The actual experimental distribution of the droplet diameter size distribution generated by pneumatic means [3] is shown in Figure 1b, for the number of drops and for the total mass carried by the drops, in both cases a broad distribution is observed. Figure 1c shows a diagram of an ultrasonic system [5], in which the aerosol is generated by the ultrasonic waves produced with a piezoelectric disc in contact with the precursor solution. The standing waves generated at/near the surface of the liquid solution result in a generation of droplets in a combination of surface waves (capillary waves) and cavitation phenomena. The first mechanism dominates at low frequencies (20–100 kHz) and the later at frequencies above 100 kHz (0.1–5 MHz). An expression that estimates well the diameter of the droplets was derived [22] as follows:

$$D = \left(\frac{\pi\sigma}{\rho f^2}\right)^{1/3} + 0.0013(We)^{0.008}(Oh)^{-0.14/n}(I_N)^{-0.28} \tag{2}$$

where

$$We = \frac{fQ\rho}{\sigma} \tag{3}$$

$$Oh = \frac{\mu}{f \cdot A_m^2 \rho} \tag{4}$$

And

$$I_N = \frac{f^2 A_m^4}{v_s Q} \tag{5}$$

In these expressions; f is the ultrasonic wave frequency, A_m is the amplitude, σ is the surface tension, ρ is the liquid density, μ is the viscosity and Q is the volumetric flow rate of the liquid and v_s is the speed of sound. Figure 1d shows the experimental distribution for the number of drops and for the total mass carried by the drops generated by ultrasound waves at three different frequencies: 0.7, 0.115, and 3 MHz. The narrow distribution of drops, as well as the control on the average size, is considered the main advantages of the ultrasonic aerosol generation systems. Once the aerosol mist is generated, it has to be transported to the material synthesis area, since in most cases the droplets in the aerosol are below 20 μm in diameter, they can be carried with a gas flow set to minimize coalescence of the drops throughout the transport process, and also to render a desired synthesis rate. Since the carrier gas will be closely present during the synthesis process, whether it is chosen to be an inert or a reactive gas becomes relevant. Thus, if the carrier gas is air, the synthesis is limited to compounds that are as stable like or even better than oxides. In the case of metals, the carrier gas has to be an inert gas which in some cases is combined with a reduction gas (N_2 and H_2, as in the case of forming gas) [1].

At the reactive zone, several parameters are determinant as to what type of material synthesis process occurs, such as temperature, droplet size and their speed. The reactive zone is, in the case of film deposition, the space near the surface of the hot substrate (a few millimeters above the surface of the substrate), or the furnace heated chamber, in the case of powder synthesis. Figure 2 shows a diagram of the different stages at which the droplet is subjected as it approaches the hot substrate for two cases a fixed droplet size and speed, different (increasing from A to D, Figure 2a) temperature of the substrate and fixed substrate temperature and speed of different droplet sizes (decreasing droplet size from A to D, Figure 2b) [23,24]. At low temperature (large initial droplet size), the solvent within the droplet is not completely vaporized and the liquid droplet hits the substrate and upon contact with it vaporizes leaving a ring-shaped dry precipitate on the substrate (process A). At low or intermediate temperature values (large or medium droplet size) the solvent is vaporized, and a dry precipitate (an amorphous precursor salt) hits the substrate surface where a pyrolysis reaction takes place (process B). At intermediate or high temperatures (medium or small droplet size) the droplet goes through all previously described stages. Near the substrate surface the dry precipitates are vaporized, propitiating a chemical vapor reaction (CVD) on the surface of the substrate (process C). Finally, for high temperature (small droplet sizes) the vaporized precipitates undergo a chemical reaction in the vapor phase before they reach the substrate surface (process D). In the case powder synthesis, similar processes occur—but for this case, the parameter that controls the occurrence of the different synthesis stages is the time of flight (time of residence) of the droplet inside the hot zone of the furnace [5].

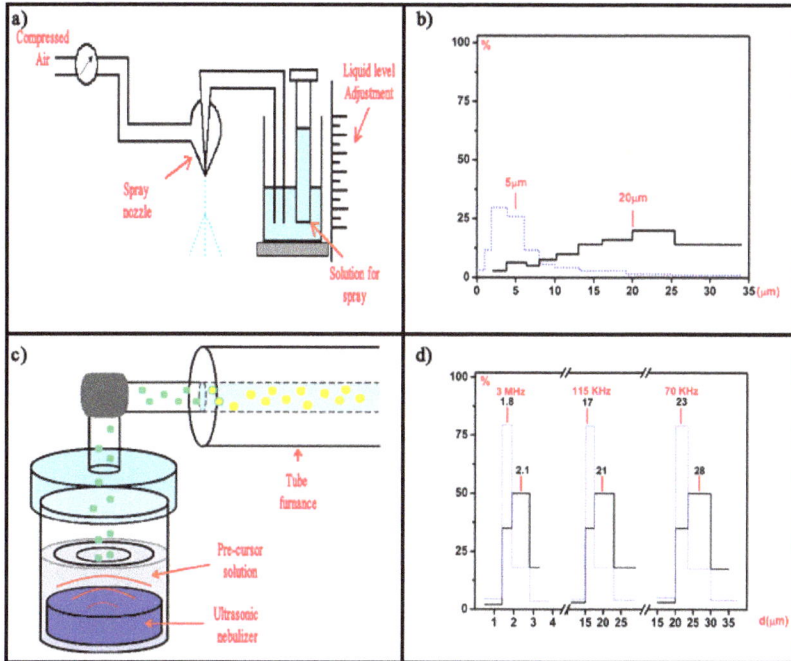

Figure 1. The most common aerosol generation systems, pneumatic and ultrasonic, and the droplet distribution by diameter size or by the amount of solution delivered: (**a**) Shows the pneumatic setup, and (**b**) the corresponding droplet distribution. (**c**) Shows the ultrasonic system, and (**d**) the droplet distribution for this system.

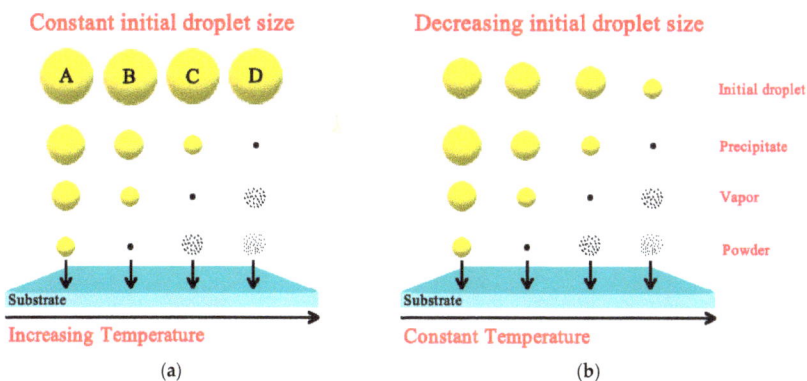

Figure 2. Diagram of the different process stages for the aerosol droplet evolution as it approaches the hot substrate for two cases: (**a**) Constant initial droplet size and increasing substrate temperature, and (**b**) constant substrate temperature and decreasing initial droplet size.

3. High-*K* Dielectric Films

Results associated to the fabrication and characterization of high-*K* dielectrics obtained by ultrasonic spray pyrolysis (USP) is shown in this section. The synthesis of high-*K* dielectric thin films by USP is considered of great importance because, as can be inferred from the last section, the technique is neither expensive nor difficult to be developed in any fair laboratory [2]. Several high-*K* dielectrics have been attempted, including aluminum oxide thin films, zirconium oxide, and yttrium oxide [16–20]. The main goal of researching high-*K* dielectrics is the preparation of metal oxides that might be of interest for the scaling and gate capacitance of some devices in the future. Literature has shown clearly the need to develop high-*K* dielectric materials [25,26] for electronic microdevices in the silicon based Complementary Metal Oxide Semiconductor (CMOS) technology. This is the case of Field Effect Transistors (FETs), one of the most important devices, because of its low power consumption and performance. However, the need of down scaling has been a very dramatic issue. Furthermore, the materials involved for these applications show a dramatic constraint in the dielectric layers that play an important role in the FETs. In them, the thickness of the SiO_2 layer needed for the gate dielectric is under 1.4 nm; so thin that the gate leakage current by direct tunneling of electrons through the SiO_2 film becomes too high. This, and other drawbacks, have resulted in a search for better suited dielectric materials than that of SiO_2. Since the tunneling current across a dielectric film decreases exponentially with increasing thickness ($\Im \propto \exp\left(-2\sqrt{\frac{2m\phi}{\hbar^2}}z\right)$, where ϕ is the barrier height for tunneling), a thicker layer of a higher dielectric constant material than SiO_2 is a possible solution. In a FET, the source-drain current depends on the gate capacitance: $C = \frac{k\epsilon_0 A}{t}$, where ϵ_0 is the permittivity of free space, k is the relative permittivity, A is the area and t is the oxide thickness. So, to solve the problem of leakage current due to tunneling, it is required to replace SiO_2 with a physically thicker layer of a higher dielectric constant material. This would preserve the capacitance value with a reduced tunneling current. With this purpose in mind, the "equivalent oxide thickness", (EOT), defined as $t_{ox} = \left(\frac{3.9}{K}\right)t_{hiK}$, where the 3.9 value is the dielectric constant of SiO_2, has been used as a figure of merit for high-*K* dielectrics to be used instead SiO_2. The requirements for choosing a new high *K* dielectric are the following: (i) Its *K* value must be high enough. (ii) The oxide should be thermodynamically stable when in contact with the Si channel. (iii) It must act as an insulator (large barrier with Si for both holes and electrons), and (iv) It should have a good electrical interface with Si. The static dielectric constant of high-*K* oxides is already known. Some oxides with its dielectric constant are listed in Table 1 [25,26].

Table 1. Static dielectric constant of a few gate dielectrics.

Oxide	*K*
SiO_2	3.9
Si_3N_4	7
Al_2O_3	9
Ta_2O_5	22
TiO_2	80
ZrO_2	25
HfO_2	25
$HfSiO_4$	11
La_2O_3	30
Y_2O_3	15
a-$LaAlO_3$	30

3.1. High-K Dielectrics Materials

Some common high-*K* metal oxides that might provide thicker dielectric layers with reduced leakage (preserving the SiO_2 equivalent capacitance values) are Ta_2O_5, $SrTiO_3$, Al_2O_3 (among others). These metal oxides' dielectric constants range from ~10 to 80, and have been employed mainly in memory capacitor applications. Among the few high-*K* materials above mentioned, Al_2O_3 is thermodynamically more stable when in contact with Si [26]. Before listing the synthesis and properties of aluminum oxide and other dielectrics prepared by the USP technique, it is important to realize the role of reagents and solvents involved in their synthesis.

3.2. The Role of the Reagents and Solvents in the USP Synthesis of High-K Dielectric Layers

The role of the reagents and solvents play a dramatic factor for achieving specific properties of films and coatings deposited by the USP technique [2]. Thus, it is important to revise the deposition process pathways, in particular, at the reactive stage (Figure 2). In the aerosol processing of materials, reactions are initiated by thermal energy. A wide number of metal-organic compounds have been used as precursors to a number of materials, via thermally induced aerosol processing. Among others, β-diketonates, carboxylates, alkoxides, and amides are frequently used. Metal β-diketonates and amides are often used as sources of metal-containing materials and frequently require reaction with an added reagent. Some metal alkoxides, β-diketonates, and amides sublimate, thus, these species are ideal and have been used for CVD like, or aerosol assisted CVD (process C in Figure 2) deposition process, that renders excellent quality layers. Metal β-diketonates have been utilized for the deposition of a large variety of materials, such as metals, metal oxides, and metal sulfides. The suggested reaction in the presence of water that produces metal oxides (MO) is as follows:

$$M(\beta - \text{diketonate})_2 + H_2O \rightarrow MO + 2H(\beta - \text{diketonate}) \tag{6}$$

Organometallic compounds have been used extensively in the gas-phase synthesis of materials, particularly CVD and gas-to-particle conversion, because these compounds are often sufficiently volatile. In addition to the appropriate choice of an appropriate source of the metal, the selection of an appropriate solvent is also important. Three basic requirements should be accomplished by a solvent for its suitability in the case of ultrasonic generation of an aerosol. Firstly, the solubility of the acetylacetonates or the organometallic compounds used. The second requirement is, because of the requirement that the aerosol droplets should arrive near the substrate surface preferably in liquid state, that the solvent used should have a relatively high boiling point. The third requirement is that the solvent should also possess a low grade of viscosity to enable proper ultrasonic excitation and aerosol generation [11]. Atomization of acetylacetonates (dissolved in organic solvents) by ultrasonic excitation has been used by G. Blandenet et al. to deposit films of Al_2O_3, Y_2O_3, ZrO_2 and other coatings on glass and stainless steel [3].

The physical properties of a few solvents used during the deposition of some high-*K* dielectric coatings in this work and in for other authors are listed in Table 2. In this review, it is highlighted the use of Dimethylformamide and a few alcohols [11].

Table 2. Physical properties of a few solvents used during the deposition of some high-*K* dielectrics.

Solvent	Boiling Point (°C)	Viscosity at Room Temperature (mPas)	Density (g/cm³)	Chemical Formula
Dimethylformamide	153.0	0.80	0.95	C_3H_7NO
Methanol (Methyl Alcohol)	65.0	0.52	0.79	CH_3OH
Ethanol (Ethyl Alcohol)	78.5	1.19	0.78	CH_3CH_2OH
Propanol (n-Propyl Alcohol)	97.4	2.25	0.80	$CH_3(CH_2)_2OH$
Butanol (n-Butyl Alcohol)	117.0	2.95	0.80	$CH_3(CH_2)_2CH_2OH$

3.3. Synthesis of Al_2O_3 Thin Films by USP Technique

Aluminum oxide thin films (Al_2O_3) have good thermal conductivity, low permeability to alkali ions, excellent hardness, high radiation resistance, high refractive index, high transparency, resistant against hostile environments and high dielectric constant [9,20,27–29]. This latter property is highly important to possibly replacement of SiO_2 as a H-K dielectric for the microelectronic devices applications. The fabrication of aluminum oxide thin films using the USP technique has been reported, at least, since the 1980s. G. Blandenet et al. deposited Al_2O_3 coatings on glass by ultrasonic spraying using aluminum isopropoxide, as source of aluminum, and butanol as solvent [3]. A lot of work has been carried out since then to get this type of films with improved characteristics. Actually, the research in these films continues up to date, using the same technique and/or a related spray pyrolysis technique. It is worth to mention the recent work of B.P. Dhonge et al. [30]; or the work of A.B. Khatibani et al., who also obtained alumina thin films by spray pyrolysis [31]. In particular, excellent quality aluminum oxide thin films have been deposited using aluminum acetylacetonate, dissolved in *N,N*-dimethylformamide as spraying solution [9,19,20]. These work shows the versatility of USP technique and how the experimental conditions of synthesis can be optimized to get the films with the required optical, structural and dielectric properties. A few highlights of these results are described below.

3.4. Experimental Details

The most appropriate reagent and solvent were found to be: Aluminum acetylacetonate ($Al(acac)_3$) as source of aluminum, and *N,N*-dimethylformamide (DMF) as solvent. Several solutions of $Al(acac)_3$ in DMF were prepared. The solutions prepared consisted in dissolving 1, 3, 5, 7, 10 and 12 g of $Al(acac)_3$ in 100 mL of DMF. The versatility of the spray pyrolysis process permits the generation of aerosol streams with different reagents and/or additives that can be supplied simultaneously during the synthesis of a thin film (for example, binary oxides of some semiconductors; such as $CuCrO_2$ have been deposited within this approach [32]). In the present case, a parallel aerosol stream of water mist to the aerosol of the $Al(acac)_3$ in DMF solution was supplied during the synthesis of the Al_2O_3 thin films. The motivation to use a water aerosol was realized by the report of J.S. Kim et al. [33], who used the addition of a water mist for fabricating thin films by the CVD method. The Al_2O_3 films were deposited on n-type silicon wafers of low and high resistivity (0.1 and 200 $\Omega \cdot$cm, respectively), and on quartz slides for the optical absorption measurements. The deposition of the films was achieved at different substrate temperatures: 450, 500, 550, 600 and 650 °C. The high deposition rate of the films led to get the film thicknesses, within a few seconds or minutes, in the range of 90–130 nm. MOS (Metal-Oxide-Semiconductor) structures were fabricated with these films by thermally evaporating aluminum contacts (1.1×10^{-2} cm^2) on top of the aluminum oxide thin film deposited on the silicon substrates [9,19,20]. The films resulted transparent in the whole visible range of the electromagnetic spectrum. The optical band gap of these films (about 5.63 eV) compared favorably to the best quality films obtained by other techniques. The films were found to be mainly amorphous in all cases. The films deposited with water mist showed a higher index of refraction, in contrast to the films deposited without water mist. These results might indicate that the films deposited by water mist show a higher specific density and were confirmed by the electrical response of the films, since the MOS structures fabricated with this type of films showed the best dielectric characteristics. The role of water during deposition process was perhaps to collect and remove the residual carbon from the acetylacetonate decomposition, reducing in this way the total amount of carbon and impurities that might remain in the oxide film [9,19,20]. 1 MHz and quasi-static capacitance versus voltage characteristic of the MOS structures were used to determine the density of interface states that was found in the range of 10^{11} eV$^{-1} \cdot$cm^{-2}. This density of interface states compared favorably to other dielectric layer used in many microelectronic applications. The current density measured by the ramp I-V characteristic curves in these MOS structures, at electric fields below 2 MV/cm, was in the range of the displacement current generated by the voltage ramp applied to the MOS structure 10^{-9} Amp/cm^2. At electric fields higher than 2 MV/cm a real current injection across the aluminum oxide (produced by Fowler-Nordheim tunneling) increases up to 10^{-6} Amp/cm^2 at approximately 5 MV/cm without any destructive breakdown of the films [9,20]. In addition, aluminum oxide thin films as thin as 30 nm were deposited by means of a pulsed spraying setup with excellent properties [19]. This last feature showed that the ultrasonic spraying is also capable of depositing extremely thin films of aluminum oxide preserving its excellent dielectric properties.

3.5. Y_2O_3 and ZrO_2 Films

Other dielectrics have also been considered. In particular, yttrium, zirconium and silicon oxides (Y_2O_3, ZrO_2 and SiO_2) deposited by spray pyrolysis. Even though SiO_2 is not a high-K dielectric, this type of film has been obtained successfully using the spray pyrolysis technique [18]. Other high-K dielectrics, such as yttrium oxide thin films, have been deposited on silicon substrates using yttrium acetylacetonate as source of yttrium, and *N,N*-dimethylformamide as solvent. For this system, a solution of H_2O-NH_4OH was sprayed in parallel during the deposition process to improve the optical, structural and electrical properties of the deposited films. In this case, the films were deposited at temperatures in the range from 400 to 550 °C. The effective index of refraction measured in the films was about 1.86, and an average deposition rate ~0.1 nm/s. A highly textured surface of the films was obtained to (400) orientation. The growth of a SiO_2 layer sandwiched between the yttrium oxide and the Si substrate was also noticed and it seemed to improve a lower interface state density, in the range of 10^{10} eV^{-1}·cm^{-2}. An effective dielectric constant up to 13, as well as a dielectric strength in the range of 0.2 MV/cm was obtained in a 100 nm thick film incorporated in a MOS structure. For this system, it seemed that the polycrystalline nature of these films results in a deterioration of the dielectric properties by reducing the threshold voltage needed for conduction current across the films [16,34]. Another high-K dielectric that has been studied is zirconium oxide (ZrO_2). ZrO_2 thin films were also deposited on silicon substrates by spray pyrolysis, in the temperature range from 400 to 600 °C. The use of zirconium acetylacetonate as source of zirconium and *N,N*-dimethylformamide was also used in this case. The films resulted with an index of refraction in the range of 2.12. The dielectric constant was about 12.5–17.5. In the best case, the films could stand an electric field up to 3 MV/cm, without presenting evidences a dielectric breakdown. Transmission electron microscopy measurements indicated that the films of ZrO_2 were constituted by nano-crystals embedded in an amorphous matrix [16].

In summary, high quality aluminum, yttrium and zirconium oxide thin films have been deposited by spray pyrolysis using acetylacetonates dissolved in *N,N*-dimethylformamide. In the case of aluminum oxide, they were obtained with excellent homogeneity to thicknesses down to 30 nm. The addition of a parallel stream of water mist into the spraying solution aerosol during the deposition process resulted in a dramatic effect over the refractive index and on the dielectric characteristics of the deposited aluminum oxide films. The density of states in the range of 10^{11} eV^{-1}·cm^{-2} and a destructive electric breakdown field larger than 5 MV/cm, were obtained on MOS structures fabricated with these films. Yttrium and Zirconium oxide thin films showed a higher dielectric constant than those of aluminum oxide, but lower dielectric strength, likely due to the polycrystalline nature of the films.

4. Luminescent Materials

Luminescent materials, in the form of powders (phosphors) and films, have been extensively studied in recent decades [35,36] because their great importance for a wide variety of applications such as: Lighting, image displays, signaling, lasers, medical applications, etc. [37,38]. They have been synthesized through a variety of physical and chemical techniques, including: Hidrothermal/Solvothermal [39–41], solid-state reaction [42], sol-gel [43,44], laser ablation [45,46], sputtering [47], Pechini Method [48], plasma electrolitic oxidation [49], conventional melt-quenching method [50,51], combustion synthesis [52], solvent evaporation method [53–55], and co-precipitation process [56]. Among these techniques, spray pyrolysis began to be used for this purpose in the mid-1980s, and it is still used today [57,58]—proving to be a practical, low cost, easy to extrapolate for large area deposition technique. In this review, an account is made on diverse luminescent materials synthesized by this technique. These materials in general involve one or more luminescent centers incorporated as dopants in a host lattice. A great variety of host lattices have been used for the synthesis (by means of spray pyrolysis) of phosphors and luminescent films, among them stand out metal oxides such as: ZrO_2, Al_2O_3, HfO_2, Y_2O_3, ZnO, In_2O_3, $ZnSiO_3$, CdO, (Y, Gd)BO_3, Gd_2O_3, $LaPO_4$, $BaMgAl_{10}O_{17}$, and some sulfur based compounds such as: ZnS, $CaSO_4$, CdS, and others. The luminescent active centers have been mainly RE (Rare Earth) and some transition metal ions. In some cases, luminescence emission has been observed to be generated by mechanisms that involve structural defects and intrinsic states in the host lattices as well. This review focuses mainly on the work done on host lattices such as: ZrO_2, Al_2O_3, HfO_2, Y_2O_3, ZnO, and ZnS with different dopants.

4.1. ZrO_2

Virtually before 1999, ZrO_2 had not been used as a host lattice to produce phosphor materials synthesized by spray pyrolysis technique. In that year, some results were reported about photoluminescence (PL) and thermoluminescence (TL) properties of ZrO_2:Tb^{3+} films deposited by a pneumatic spray pyrolysis (PSP) system [59]. These films excited by 275 nm exhibited four peaks at 487, 542, 582, and 619 nm—typical of electronic transitions in the Tb^{3+} ions. The TL glow curve displayed two peaks at 112 °C and 270 °C for the ZrO_2:Tb^{3+} films exposed to 260 nm UV radiations. In addition, the TL response was linear in the range of 40 to 240 mJ·cm^{-2} spectral UV irradiance. These results exhibited that ZrO_2:Tb^{3+} films had appropriate characteristics for their use as a UV dosimeter as well as PL phosphor. In a later investigation (2001) on this material [60], a deeper analysis was made on the thermoluminescence mechanisms. Two important parameters in TL studies such as activation energy (E) and the frequency factor (S) were investigated. In this contribution, the Lushchik and Chen methods were used to determine the kinetic parameters which showed second order kinetics for both the first and second glow TL peaks.

Furthermore, in 2001, PL and cathodoluminescence (CL) feature of ZrO_2:Tb^{3+} films, deposited by the PSP, technique was reported [61]. In this case different deposition parameters, such as substrate temperatures, doping concentrations, and the flow of the precursor solution, were studied. Substrate temperatures higher than 400 °C rendered a polycrystalline material with metastable tetragonal or cubic phases. With increasing deposition temperatures, the PL and CL emission intensities (excited with 250 nm light) also increased. The PL and CL emission spectra showed the characteristic peaks associated with the electronic transitions of Tb^{3+} ions. Concentration quenching for the PL and CL emissions occurred at doping concentration greater than 1.96 and 1.17 at.%, respectively. Similar studies were conducted on ZrO_2:Eu^{3+} films [62]. Depending on the substrate temperature, these films were amorphous or polycrystalline (tetragonal-cubic phase). A strong red emission was observed which was generated by the $^5D_0 \rightarrow {^7F_2}$ transition typical of the Eu^{3+} ions. From those studies, it became clear that zirconia was a suitable host lattice for RE ions.

For the first time, a study on luminescent emissions from ZrO_2: Mn^{2+} films deposited by the USP technique was reported in 2002 [63]. These films were deposited at substrate temperatures ranging from 250 to 500 °C. The PL and CL (7 KeV) emission spectra showed a broad band (450–750 nm) centered at 650 nm (red), which is associated with the electronic transitions $^4T_1(^4G) \rightarrow {^6A_1(^6S)}$ of the Mn^{2+} ions. A decrease of the luminescence, as a function of the doping concentration, substrate temperature and electron accelerating voltage was observed. The maximum emission intensity was observed for films deposited at 250 °C, EDS measurements showed that these films had a high amount of incorporated chlorine (from the precursors in the spraying solution), which acts as a co-activator for the red emission. As the deposition temperature increased, the amount of chlorine in the film (as well as the red luminescence emission intensity) decreased. The presence of chlorine was necessary for the red luminescence emission to occur. CL spectra obtained at higher electron accelerating voltages (10 KeV) from samples deposited at 500 °C showed, instead of the red emission, a wide band centered at 590 nm (yellow)—which is also characteristic of Mn^{2+} ions.

ZrO_2:Eu^{3+} phosphors consisting of spherical, dense and sub-micrometer size particles were successfully synthesized by the USP technique in 2005 [64]. The X-ray diffraction (XRD) measurements indicated that the crystallinity of these powders increased with increasing postdeposition annealing temperature. Several characterization techniques were used to study this material: Including PL emission spectra, and decay time measurements. The excitation spectrum showed a band centered at 248 nm corresponding to a charge transfer transition from Eu-O generated electronic states in the ZrO_2 host matrix. The emission spectra exhibit the typical (red) bands of Eu^{3+} ions. The optimal concentration of Eu^{3+} ions was 10 at.% and it was observed that the spherical morphology of the particles improves the intensity of the PL emission.

A research work on $ZrO_2:Pr^{3+}$ films was published in 2007 [65]. In this case, PL and CL properties were studied as a function of growth parameters such as the substrate temperature and the Pr^{3+} ions concentration. XRD studies indicated a tetragonal crystalline structure for zirconia as the substrate temperature was increased. The PL spectra exhibited bands centered at 490, 510, 566, 615, 642, 695, 718, 740 and 833 nm; associated with the electronic transitions $^3P_0 \rightarrow {}^3H_4$, $^3P_0 \rightarrow {}^3H_4$, $^3P_1 + {}^1I_6 \rightarrow {}^3H_5$, $^1D_2 \rightarrow {}^3H_4$, $^3P_0 \rightarrow {}^3H_6$, $^1D_2 \rightarrow {}^3H_5$, $^1D_2 \rightarrow {}^3H_5$, $^3P_0 \rightarrow {}^3F_{3,4}$, and $^1D_2 \rightarrow {}^3F_2$ of the Pr^{3+} ions. As the substrate temperature was increased, an increasing intensity of the PL emission was observed. Also, a quenching of the PL and CL emissions, with increasing doping concentration, was detected. Interestingly the CL spectra, as a function of the electron accelerating voltage, showed an evolution of the highest peak: For low electron accelerating voltages (4 kV) the red emission (615 nm) is the maximum, and for high voltages (15 kV) the most intense band is the blue (around 490 nm).

The cathodoluminescence properties of $ZrO_2:Er^{3+}$ films were reported in 2014 [66]. These films were deposited at different temperatures from 400 °C up to 550 °C. As substrate temperatures are increased, the films showed a tetragonal phase. CL emission spectra showed bands centered at 524 (green), 544 (green) and 655 (red) nm associated with the electronic transition $^2H_{11/2} \rightarrow {}^4I_{11/2}$, $^4S_{3/2} \rightarrow {}^4I_{15/2}$, and $^4F_{9/2} \rightarrow {}^4I_{15/2}$ of Er^{3+} ions. The highest emission intensity is achieved in samples deposited at 500 °C doped with 5 at.% of Er^{3+} ions. Also, the CL emission intensity increases as the substrate temperature and electron accelerating voltage values increase.

Investigations on $ZrO_2:Dy^{3+}$ and $ZrO_2:Dy^{3+}+xLi^+$ films were published in 2015 [67]. XRD measurements, as a function of the deposition temperature, indicated a meta-stable tetragonal crystalline structure of the zirconia. PL and CL features of the films were studied as a function of synthesis parameters such as the substrate temperature and the Dy^{3+} and Li^+ concentrations. All luminescent emission spectra showed peaks located at 485 (blue), 584 (yellow), 670 (red) and 760 nm; which correspond to electronic transitions $^4F_{9/2} \rightarrow {}^6H_{15/2}$, $^4F_{9/2} \rightarrow {}^6H_{13/2}$, $^4F_{9/2} \rightarrow {}^6H_{11/2}$, and $^4F_{9/2} \rightarrow {}^6H_{9/2}$, of Dy^{3+}, respectively. The Li^+ incorporation in the $ZrO_2:Dy^{3+}$ films produced an improvement in the intensity of the luminescent emission, presumably because it acts as a charge compensator and because it contributes to improving the crystalline structure of the host lattice. The CIE color coordinates (0.3475, 0.3609) of these films were found within the warm white light emission region. These spectroscopic characteristics allowed to propose this material for application in solid-state lighting (SSL), especially for white lighting emission applications. It is observed that, as the concentration of Li^+ ions increases, they come closer to the perfect white area of the CIE color coordinates (0.3333, 0.3333).

Moreover, in 2015 a work on ZrO_2, $ZrO_2:Dy^{3+}$ and $ZrO_2:Dy^{3+} + Gd^{3+}$ films was published [68]. The synthesis and the characterization conditions were carried out as described in Reference [67]. The relative concentrations of Dy^{3+} and Gd^{3+} ions were varied; the emission spectra of these films exhibited bands in the blue and yellow regions. The incorporation of Gd^{3+} ions in $ZrO_2:Dy^{3+}$ films generated a remarkable increase in the intensity of the luminescent emission (approximately 15 times). In principle, the host lattice absorbs the excitation energy which is transferred to the Gd^{3+} ions which in turn transfers it to the Dy^{3+} ions. The CIE chromaticity diagram exhibited a cold-white emission (Dy^{3+}-Gd^{3+} doped samples) and a warm-white emission (Dy^{3+} doped samples), which shows the potential of these films for generating white light coatings for solid state lighting (SSL) applications.

The PL and structural properties of co-doped ZrO_2: $Eu^{3+} + Tb^{3+}$ films, were also reported in 2015 [69]. The PL spectra showed the typical emission bands associated with the Tb^{3+} and Eu^{3+} ions, as well as a broad emission, peaked at 440 nm associated to radiative transitions within the ZrO_2 host lattice. These films displayed multicolored emissions depending of the ratio Eu^{3+}/Tb^{3+} and the excitation wavelength. The observed colors were: Blue (from the host lattice), green (from the $ZrO_2:Tb^{3+}$ films), red-orange (from the $ZrO_2:Eu^{3+}$ films), yellow (from the $ZrO_2:Eu^{3+} + Tb^{3+}$ films, excited with 288 nm) and bluish-white and yellowish white (from the $ZrO_2: Eu^{3+} + Tb^{3+}$ films, excited with 368 or 380 nm). The CIE coordinates of the double-doped $ZrO_2:Tb^{3+}$ (10 at.%) + Eu^{3+} (5 at.%) films lie in the white light region of the chromaticity diagram and show good potential for lighting devices and photonic applications.

4.2. Al_2O_3

A pioneering work on luminescent $Al_2O_3:Tb^{3+}$ films appeared in 1992 [70]. The films were deposited by the PSP technique on either plain or conductive oxide coated glass substrates at deposition temperatures in the range of 270–450 °C. PL emission from these films showed well-defined peaks at 490 and 550 nm, which were associated to the electronic transitions corresponding to Tb^{3+} ions. The relative emission intensity was strongly dependent on the type of substrate, the deposition temperature and the amount of Tb^{3+} ions incorporated in the films. Two years later, an investigation on $Al_2O_3:CeCl_3$ films was published in 1994 [71]. PL spectra (excited with 300 nm light) showed a broad emission formed by two overlapping peaks at 365 and 395 nm. It was suggested that these bands originate from the 5d to 4f electronic energy levels of Ce in the $CeCl_3$ molecule. The PL emission intensity of these peaks was strongly dependent on the doping concentration and the substrate temperature. The films with greater intensity were those deposited at the lowest temperature, where there is a greater amount of $CeCl_3$ incorporated in the films. As the temperature increases, the concentration of $CeCl_3$ molecules decreases and so does the PL emission intensity—therefore, the presence of this molecule is essential for an optimal emission of blue light. Also, a quenching of the PL is observed for $CeCl_3$ concentrations higher than 1 at.%. Another research on $Al_2O_3:Eu^{3+}$ films was published in 2000 [72]. These films were deposited by the USP technique at substrate temperatures from 300 to 540 °C and the Eu^{3+} doping concentration was varied. All films were amorphous in structure and the PL spectra were measured as a function of substrate temperature and doping concentration. The excitation spectrum showed an intense peak centered at 395 nm. All the PL emission spectra (excited by 395 nm) showed bands located at 587, 600, 612, and 648 nm—typical of the electronic transitions in Eu^{3+} ions. It was observed a concentration quenching of the PL emission intensity at values of above 1.5 at.% in the films. Thus, it was shown that Al_2O_3 is a suitable host lattice to support RE ions (such as Eu^{3+}) to generate strong PL emissions.

In 2003, a new research in $Al_2O_3:Tb^{3+}$ films was published [73]. In this case, the transparency of the films was up to 88% on the 400 to 700 nm range. These was possible because the use of organic source reactive for both aluminum and terbium (acetylacetonates) that were dissolved in dimethylformamide and sprayed, deposited at temperatures up to 600 °C. These films were mostly amorphous in the range of deposition temperatures studied with an average roughness of 14 Å or less; which was perfect for the design and development of microdevices integrating this type of films. PL and CL spectra, studied as a function of the deposition parameters such as doping concentrations and substrate temperatures, were typical of the transitions among the electronic energy levels of the Tb^{3+} ions. Thus, from this work, it is clear that the use of acetylacetonates as precursors, generates the formation of high transmittance films with low roughness, as described in the dielectric section thin films, in contrast to those films synthesized from chlorides, nitrates or acetates (dissolved in water) which are, in general, very rough and opaque.

An energy transfer mechanism between Ce^{3+} and Mn^{2+} ions in alumina films was reported in 2005 [74]. Blue and red light emitting $Al_2O_3:Ce^{3+}:Mn^{2+}$ films, under ultraviolet light excitation, were investigated in this case. The blue emission is due to transitions from the excited state 5d to the split ground state ^2F of the Ce^{3+} ions. The usually weak Mn^{2+} ions red emission, attributed to intra 3d transitions, was enhanced by an efficient energy transfer from the Ce^{3+} ions. The energy transfer mechanism was an electric dipole–quadrupole interaction with a quantum efficiency estimated to be near to 100%, which makes these films interesting phosphors for the design of microdevices based on luminescent layers in flat-panel displays. Other studies on this type of amorphous $Al_2O_3:Ce^{3+}:Mn^{2+}$ films were also published [75,76]. However, in this case, the precursors were $AlCl_3$, $CeCl_3$ and $MnCl_2$ dissolved in deionized water (Ce: 10 at.%; Mn: 1, 3, 5, 7 and 10 at.%), deposited at a substrate temperature of 300 °C. The chemical composition and the profile distribution of the dopant ions across the films were determined by Rutherford backscattering (RBS). A homogeneous depth profile of both Ce^{3+} and Mn^{2+} ions was found within the films, and the overall measured quantities were as expected from the solution concentrations. Chlorine, which plays a significant role in luminescent properties, was detected in important quantities, something that was expected due to the low deposition temperatures used in this case. The red emission from manganese-doped samples was strongly enhanced with the co-doping with Ce due to the efficient energy transfer mechanism from Ce^{3+} to Mn^{2+} ions. From XPS analysis, it was determined that a considerable amount of Mn ions remains linked to chlorine, while Ce is mostly in an oxidized state.

In 2010, alumina was used to host three ions (Tb^{3+}, Ce^{3+}, and Mn^{2+}) to generate white light when excited by ultraviolet light [77]. These amorphous films were also deposited at 300 °C. Sensitization of Tb^{3+} and Mn^{2+} ions by Ce^{3+} ions gave rise to blue, green and red luminescent emission when the film was excited with UV radiation. The overall efficiency of such energy transfer was about 85% upon excitation with 312 nm light. Energy transfer from Ce^{3+} to Tb^{3+} ions through an electric dipole–quadrupole interaction mechanism appeared to be more probable than the electric dipole–dipole one. A strong white light emission from the $Al_2O_3:Ce^{3+}$ (1.3 at.%):Tb^{3+} (0.2 at.%):Mn^{2+} (0.3 at.%) films under UV excitation was obtained. The high efficiency of energy transfer from Ce^{3+} to Tb^{3+} and Mn^{2+} ions, resulted in a cold white light emission ($x = 0.30$ and $y = 0.32$). Thus, these films resulted interesting material for the design of efficient UV pumped phosphors for white light generation which could be integrated in light emitting microdevices.

Similarly, alumina co-doped with Dy^{3+} and Ce^{3+} ions was reported in 2011 [78]. The PL properties of these films were studied through excitation, emission spectra measurements and decay time spectroscopy. These films emitted a combination of blue and yellow colors through an efficient energy transfer (77%) from Ce^{3+} to Dy^{3+} ions. It was inferred that such energy transfer was non-radiative, taking place between Ce^{3+} and Dy^{3+} clusters, through a short-range interaction mechanism. Ce^{3+} doped single films emitted in the violet-purplish-blue region; whereas co-doped films the presented a cold-white light emission. The PL properties of tri-doped $Al_2O_3:Ce^{3+}:Dy^{3+}:Mn^{2+}$ films were published in 2012 [79]. Nonradiative energy transfer from Ce^{3+} to Dy^{3+} and Mn^{2+} was reported upon UV excitation at 278 nm. From lifetime data, it was deducted that the energy transfer was nonradiative in nature. Simultaneous emission of all co-dopant ions in the blue, yellow and red regions, resulted in white light emission with CIE 1931 chromaticity coordinates, $x = 0.34$ and $y = 0.23$, with a color temperature of 4900 K. Thin films as these might contribute to the development of materials that, pumped with AlGaN-based LEDs, could generate white light emission.

Also, in 2012, a study on the PL characteristics, under continuous and pulsed excitation of Eu-doped alumina films was reported [80]. It was determined that localized states in the undoped Al_2O_3 host lattice, excited with 250 nm radiation, emit a violet color (broad band centered at 415 nm) associated to a radiative recombination process involving F centers. When Eu^{3+} ions were incorporated into these films, a charge transfer mechanism to these ions from the localized states seems to occur predominantly. The Eu^{3+} related emission, generated in this way, results intensified and luminescence decay time extended as compared to that obtained when the excitation is achieved through an inter-electronic energy level transition in the Eu^{3+} ion, excited by 395 nm radiation.

Subsequently, in 2013, a contribution on the white light emission from $Al_2O_3:Ce^{3+}:Tb^{3+}:Mn^{2+}$ and $HfO_2:Ce^{3+}:Tb^{3+}:Mn^{2+}$ films was published [81]. These oxide films doped with $CeCl_3/TbCl_3/MnCl_2$ were deposited at 300 °C. XRD measurements exhibited a very broadband typical of non-crystalline materials. Non-radiative energy transfer from Ce^{3+} to Tb^{3+} and Mn^{2+} ions is observed upon UV excitation at 280 nm; the energy transfer could take place in Ce^{3+}-Tb^{3+} and Ce^{3+}-Mn^{2+} clusters through an electric dipole-quadrupole interaction mechanism. This energy transfer gives place to a simultaneous emission of the donor and acceptor ions in the blue, green, yellow and red regions, resulting white light emission. The chromaticity coordinates for Al_2O_3:Ce^{3+}:Tb^{3+}:Mn^{2+} films and color temperatures were: (0.30, 0.32) and 7300 K (cold-white color). The chromaticity coordinates for HfO_2:Ce^{3+}:Tb^{3+}:Mn^{2+} films and color temperatures were (0.32, 0.37) and 6000 K (warm-white color).

Another study on PL emission (white emission) from single and double layered Al_2O_3:Ce^{3+}:Tb^{3+}:Eu^{3+} films was presented in 2013 [10]. These films were deposited using acetylacetonates (dissolved in dimethylformamide) as precursors. Eu^{3+} and Tb^{3+} doped films showed the typical emissions of these trivalent ions (red and green, respectively). Ce doped films showed two broad bands associated with the 5d to 4f transitions of the Ce^{3+} ion, centered at ~400 and 510 nm. As expected from films deposited with organic precursors, these films had low surface roughness (lower than 3 nm) and thicknesses between 50 and 260 nm. The double layer stacks involved first an Eu^{3+} doped film followed by a second Ce^{3+}-Tb^{3+} co-doped layer. The films were transparent in the visible region, with an optical bandgap of approximately 5.63 eV. The PL of these stacks was an overlap of the emissions corresponding to all the dopants when excited with 300 nm light, resulting in an intense white light emission, which would be suitable for the design of electroluminescent microdevices.

The PL characteristics of Eu^{3+} doped alumina films co-doped with Bi^{3+} and Li^+ were published in 2015 [82]. In this case, the incorporation of Bi^{3+} and Li^+ ions as co-dopants in Al_2O_3:Eu^{3+} films and its effect on the luminescence characteristics of this material were described. Both Bi^{3+} and Li^+ do not introduce new luminescence features but affect the luminescence intensity of the Eu^{3+} related emission spectra as well as the excitation spectra. The introduction of Bi^{3+} generates localized states in the aluminum oxide host that result in a quenching of the luminescence intensity, while Li^+ and Bi^{3+} co-doping increases the luminescence intensity of these films. It was found that the Eu^{3+} ions emission intensity in these films, when Bi^{3+} ions were added together with Li^+, produce an increase of 62% in the emission intensity. It was suggested that the role of Li^+ co-doping was to redirect the energy paths back to the Eu^{3+} ions from the Bi^{3+} ions. Analysis of time decay measurements of the Eu^{3+} related emission in the amorphous alumina films indicated the presence of two type of sites in the short-range surroundings of the Eu^{3+} ions that could be correlated with those around this ion in α or γ Al_2O_3 crystalline phases.

4.3. HfO2

Luminescent HfO_2:Mn^{2+} films (deposited by Ultrasonic Spray Pyrolysis technique) were reported for the first time, in 2004 [83]. The deposited films were amorphous at deposition temperatures up to 300 °C; for higher temperatures a polycrystalline material was obtained with a monoclinic HfO_2 phase. The cathodoluminescence (CL) spectra showed blue–green and red bands associated with the electronic transitions $^4T_1(^4G) \rightarrow {}^6A_1(^6S)$ of the Mn^{2+} ions. A dependence of the CL emissions, as a function of the doping concentration, substrate temperature and electron accelerating voltage was reported. It was determined that both amorphous and polycrystalline hafnium oxide make efficient host for Mn^{2+} ions, and that the relative content of chlorine in the processed films have an important role on the luminescent emission intensity of the studied materials.

USP deposited HfO_2:$CeCl_3$ films luminescent properties were published in 2007 [84]. These films were deposited from hafnium dichloride oxide and $CeCl_3$ dissolved in deionized water (18 MΩ/cm). The PL characteristics of the HfO_2:$CeCl_3$ films were studied as a function of doping concentrations and substrate temperature. XRD measurements showed the monoclinic phase of HfO_2 for samples deposited at deposition temperatures higher than 400 °C. These films showed a violet–blue PL emission that could easily be seen with the naked eye in normal room light. Also, PL emission and excitation spectra evidence the presence of two different Ce^{3+} centers in HfO_2. A complete concentration quenching of the luminescence of one of the two centers is observed at high concentration of $CeCl_3$ (15 at.% in the start solution), which suggests a fast energy transfer from the high-energy to the low energy centers. Finally, it was confirmed that HfO_2 is an adequate host matrix for rare earth ions as active centers to generate strong violet–blue PL emissions.

Also, in 2007, a work on PL properties of HfO_2:Tb^{3+} films was published [85]. The PL properties of these films were studied as a function of deposition temperature and Tb^{3+} ions concentration. The films were deposited the USP technique from aqueous solution of Hafnium and Terbium chlorides. Results showed that crystalline structure of HfO_2:Tb^{+3} films depends on the deposition temperature. PL excitation spectrum showed a wide band centered at 262 nm while the PL emission spectra showed bands centered at 488, 542, 584 and 621 nm, which correspond to the electronic transitions: $^5D_4 \rightarrow {}^7Fj$ (j = 3, 4,5, 6) typical of trivalent terbium ions. The dominant emission intensity corresponds to the green color (542 nm), which depended on the terbium concentration incorporated in the host lattice; the optimum doping concentration was 5 at.% Tb^{+3} in the spraying solution.

The PL and CL characteristics of HfO_2:Sm^{3+} films were published in 2008 [86]. These films were deposited by the USP technique on Corning glass substrates at deposition temperatures ranging from 300 to 550 °C using chlorides as precursor materials. Scanning electron microcopy (SEM) micrographs revealed rough surfaces morphology with spherical particles. The PL and CL spectra exhibited four main bands centered at 570, 610, 652 and 716 nm, which are due to the well-known intra-4f transitions of the Sm^{+3} ions. It was found that the overall emission intensity rose as the deposition temperature was increased. Moreover, a concentration quenching of the emission intensity was observed for doping concentration higher than 0.7 at.% as measured by EDS. These films showed good adherence to the substrate and a high deposition rate of up to 2 μm per minute. In addition, The CL emission intensity was found to increase as the electron accelerating voltage was raised.

Also, in 2008, HfO_2 films doped with $CeCl_3$ and/or $MnCl_2$ were deposited at 300 °C by the USP technique [87]. The XRD results revealed that the films were predominantly amorphous. HfO_2: $CeCl_3$ showed a violet-blue emission. The weak green–red emissions of Mn^{2+} ions was enhanced through an efficient energy transfer from Ce^{3+} to Mn^{2+} ions in the co-doped films. Spectroscopic data indicated that this energy transfer was nonradiative in nature and it could occur in Ce^{3+} and Mn^{2+} clusters through a short-range interaction mechanism. The efficiency of this energy transfer increases with the Mn^{2+} ion concentration, so that an efficiency of about 78% is achieved for a 5 at.% of $MnCl_2$ concentration. The HfO_2:$CeCl_3$:$MnCl_2$ films are interesting phosphors for the design of luminescent layers emitting simultaneously in the three primary colors: Violet-blue, green and red.

The HfO_2 host lattice was also used to house, simultaneously, ions such as Ce^{3+}, Tb^{3+} and Mn^{2+} to generate cold white light [88]. These films were either doubly doped with $CeCl_3$ and $TbCl_3$ or tri-doped with $CeCl_3$, $TbCl_3$, and $MnCl_2$ and deposited at 300 °C. In the doubly doped films, energy transfer from Ce^{3+} to Tb^{3+} ions could take place in Ce^{3+}-Tb^{3+} clusters through an electric dipole-quadrupole interaction; the efficiency of this transfer was about 81% upon excitation with 270 nm light. In the triply doped films, both Tb^{3+} and Mn^{2+} ions, can be sensitized by Ce^{3+} ions. The efficiency of energy transfer from Ce^{3+} to Tb^{3+} and Mn^{2+} ions was enhanced by increasing the Mn^{2+} concentration, up to about 76% for the films with the highest Mn^{2+} ions content (1.6 at.%). The simultaneous emission of these ions under UV excitation resulted in white light luminescence.

The PL and TL properties of HfO_2 films were investigated [89], these films were synthesized from hafnium chloride as raw material in deionized water as solvent and were deposited at temperatures from 300 to 600 °C. SEM images showed that the film's surface resulted very rough with semi-spherical promontories. UV irradiation was used in order to perform the thermo-luminescent (TL) characterization of these films; the 240 nm wavelength irradiation induced the best response. The PL spectra showed emission bands, centered at 425, 512 and 650 nm, associated to impurities such as chlorine and/or structural defects. As the substrate temperature was raised, a higher intensity of the band centered at 425 nm was observed. The TL experimental results showed that HfO_2 films could be useful in UV radiation dosimetry applications, using the TL method mainly in the interval of 200–400 nm; indicating an advantage over other ultraviolet dosimeters currently used.

An investigation on the luminescent properties of HfO_2 films co-doped with Ce^{3+} and several concentrations of Dy^{3+} was presented in 2011 [90]. The deposition temperature was 300 °C. PL emissions from Dy^{3+} ions centered at 480 nm (blue) and 575 nm (yellow) associated with the $^4F_{9/2} \rightarrow {}^6H_{15/2}$ and $^4F_{9/2} \rightarrow {}^6H_{13/2}$ electronic transitions, respectively, were observed upon UV (280 nm) excitation via a non-radiative energy transfer from Ce^{3+} to Dy^{3+} ions. Such energy transfer via an electric dipole–quadrupole interaction appeared to be the most probable transfer mechanism. The efficiency of this transfer increases up to 86 ± 3% for the film with the highest Dy^{3+} content (1.9 ± 0.1 at.% as measured by EDS). The possibility of achieving the coordinates of ideal white light with increasing the concentration of Dy^{3+} ions was demonstrated.

The PL, CL, and TL characteristics of $HfO_2:Dy^{3+}$ films were also reported in 2014 [91]. The films were deposited at temperatures ranging from 300 to 600 °C, using chlorides as precursor reagents. XRD diffraction studies showed the presence of HfO_2 monoclinic phase in the films deposited at substrate temperatures greater than 400 °C. The surface morphology of films showed a veins shaped microstructure at low deposition temperatures, while at higher temperatures the formation of spherical particles was observed. The PL (excitation = 248 nm) and CL spectra of the doped films showed the highest emission in the band centered at 575 nm (yellow) corresponding to the transitions $^4F_{9/2} \rightarrow {}^6H_{13/2}$, which is a typical transition of Dy^{3+} ions. Regarding the TL behavior, the glow curve of $HfO_2:Dy^{+3}$ films exhibited spectrum with one broad band centered at about 150 °C. The highest intensity TL response was observed on the films deposited at 500 °C. A concentration quenching was observed and the optimum $DyCl_3$ concentration was 1 at.% in the initial solution. It was also determined that substrate temperature for the sample with maximum PL emission intensity was 600 °C. The PL (yellowish-white emission) is intense since it can be observed by the naked eyes with normal ambient illumination.

HfO_2 films co-doped with Tb^{3+} or Eu^{3+} ions using acetylacetonates as precursors, were studied [92]. The films presented transmittance values in the visible region $\cong$90% and surface roughness less than 3.9 nm. These films were polycrystalline with a monoclinic phase for films deposited at substrate temperatures higher than 500 °C. The luminescent emissions (PL and CL) were typical of Tb^{3+} and Eu^{3+} ions with a luminescence concentration quenching observed for both Tb^{3+} and Eu^{3+} ions at 5 and 10 at.%, respectively. The peak PL and CL emission intensities for single doped films were observed for $HfO_2:Tb^{3+}$ (5 at.%) and $HfO_2:Eu^{3+}$ (10 at.%) films deposited at 500 °C. The refractive index observed in these films was between 1.97 and 2.04 and an optical band gap of 5.4 eV. The PL decay time measurements was measured on some $HfO_2:Tb^{3+}$, Eu^{3+} samples. QE around 35% and 25% were obtained using excitation wavelengths of 204 nm for Tb^{3+} and 215 nm for Eu^{3+}, respectively. HfO_2 films co-doped with Tb^{3+} and Eu^{3+} ions were synthesized at substrate temperatures from 400 to 600 °C using chlorides as reactive source materials [93]. These films became polycrystalline at 600 °C exhibiting the HfO_2 monoclinic phase. Tuning by the means of the excitation wavelength and the relative concentration of the co-dopants, PL spectra with several shades, from blue to yellow (including white light) were obtained due to the combined emissions of Tb^{3+} (green), Eu^{3+} (red) ions and the host lattice (HfO_2) violet-blue emission. The best white light emission (x = 0.3343, y = 0.3406) was obtained with 382 nm excitation light and 1.35 and 0.88 at.% of Tb and Eu in the films, respectively. The CL emission spectra for these films also showed emissions from green to red (including yellow, orange, and other intermediate emissions depending on the relative content of Tb and Eu in the film). Quantum efficiency values between 47% and 78% were obtained for these films, depending on the excitation wavelength and co-doping concentrations.

4.4. Y_2O_3

The first publication on $Y_2O_3:Eu^{3+}$ particles (synthesized by the spray pyrolysis process) was registered in the year 2000 [94]. These particles were prepared from high solution concentrations which had a more hollow and porous structure than those prepared from low-concentration solutions. The PL spectra showed a prominent peak at 612 nm (pure red color). The colloidal seed-assisted spray pyrolysis introduced in this paper was found to be applicable to the control of morphology of phosphor particles when the stock solution concentration was high. For the colloidal seed-assisted spray pyrolysis, the stable colloidal solution should be used for homogeneity of phase and morphology of the phosphor particles. The colloidal solution of Y and Gd hydroxy carbonate sol obtained by the liquid phase reaction method, using urea, was appropriate for the preparation of Y_2O_3:Eu^{3+} particles of filled and non-porous structure at high concentration of the precursor solution. The fine particles size prepared from the colloidal solution compared to those of the aqueous solution also revealed that the particles prepared from colloidal solution are much less hollow.

CL of USP deposited Y_2O_3 thin films doped separately with Eu^{3+}, Tb^{3+} and Tm^{3+} were reported in 2001 [95]. CL spectra for films doped with Eu^{3+}, Tb^{3+} or Tm^{3+} ions presented red, green, and blue light emissions, respectively. The blue emission of Y_2O_3:Tm^{3+} films had dominant peak at 476 nm. The CL intensity of these films depended strongly on annealing conditions and thulium doping concentration, presenting a maximum luminance of 30.4 cd/m^2. For the Eu^{3+}-doped films, a luminance of 255 cd/m^2 was obtained with a dominant peak centered at 604 nm. The luminance for the Tb^{3+}-doped film was 72 cd/m^2 with a dominant peak at 547 nm.

The role of LiCl added as flux on the luminescence properties of USP synthesized Y_2O_3:Eu^{3+} phosphors was investigated in 2002 [96]. The maximum PL intensity was obtained for phosphors prepared at 1300 °C from solution with LiCl flux, their intensity was 50% higher than that of phosphors prepared from solution without flux. The PL intensities of phosphors prepared at 700 and 900 °C from flux solution were 200% and 134% of those phosphors processed from solutions without flux at the same synthesis temperatures. LiCl flux played the role of enhancing the luminescence of Eu^{3+} ions into Y_2O_3 host lattice by reducing defects in the phosphor particles.

Furthermore, in 2002, a study on spherical particles of Y_2O_3:Eu^{3+} was published [97]. Y_2O_3:Eu^{3+} luminescent particles of spherical shape, filled morphology, and high brightness were prepared by combination of colloidal seed assisted spray pyrolysis and flux-added spray pyrolysis. Y_2O_3:Eu^{3+} particles processed from Y colloidal solution with 5 at.% LiCl/KCl flux showed completely spherical shape, filled morphology, high crystallinity, and significantly improved PL emission intensity, which was 30% higher than that of particles prepared by general spray pyrolysis.

Another study on Y_2O_3:Eu^{3+} powders was published in 2005 [98]. These powders were synthesized by spray pyrolysis process and annealed at several temperatures, in the range 900–1400 °C, to achieve crystallized luminescent materials. The microstructure and macrostructure of these powders were investigated by high resolution SEM images and XRD measurements. The luminescent properties were measured under VUV excitation (254 nm). The results of this work allowed to understand the influence of the phosphors' microstructure on PL characteristics. The spray pyrolysis powder PL efficiencies excited at 254 nm were lower than that of the commercial phosphor but under a 600 mbar Ne–Xe plasma excitation (this measurement provides a characteristic close to the working conditions in plasma display panels); the powder the brightness was equal that of the commercial phosphor. The results allowed differentiating the microstructure and macrostructure influence on luminescence. Eventually, a suitable phosphor powder for plasma display panels less dense than the commercial one has been prepared by spray pyrolysis.

A control of the morphology of Y_2O_3:Eu^{3+} phosphor particles in the spray pyrolysis process was attempted by using citric acid and polyethylene glycol (PEG) as additives in the spray precursors [99]. Three different morphologies of phosphor particles were obtained: Smooth spheres, rods, and flakes (with the presence of PEG with different molecular weights or without the presence of PEG, respectively). It was shown that the spherical Y_2O_3:Eu^{3+} particles, obtained through a two-step spray pyrolysis process, had higher PL intensity than those with other morphologies.

In a similar work to the previous ones, also published in 2005, it was demonstrated that the densified particles of Y_2O_3:Eu^{3+} remarkably improved the intensity of PL emissions [100]. High luminous Y_2O_3:Eu^{3+} phosphor particles with spherical shape were synthesized by Spray Pyrolysis technique. A simple but effective preparation strategy for enhancing the PL intensity of these particles was implemented. The yttrium nitrate solution was modified using an organic additive, then non-hollow particles were reached, but they were very porous, and the PL intensity was not improved. To solve this disadvantage, a drying control chemical additive (DCCA) was used as a secondary additive. It was found that the surface area was greatly reduced, and the crystallite size was increased by the use of DCCA. As a consequence, densified Y_2O_3:Eu^{3+} particles showed great improvement in their PL emission intensity.

The luminescent characteristics of Y_2O_3:Eu^{3+} (5 and 10 at.%) submicron particles, synthesized from the pure nitrate solutions at 900 °C, was also reported in 2010 [101]. The synthesis conditions (gradual increase of temperature within triple zone reactor and extended residence time) assured formation of spherical, dense, non-agglomerated particles with a crystallite size about 20 nm with a cubic Y_2O_3 crystalline phase. PL emission spectra were studied under excitation with 393 nm and together with the decay lifetimes for Eu^{3+} ion $^5D_0$ and $^5D_1$ levels revealed the effect of nanocrystalline nature on the luminescent properties of the powders. The PL emission spectra showed typical Eu^{3+} $^5D_0 \rightarrow {}^7F_i$ (i = 0, 1, 2, 3, 4) electronic transitions with dominant red emission at 611 nm, while the lifetime measurements revealed the quenching effect with the rise of dopant concentration and its more consistent distribution into host lattice due to the thermal treatment. The nanostructured Y_2O_3:Eu^{3+} phosphors possess favorable morphological properties for applications as red phosphor in optoelectronic microdevices, for example for luminescent displays.

Y_2O_3 powders doped with Yb^{3+} and co-doped either with Tm^{3+} or Ho^{3+} were synthesized and reported in 2012 [102]. These powders were processed at 900 °C using 0.1 M nitrates precursor solution and a cubic structure with space group Ia-3 was confirmed for all samples. Spherical particles with average size about 400 nm were generated with certain degree of porosity which alters their morphology during additional thermal treatment. The up-conversion emission spectra after excitation with 978 nm, as well as emission lifetimes and up-converted emission intensity dependence on excitation power were investigated. Dominant green ($^5F_4$, $^5S_2 \rightarrow {}^5I_8$) and blue ($^1G_4 \rightarrow {}^3H_6$) emissions were found for Ho^{3+} and Tm^{3+} samples, respectively. The enhanced emission intensities and lifetime in thermally treated samples were correlated with morphological and structural changes observed.

The enhancement of the PL emission intensity from Y_2O_3:Er^{3+} thin films with Li^+ as co-dopant was published in 2013 [103]. These films were deposited using 0.03 M of yttrium acetylacetonate, dissolved in *N,N*-dimethylformamide. The doping of the films with Er was achieved by adding erbium (III) acetate in the solution at 1.5% in relation to the Y content. The co-doping with Li was achieved adding lithium acetylacetonate to the spraying solution; the Li contents studied were 0, 0.5, 1, 2, 3, 3.5, and 4 at.% in relation to the Y content. The films were deposited at 500 °C on (1 0 0) silicon wafers. These films were polycrystalline with a pure Y_2O_3 cubic phase. The typical Er^{3+} related emission spectra showed an intensity increase by a factor of ~4–5 times with the addition of 2% of Li^+. This behavior is attributed to the distortion of the local crystalline field induced by the incorporation of Li^+ ions. The addition of Li^+ reduces the intensity of the diffraction peaks after 1%, and shifts the main diffraction peak toward large angles for Li^+ doping less than 3%. The distortion of the crystalline field leads to an increment of the efficiency of intra-4f transitions by permitting the otherwise parity forbidding transitions and reducing alternative nonradiative processes. These results showed that the low-cost ultrasonic spray pyrolysis technique was a simple way to obtain rare earth doped metallic oxide films co-doped with Li^+ ions as a strategy to improve their PL emission intensity.

The enhancement of the PL emission from Y_2O_3:Er^{3+} films, with the incorporation of Li^+ ions, was reported in 2014 [104] for both visible and IR characteristic emissions of Er^{3+} ions. The presence of Li^+ ions in the USP deposited films was inferred from Fourier transform infrared (FTIR) spectroscopy and also measured by Ion Beam Analysis (EBS), in which the high energies α particle yield from the $^7Li(p,\alpha)^4He$ nuclear reaction was used to determine the content of Li^+ inside the films. The average content of Li^+ inside the films, as determined by EBS, increases from 0 up to 18.5 at.% for un-doped to 4 at.% Li^+ co-doped samples. The Li-C-O absorption band in the IR region was directly proportional to the Li^+ content inside the films and a calibration curve was generated with the EBS analysis. In a related work [105], the effect of Li$^+$ co-doping on PL time decay characteristics of Y$_2$O$_3$:Er^{3+} was reported for films deposited at 500 °C. The Er^{3+} content, in this case, was fixed at 1.5 at.% while the Li$^+$ content in the spraying solution was varied from 1 to 4 at.% in relation to Y^{3+} content. The addition of Li$^+$ content up to 2 at.%, besides resulting in an increase of the luminescence emission intensity, modified the luminescence time decay behavior as well. A simple model in which charge transfer from localized centers to the Er^{3+} ions was proposed to describe the temporal evolution of the PL emission. The introduction of Li$^+$ ions in Y$_2$O$_3$:Er^{3+} had an impact on the charge transfer (CT) process and on the total number of Er^{3+} ions contributing to the PL emission. The PL time decay characteristics of Y$_2$O$_3$:Er^{3+} films under 207 nm or 414 nm excitation light were analyzed with a simple model in which, in addition to the radiative recombination sites associated with Er^{3+} ions, a CT process from localized states was considered.

Luminescent and structural characteristics of Y$_2$O$_3$:Tb^{3+} thin films deposited from β-diketonates as precursors on c-Si substrates, at temperatures in the 400–550 °C range, were reported in 2014 [106]. The PL and CL spectra intensity depended strongly on substrate temperature, the thickness of the films and the Tb^{3+} doping concentration. Y$_2$O$_3$:Tb^{3+} thin films exhibited one main band centered at 547 nm due to the $^5D_4 \rightarrow {}^7F_5$ electronic transition of the Tb^{3+} ion. A concentration quenching of the luminescence intensity was observed. At high temperatures the cubic crystalline phase of Y$_2$O$_3$ was observed as well as a reduction of organic residues. Also, at elevated temperatures, a low average surface roughness was obtained in the films with a high transmittance in the visible region.

PL and CL from Y$_2$O$_3$ doped with Tb^{3+} and Eu^{3+} ions films results were published in 2015 [107]. The deposition conditions were similar to those of the work described above. The optical and structural characterization of these films was carried out as a function of substrate temperature and Tb^{3+} and Eu^{3+} concentrations. Films deposited above 450 °C exhibited the typical PL bands associated with either Tb^{3+} or Eu^{3+} intra electronic energy levels transitions. The most intense PL and CL emissions were found for dopant concentration of 10 at.% for Tb^{3+} and at 8 at.% for Eu^{3+} ions in spraying solution. Higher substrate temperatures improved the crystallinity of Y$_2$O$_3$ films, and showed a low average surface roughness (62 Å for Y$_2$O$_3$:Tb^{3+}, and 25 Å for Y$_2$O$_3$:Eu^{3+} thin films). The films reported in this work were dense, and showed high refraction index (1.81), as well as a high optical transmittance in the UV-Vis range (about 90%) of the electromagnetic spectrum. These results suggest the possibility of applying those films in electroluminescent microdevices.

Recently, in 2017, an investigation on luminescent (PL and TL) Y$_2$O$_3$:Sm^{3+}, Li nanostructured thin films was presented [108]. XRD measurements confirmed the cubic structure of Y$_2$O$_3$ thin films. Li ions were successfully incorporated into the Y$_2$O$_3$ host lattice and it served as a sensitizer for better crystallization. The crystallites sizes are found to be ~50 nm. Surface morphology appeared as carved sculptures of particles with agglomeration. Optical absorption spectrum exhibited a prominent absorption peak at 270 nm and the corresponding energy gap was found to be ~5.53 eV. A broad PL emission was observed in the range 560–690 nm with peaks at 595, 608 and 622 nm corresponding to characteristic electronic transitions in the Sm^{3+} ions. These films were irradiated with γ-rays in a dose range 187–563 Gy; TL glow curve is deconvoluted into three peaks with temperature maxima at 400, 460 and 580 K. The activation energy and frequency factor of these TL glows were found to be in the order of ~0.58 eV and ~10^6 s^{-1}, respectively. Trap depths for the three luminescent centers were calculated and dose response was found to be linear in the range of 422–469 Gy.

4.5. ZnO

Zinc oxide (ZnO) is one of the most studied materials due to the various areas in which it is used. This material in the form of films and powders has also been frequently synthesized by the spray pyrolysis technique. One of the first studies on luminescent films deposited by the PSP technique of this material was on ZnO:TbCl$_3$ films published in 1987 [8]. Both intrinsic and ZnO:TbCl$_3$ films were deposited at atmospheric pressure, using air as the carrier gas. The substrate temperature during deposition was varied from 270 to 400 °C. The solution flow rate was changed in the range of

4–16 cm^3/min and the carrier gas flow rate was kept constant throughout the deposition process at 10 l/min; the deposition time was 10 minutes in all cases and the TbCl$_3$ concentration was 10 at.%. These films were polycrystalline with a hexagonal wurtzite structure. The PL spectra from un-doped films showed a peak centered at 510 nm [109], while ZnO:TbCl$_3$ films showed a peak at 550 nm associated to electronic transitions in the Tb^{3+} ions. Later in a follow up study about these films [110] it was reported that the light emission of the ZnO:TbCl$_3$ decreased with time of exposure of the sample to the excitation radiation. The phenomenon was interpreted in terms of a simple model in which a competitive process of hole trapping and photo-detrapping occurred at a radiative recombination center generated by the presence of TbCl$_3$.

The luminescence of undoped ZnO films, deposited from zinc nitrate solution, was also published in 1998 [111]. The films had a polycrystalline hexagonal wurtzite type structure with no preferred orientation. Green and orange PL (excited by 320 nm light) with emission intensity strongly dependent on the deposition and annealing temperatures was reported. The best green (broad band peaked at 510 nm) luminescent films had a porous structure while orange (band peaked at 640 nm) films consisted of close-packed grains with diameters of up to more than 1 micrometer. Green and orange PL bands resulted from oxygen-poor and oxygen-rich states, respectively, in ZnO. In the case of the green films, the vacancies did not appear to penetrate deeply into the crystallites.

The effect of Li ions incorporation on the luminescence of ZnO films was reported in 1990 [112]. The spraying solution was 0.1 M zinc acetate in isopropyl alcohol and deionized water mixed in equal proportions. Lithium chloride was added to the spraying solution at a concentration of 10 at.%. All deposited films exhibited a hexagonal polycrystalline structure. The optical transmission depended on the deposition temperatures (Ts = 340–330 °C) which showed an absorption edge shifting to longer wavelengths with higher Ts. The PL spectra of samples deposited at low Ts showed two emissions located at 420 nm and 500 nm, associated with blue emission from the Pyrex glass substrate and the blue green emission typical of un-doped zinc oxide, respectively. Films deposited at high Ts showed an emission peak centered at 555 nm apparently associated with the localized states generated by incorporation of Li ions in the ZnO films.

The photoluminescence from PSP deposited indium doped ZnO films was reported in 1992 [113]. This study was carried out as a function of the substrate temperature and solution flow rate. Deposition solution was 0.1 M zinc acetate in three parts of isopropyl alcohol mixed with one part of deionized water. Indium doping was achieved by adding InCl$_3$ to the spraying solution in a concentration of 2 at.%. The substrate temperature was varied from 260 to 320 °C. These films were polycrystalline with a hexagonal crystalline structure; high solution flow rates resulted in larger disorder on the orientation of the polycrystallites. The PL spectra from films deposited at low substrate temperature or with high solution flow rate showed a broad peak centered at 530 nm which was associated with (In$_{Zn}$ Vz)-luminescent centers.

The green photoluminescence efficiency and free-carrier density in ZnO phosphor powders were investigated in 1997 [114]. An aqueous zinc nitrate solution (10 at.% Zn) was utilized in the synthesis of all powders at processing temperatures from 700 to 900 °C. Electron paramagnetic resonance, optical absorption, and photoluminescence spectroscopy were combined to characterize ZnO powders. Green PL emission was generated and a good correlation between the 510 nm green emissions with the density of paramagnetic isolated oxygen vacancies was observed. Also, both quantities increase with free-carrier concentration n$_e$, as long as n$_e$ < 1.4 × 10^{18} cm^{-3}. At higher free-carrier concentrations, both quantities decrease. A model is proposed involving the isolated oxygen vacancy as the luminescence center. It was also shown that a free-carrier depletion layer, which forms at the surface of the powder particles, and the overall free-carrier concentration of the particles have a large impact on the green emission intensity of the ZnO powder.

PL from ZnO and ZnO:Li films, reported in 1997 [115], showed the well-known blue-green emission typical of ZnO for the undoped films. The Li-doped films PL emission was a broad band composed of four overlapping peaks at 508, 590, 604 and 810 nm (the excitation wavelength was

365 nm); the PL excitation spectra indicated that the excitation mechanism is primarily due to electron-hole pair generation across the ZnO energy bandgap. The decay time measurements of the PL showed that the lifetime of the luminescence emission was 187 ns. The dependence of the luminescent intensity with temperature showed an activation energy of 0.057 eV for competitive non-radiative transitions. These results were indicative that the lithium was atomically incorporated giving rise to a donor level in the ZnO.

PL dependence on the deposition temperature, film thickness, and post heat treatment of ZnO films, deposited from 0.4 M solution of zinc acetate dihydrate in a mixture of deionized water and isopropyl alcohol, was reported in 2000 [116]. Chlorine free ZnO films were obtained using zinc acetate as a precursor with the (002) oriented wurtzite structure in the substrate temperature range 250–350 °C. For films with the same thickness, the intensity of green emission decreased with an increment of the O/Zn ratio as determined by XPS. The green emission intensity was gradually enhanced with increasing film thickness. Increasing deposition temperatures resulted in a reduction of the O/Zn ratio and an increment of the intensity of the green PL emission. Also, as the annealing temperature was increased, the O/Zn ratio decreased, and the green emission was consequently enhanced.

The CL from ZnO and ZnO:F (5 at.%) films deposited from $ZnCl_2$ precursor solutions and fluorine doped by adding NH_4F to spraying solution was reported in 2002 [117]. The optimal substrate temperature was 450 °C presenting a hexagonal close packed structure. The CL spectra of both ZnO and ZnO:F films exhibited near-ultra-violet band peaked at λ = 382 nm, but they differ on the visible emissions; the undoped ZnO films emitted an intense blue-green light at 520 nm and a red emission at 672 nm, the fluorine doped samples presented a new band emission centered at 454 nm and no blue-green emission. This emission was interpreted as coming from a lattice modification of the Zn^{2+} environment in the crystal that could be due to a total anionic substitution process of O by F species.

Luminescent properties of ZnO and ZnO:Sn (6 at.%) films were studied through cathodoluminescence as well, in 2003 [118]. The spraying solutions (0.05 M) were prepared from Zn and Sn chlorides dissolved in deionized water. The substrate temperature was fixed at 450 °C. Luminescence films had a polycrystalline hexagonal wurtzite type structure. The CL measurements of the undoped films showed three bands centered at 382, 520 and 672 nm. Incorporation of tin ions extinguishes the blue–green band (520 nm) while appears a blue light at 463 nm and increases the value of the band-gap transition. CL imaging of ZnO films showed that the luminescence was located at defined sites giving rise to a grain-like structure inherent to the surface morphology. The presence of Sn inside the films led to great luminescent spots, attributed to large grain sizes.

The photoluminescent properties of Eu^{2+} and Eu^{3+} ions in ZnO phosphors were reported in 2004 [119]. These particles were synthesized from a zinc acetate solution and europium nitrate as the europium ions source. The crystal structure (zincite) of samples depended on the europium ions and the synthesis temperature. It was identified the coexistence of Eu^{2+} and Eu^{3+} ions in the as prepared ZnO samples. With addition of a 0.5 mol% concentration of europium ions, only the Eu^{2+} ion was detected inside the samples, while both Eu^{2+} and Eu^{3+} ions existed in samples using 1 mol% or higher concentration of europium ions. Changing the excitation wavelength, it was observed that both the blue and red PL can be obtained. The reduction of the Eu^{3+} to Eu^{2+} ions occurred in the particles prepared by the addition of a low concentration of europium ions. This reduction changed the color of PL from red to blue. Blue PL can be enhanced by increasing the synthesis temperature. At a high concentration of europium ions, the Eu^{3+} created the Eu_2O_3 component forming a $ZnO–Eu_2O_3$ composite.

The origin of the well-known blue-green emission of ZnO thin films was discussed on the basis of variation of the properties induced by different treatment of these films, such as ion beam irradiation (120 MeV Au ions and 80 MeV Ni ions were used for ion beam irradiation), and doping (Indium) [120]. PL studies of untreated thin films showed only one emission at 517 nm at room temperature while the irradiated films showed a decrease in this emission intensity. Indium doping also reduced the intensity of this emission; but additional emissions (centered at 407, 590 and 670 nm) were observed in these thin films. It was proposed that the blue-green emission was due to the transitions from the bottom of the ZnO conduction band to the level associated with an oxygen antisite (O_{Zn}).

Photoluminescence from Er-doped ZnO films were reported in 2008 [121]. These ZnO:Er films were deposited on (1 0 0) MgO wafers at 550 °C; the concentration of Er ions in the deposition solution (from Zn and Er acetates in methanol at 0.1 M) varied from 1.0 to 3.0 at.%. The films were polycrystalline with a dominant [002] preferential orientation. The near-ultraviolet (n-UV) PL from undoped ZnO films, n-UV peaks at 3.375 and 3.360 eV were observed at 18 K, which were proposed to be originated by free excitons and donor-bound excitons, respectively. The peaks from the free exciton transitions disappeared at room temperature. However, Er-doping enhanced the room temperature n-UV emission of ZnO:Er films. ZnO:Er (2.0 at.%) films showed n-UV peaks which were ~15 times stronger than those of undoped ZnO films.

Also, the luminescence of ZnO and ZnO:Ag nanocrystalline films deposited on Si (1 0 0) substrates from aqueous solution prepared by Zinc acetate dehydrate and Silver nitrate (6 at.%) was reported in 2008 [122]. Intrinsic samples deposited at 500 °C with spray rate of 0.15 mL/min presented the best near-band edge near-ultraviolet emission at 378 nm observed within a set of samples deposited at different deposition temperature and spray rates. The PL intensity ratio of the n-UV emission to the deep-level emission had a largest value of 470 and the full-width at half-maximum of n-UV peak had a smallest value of 10 nm (87 meV). In addition, the n-UV emission intensity of ZnO:Ag films (with the Ag:Zn atomic ratio = 3% in the precursor solution) is markedly enhanced and the ratio to the deep-level emission, increased to at least 700. However, a silver phase was detected and the n-UV luminescence became weak for ZnO:Ag films after the annealing at 700 °C in air for 1 h.

The electrical resistivity and the photoluminescence of zinc oxide films were correlated and reported in 2009 [123]. ZnO thin films were deposited, in this case, using zinc acetate dehydrate dissolved in methanol, ethanol, and deionized water within the substrate temperature range 320–420 °C. PL measurements showed that the as-grown ZnO thin films exhibited ultraviolet and green emission bands when excited by an Hg arc lamp using 313 nm as the excitation source. A red-shift in the near band edge was observed with the increase in the deposition temperature and was attributed to the compressive intrinsic stress present into the films. It is confirmed that oxygen vacancy (VO) is the most important factor that causes the broad visible emission. Furthermore, the visible emission and electrical resistivity of ZnO thin films are found to be a function of porosity. Additionally, it has been found that the intensity of the green emission at ~2.5 eV increased when ZnO films were deposited at 320 °C. The reason might be that the intrinsic stress, surface-to-volume ratio and porosity were incremented at low substrate temperatures. The resistivity presented similar behavior as the intensity of the green emission. A new luminescence mechanism based on the recombination related to oxygen vacancies in Zn-rich or stoichiometric conditions, was proposed.

Another study about ZnO:Li films was reported [124] for thin films deposited on borosilicate glass substrates; the deposition temperature was kept at 250 °C. The spraying solution was 0.2 M zinc acetate in a mixture of equal proportion of isopropyl alcohol and deionized water. Lithium doping was achieved by adding required amount of lithium acetate to the spraying solution. The spray time was 2 min with solution flow rate of 18 $cm^3 \cdot min^{-1}$ and gas flow rate of 15 $L \cdot min^{-1}$. The polycrystalline nature of the films was confirmed from XRD and TEM studies. A two-dimensional fringe moiré pattern with spacing of 1.2 nm was observed for the Li doped thin films. Lithium doping increased the roughness of the surface, thus making the film more passivated. Lithium was founded to play a key role in the excitonic as well as visible PL of ZnO films.

The effect of introducing Yb ions into ZnO films was reported in 2011 [125]. Yb-doped ZnO thin films were deposited on glass substrates at 350 °C during 77 min with a flow rate of the solution fixed at 2.6 mL/min; the molar ratio of Yb in the spray solution was varied in the range of 0–5 at.%. XRD measurements showed that the undoped and Yb-doped ZnO films exhibit the hexagonal wurtzite crystal structure with a preferential orientation along [002] direction. All films exhibited a high transmittance. The PL measurements showed a band at 980 nm that is characteristic of Yb^{3+} transition between the electronic levels $^2F_{5/2}$ and $^2F_{7/2}$. This was an experimental evidence for an efficient energy transfer from ZnO matrix to Yb^{3+} ions. These films showed low resistivity and high carrier mobility which makes of interest to photovoltaic devices; all ZnO:Yb thin films were n-type semiconductor. Also, ZnO:Yb^{3+} films had potential as candidates for photons down conversion process.

An investigation of structural, optical and luminescent properties of sprayed N-doped zinc (NZO) oxide thin films was reported in 2012 [126]. The precursor solution (0.1 M of zinc acetate and *N,N*-dimethylformamide) was sprayed onto the preheated corning glass, and fluorine doped tin oxide substrates held at optimized substrate temperature of 450 °C. Influence of N doping on structural, optical and luminescence properties were studied. These films were nanocrystalline having hexagonal crystal structure. Raman analysis depicted an existence of N-Zn-O structure in NZO thin film. XPS spectrum of N 1s showed the 400 eV peak terminally bonded, well screened molecular nitrogen (γ-N_2). Lowest direct band gap of 3.17 eV was observed for 10 at.% NZO thin film. The UV, blue and green deep-level emissions in PL of NZO films were due to Zn interstitials and O vacancies. The intensity of UV emission band increased with the concentration of activated nitrogen impurities. Shifting of PL peak from 393 to 388 nm seemed to be associated with free electron to neutral acceptor transition or some LO phonon replicas, followed by free electron-acceptor transitions.

The effect that Ga has on the properties of ZnO films deposited with an aqueous solution of 0.1 M zinc acetate and gallium nitrate on corning glass substrates was reported in 2012 [127]. XRD study depicted that the films were polycrystalline with hexagonal crystal structure and strong orientations along the (0 0 2) and (1 0 1) planes. Presence of E^{high}_2 mode in Raman spectra indicated that the gallium doping does not affect the hexagonal structure. The ZnO:Ga thin films were adherent, compact, densely packed with hexagonal flakes and spherical grains. Optical transmittance was high (about 80%). PL spectra showed violet, blue and green emission in these films. The specific heat and thermal conductivity study showed that the phonon conduction behavior was dominant in these films. XPS analysis confirmed that the majority Zn atoms remain in the same formal valence state of Zn^{2+} within an oxygen-deficient ZnO host lattice. The presence of zinc and oxygen vacancies was confirmed from PL results. The potential use of these films for optoelectronic microdevices was considered possible.

Optical and structural characteristics of ZnO:Al microrod films, obtained using different solvents (methanol and propanol), were published [128]. Zinc chloride at 0.1 M concentration in methanol and propanol was used as spraying solution. The doping was achieved by the addition of Alq3 (tris(quinolin-8-olato) aluminum(III)) dissolved in chloroform with a concentration of 7 at.% Al; a 50 nm/min deposition rate on glass substrates, at 500 °C and a spray rate of about 5 mL/min, was achieved. Both undoped ZnO and ZnO:Al films were composed of microrods with hexagonal crystal structure and a (0 0 2) preferential orientation. SEM images revealed a quasi-aligned hexagonal shaped microrods with diameters varying between 0.7 and 1.3 micrometers. Optical studies showed that microrods had a low transmittance (~30%) and the band gap increased from 3.24 to 3.26 eV upon Al doping. PL measurements showed the two emission bands usually present in ZnO PL spectra: One sharp and intense peak at ~383 nm and one broadband ranging from 420 to 580 nm.

The lithium effect on the blue and red emissions of ZnO:Er thin films was reported in 2013 [129]. These films were successfully deposited on heated (at 450 °C) glass substrates. The spraying solution was 0.05 M zinc chloride; erbium doping was achieved by adding $ErCl_3$ in concentrations of 2, 3, 5, and 7 at.%. Lithium was obtained from Li_2SO_4 in concentrations of 3, 5 and 7 at.%. This study was an investigation of the Li effect on the enhancement of CL emission intensity on Er-mono doped ZnO films. The Li–Er co-doped ZnO films showed a higher CL intensity of blue and red emissions than the Er-mono doped ZnO films. This behavior was attributed to the modification of the local symmetry of the Er^{3+} ions, which increases the probabilities for the radiative intra 4f transition of the Er^{3+} ions to occur. These results suggested that optimized Er–Li-codoped ZnO films could be used in data storage devices.

The blue luminescence of ZnO:Zn nanocrystals prepared from zinc acetate dihydrate aqueous solutions (0.05 M), and air as carrier gas with 1, 3, and 5 L/min flow rate was also reported in 2013 [130]. The temperature of the tubular reactor was set at 500, 600, and 700 °C. The crystal sizes were about 14–22 nm with a zincite structure; the observed morphology was partially spherical with other particles of irregular shape. The highest PL intensity, peaked at 450 nm (excitation wavelength of 250 nm), was obtained from samples prepared using 5 L/min carrier gas at 700 °C. These PL emission was associated to oxygen vacancy in the ZnO:Zn nanocrystals.

PL emission from ZnO:Ag films, formed by nanorods (NRs) as a function of the measurement temperature (10–300 K), was published in 2014 [131]. These films were deposited on soda-lime glass substrates at the deposition temperature of 400 °C and different deposition times (3, 5, and 10 min). The spraying solution (0.4 M) was prepared from zinc acetate and silver nitrate dissolved in in a mix of deionized water, acetic acid and methanol, a constant [Ag]/[Zn] ratio of 2 at.% was used for ZnO: Ag films deposition. The de position time variation permitted modifying the ZnO phase from the amorphous to crystalline, to change the size of ZnO:Ag NRs and to vary the PL emission spectra. PL spectra, versus temperature, revealed that the band related to the acceptor AgZn (LO phonon replicas of an acceptor bound exciton (2.877 eV)), and its second-order diffraction peak (1.44 eV) disappeared in the temperature range of 10–170 K with the formation of free exciton (FX). The PL intensity of defect related PL bands decreases monotonously in the range 10–300 K with the activation energy of 13 meV. The PL band (3.22 eV), related to the LO phonon replica of free exciton (FX-2LO) and its second-order diffraction peak (1.61 eV) increased in the range 10–300 K. FX related peak dominated in PL spectra at room temperature testifying the high quality of ZnO:Ag films deposited by the ultrasonic spray pyrolysis process.

A study on the role of substrates on the structural, optical, and morphological properties of ZnO films (nanotubes) was also reported in 2014 [132]. The role of substrate on the properties of ZnO films was investigated; these films were deposited onto glass, ITO coated glass and sapphire substrate and annealed at 400 °C for 3 hours. Aqueous solution (0.1 M) of zinc acetate was used to deposit these films at 350 °C. In the characterization XRD, SEM, Atomic force microscopy (AFM), and PL measurements were employed. XRD measurements showed that the ZnO films deposited on sapphire and ITO substrates exhibited a strong c-axis orientation of grains with hexagonal wurtzite structure. Extremely high UV emission intensity was observed in the film on ITO. The different luminescence behavior was discussed, which would be caused by least value of strain in the film—it is well known that the visible emission of ZnO thin films is due to the lattice defects that form deep energy levels in the bandgap. Films grown on different substrates revealed differences in the morphology. ZnO films on ITO and sapphire substrates revealed better morphology than that of the films deposited on glass. AFM images of the films prepared on ITO showed uniform distribution of grains with large surface roughness, suitable for application in dye sensitized solar cells. It was concluded that the nature of substrate had significant effect on the crystal structure, PL spectra, and morphological characteristics of the deposited ZnO films.

A comprehensive review on the structure, optical, and luminescence properties of ZnO:RE nanophosphors, including up-conversion (UC) and down-conversion (DC) and/or down shifting PL, was published in 2017 [133]. Some of ZnO:RE nanophosphors reviewed were synthesized by spray pyrolysis technique. The interest on RE doped ZnO for UC and DC nanophosphors has been motivated by the potential application of these materials in light emitting microdevices and photovoltaic cells. The two characteristic emissions observed in ZnO at the ultraviolet and visible regions are related, respectively, to excitonic recombination and intrinsic defects. XPS data demonstrated a correlation between the visible emission and intrinsic defects in these phosphors. In the case of the DC or down shifting processes, there was simultaneous emission related to intra-f level transitions of the RE ions and defects associated transitions in the ZnO host lattice. These emissions were mainly dependent on the synthesis process, annealing temperature, and RE ion concentration; only f $\rightarrow$ f transitions of RE ions were observed in the case of the UC process. These down and up conversion RE doped ZnO phosphors were evaluated for a possible application in solid state lighting devices and photovoltaic cells.

Also, in 2017 a study on the morphological, structural and optical properties (PL and CL) of ZnO thin films formed by nanoleafs or micron/submicron cauliflowers was reported [134]. Precursor solution was composed of zinc acetate dihydrate in deionized water (resistivity: 18 MΩ·cm); solution concentration was 0.002–0.064 mol·dm^3 and reactor temperature was varied from 300 to 450 °C, in 50 °C steps. These films formed by nano and microstructures with hexagonal crystal phase were successfully synthesized on aluminum or silicon substrates. The morphology showed the presence of three types of particles: Nano-leafs, single microparticles, and particles formed by the aglomeration of microparticles. The largest zone was formed by nanoleafs with a width of 25 nm and a length 200 nm long regardless of the roughness of the substrate. Moreover, the energy bandgap (3.26 eV) was invariant to changes in synthesis parameters. The optical measurements showed no considerable differences between the luminescence properties of films formed by nanoleafs and cauliflower particles. Deconvolution of PL emission spectra made it possible to elucidate the existence of oxygen vacancies, interstitial oxygen, zinc vacancies and interstitial zinc, structural defects in nanoleafs, and micro-cauliflowers. Defects such as these play an important role into PL and CL emissions of ZnO because electronic transitions associated to these defects originated almost the 100% of these emissions.

In 2018, a paper on the enhancement of visible luminescence and photocatalytic activity of ZnO:Cu thin films was published [135]. ZnO thin films doped with copper (0–4 at.%) were deposited on glass substrates maintaining a substrate temperature of 400 $\pm$ 10 °C. 0.4 M solution of zinc acetate and cupric acetate dissolved in a mixture of methanol, deionized water and acetic acid was used as the precursor for the deposition of these thin films. Hexagonal crystallinity (wurtzite) of the films improved at lower doping concentrations due to the easy fitting of Cu dopants in the Zn host lattice sites and preferred highly textured growth along the (0 0 2) plane. Higher doping concentration deteriorated the crystallinity and the optical transmission. EDX measurements confirmed the incorporation of Cu in the doped films. Optical energy gap red-shifted with the addition of Cu contents due to the exchange interactions and difference in iconicity of Zn and Cu. Cu doped films exhibited strong PL visible emission due to the modulation of the band structure and subsequently new levels acting as emission centers were formed in the forbidden bandgap of ZnO films. The addition of Cu ions increases the concentration of zinc interstitials, as well as zinc and oxygen vacancies which cause more intense emission in the visible region. ZnO:Cu thin films exhibited very good photocatalytic activity due to the efficient trapping of photo-generated electrons thereby suppressing the electron-hole recombination and higher doping level slightly decreased the degradation efficiency because excess dopants may act as recombination centers.

The effect of fluorine and boron co-doped ZnO thin films on the structural and luminescence properties was published in 2018 [136]. Fluorine and boron co-doped zinc oxide (ZnO:B:F) thin films were deposited on the corning glass substrates at 400 $\pm$ 5 °C: The spraying solution was prepared by mixing zinc acetate, boric acid, and ammonium fluoride, dissolved in methanol and deionized water with a ratio of 3:1. After characterization it was found that ZnO:B:F films had high average optical transmittance; XRD patterns indicated that the obtained ZnO:B:F films had a hexagonal wurtzite type structure with (0 0 2) preferential orientation. The crystallite sizes were in the 18–40 nm range. Green emission and UV emission band are observed in PL spectra of ZnO and ZnO:B:F. Undoped ZnO films exhibited only one peak around 390 nm associated with near band ultra violet emission. It is well known that the UV emission peak usually originates from the near band-edge emission from the recombination of free exciton. Also, it was considered that the intensity ratio of UV to visible emission is commonly considered as a sign of perfect crystal quality and low defect concentration. A green emission peak was observed for ZnO:B:F films; the intensity of this peak centered at 520 nm increased while the B–F concentrations increased. The observed green emission is also due to the impurity levels related the oxygen vacancy (Vo) in ZnO:B:F films. The electrical resistivity, carrier concentration and Hall mobility also were measured. The highest Hall mobility of 13.22 cm^2 v^{-1}·s, and the lowest electrical resistivity of 3.13×10^{-4} $\Omega \cdot$cm, were obtained at the optimal boron-fluorine co-doping concentration of 5 at.%. All of the results were appreciated in point of view of optoelectronic industry and photovoltaic solar cell applications and it was concluded that B-F co-doping has a positively effect on electrical properties.

4.6. ZnS

A research on the luminescence of ZnS, ZnS:TbC1$_3$ and ZnS:SmC1$_3$ films, deposited by the Pneumatic SP technique, was first reported in 1988 [137]. The ZnS films were deposited using a spraying solution obtained by mixing in equal proportions solutions of 0.1 M of Zn acetate and 0.1 M of dimetylthiourea ($C_3H_8N_2S$), both dissolved in three parts of isopropyl alcohol and one part of deionized water. The doped films were prepared by adding TbCI$_3$; or SmCI$_3$, to the spraying solution at a 10 at.% concentration; the substrate temperature, during the deposition, was either 300, 330, 360, or 375 °C. The solution flow rate was 14 cm^3/minute and the nozzle substrate distance was 30 cm in all cases. The doped films exhibited strong PL emission with a blue dominant peak at about 460 nm. This peak is characteristic of chlorine-doped ZnS phosphors. The films had poor crystallinity with a cubic crystalline structure. The optical transmission (T about 80%) characteristics of these films showed an absorption edge shifted to shorter wavelengths compared with those of the undoped films. Photoluminescent characteristics of In-, Al-, and Cu-doped ZnS films were reported in 1989 [138]. These films were deposited from a spraying solution formed by 0.1 M $(CH_3COO)_2$ + 0.1 M $C_3H_8N_2S$ dissolved in a mixture of three parts of isopropyl alcohol plus one part of deionized water. Doping was reached adding InC1$_3$, A1C1$_3$, or CuCl to the starting solution. The substrate temperatures were either 270, 300 or 330 °C. The substrates were pyrex glass slides, pyrex glass coated with In$_2$O$_3$ and silicon oxide. All films showed polycrystalline features which could be associated to a wurtzite structure of ZnS. Also, the presence of chlorine was detected into the films in quantities that depended on the deposition parameters. The PL spectra measured at room temperature displayed different emission peaks for each one of the impurities. The PL spectra from the Al-doped ZnS films showed a peak centered at 470 nm. The In-doped ZnS films showed a peak about 545 nm and the PL spectrum from the Cu-doped films exhibited a peak at 570 nm. The shape and intensity of the PL spectra do not depend strongly on the type of substrate.

The luminescent properties of ZnS:Mn films deposited by the pyrolysis spray technique on glass substrates at atmospheric pressure using air as a carrier gas were reported, for the first time, in 1992 [139]. The spraying solution in this case consisted of 0.1 M of Zn acetate and 0.1 M of dimethylthiourea in a mixture of three parts of isopropyl alcohol and one part of deionized water. The Mn doping was achieved by mixing MnCl$_2$ (0–20 at.%) in the spraying solutions; the deposition temperature was varied between 340 and 500 °C in steps of 20 °C. All films resulted polycrystalline with a wurtzite (hexagonal) structure. The PL spectra show, besides the characteristic light emission associated with Mn (yellow at 590 nm) in a ZnS host lattice, a peak associated with the self-activated emission (blue at 490 nm) observable at low substrate temperatures and/or long deposition times. The presence of chlorine impurities in the films was suggested to be associated with this emission. The Mn related luminescence showed a quenching effect with the Mn concentration (at concentrations higher than 3 at.% Mn in the spraying solution). The light emission at this center had an activation energy of 0.71 ± 0.05 eV with the deposition temperature. This energy was proposed to be related with the energy required for the Mn atoms to find a proper site during the growth process to form a Mn^{2+} center. These films were incorporated in a Metal-Insulator-active layer- Insulator-Metal (M-I-S-I-M) structure and their electroluminescent features were reported in 1995 [14]. These alternating current electroluminescent thin film structures were prepared using, for the first time, high-quality SiO$_2$ insulating thin films and spray pyrolyzed ZnS:Mn^{2+} as the active layer. The structures prepared with 60 nm thick insulating films showed threshold voltages of 30 V (rms) and saturation voltages of about 56 V (rms). The electroluminescent emission spectra presented a peak centered at 590 nm (yellow emission) associated with the Mn^{2+} center. The brightness-voltage characteristics were typical for a structure of the M-I-S-I-M type. The external efficiency calculated from the charge-voltage characteristics had a value of 1.8 Lumen/Watt.

Spray pyrolysis synthesis of ZnS nanoparticles (sub-10 nm) from a single-source precursor was published in 2009 [140]. Here, it was reported the synthesis of cubic ZnS nanoparticles from a low-cost single-source precursor in a continuous spray pyrolysis reactor. In this study, a single-source precursor: Diethyldithiocarbamate, $[(C_2H_5)_2NCS_2]_2Zn$, dissolved in toluene was used to synthesize ZnS nanoparticles. The furnace setpoint temperature was typically 600–800 °C. In this method, the evaporation and decomposition of precursor and nucleation of particles occur sequentially. XRD indicated a Cubic ZnS (zinc blende) for the synthesized particles. High Resolution Transmission Electron Microscopy (HRTEM) images showed ZnS particles with diameters ranging from 2 to 7 nm were. As-synthesized ZnS nanoparticles (excited at 350 nm) exhibited blue photoluminescence near to 440 nm had quantum yields up to 15% after HF treatment. This demonstrated a potentially general approach for continuous low-cost synthesis of semiconductor quantum dots, and applications in solar cells, lasers and displays. Also, ZnS nanoparticles can be applied as phosphors, probes for bio-imaging, emitters in light emitting diodes and photocatalysts.

5. Conclusions

This review describes some of the very extensive research work about the spray pyrolysis technique, which without doubt, is an extraordinarily flexible and practical materials synthesis method. It is a low-cost, non-vacuum required, way to synthesize materials in the form of powders and films deposited over a wide variety of substrates, and can be easily adapted for large area deposition and industrial production processes. The present work has been limited to review several luminescent materials and those with high-*K* dielectric properties, most of them metal oxides, synthesized by this process. Concerning the dielectric materials, it has been focused on the work carried out in high-*K* dielectric films of aluminum oxide, yttrium oxide, and zirconium oxide, developed for application on MOSFET technology devices. Through the works reviewed, the spray pyrolysis technique has been proved to be a technique capable of producing films as thin as 30 nm on silicon wafers with an outstanding dielectric and optical qualities, which could also be considered for design and development of sensors and other multilayered microdevices—such as planar waveguides and resonant optical structures. In the case of luminescent materials (PL, CL, TL, Up-conversion), the information reviewed shows that metal oxides (ZrO_2, HfO_2, Al_2O_3, Y_2O_3, ZnO) and ZnS doped with rare earth and transition metal ions with specific luminescent characteristics could be tailored according to light emitting devices, and to many other applications requirements.

Author Contributions: Writing-Review & Editing, C.F.; Writing-Review of High K dielectrics section, M.A.A.-F.; Writing-Review of luminescent materials section, M.G.-H.

Funding: This research was funded by CINVESTAV IPN and CONACyT-Mexico grant number (CB-2015/253342).

Acknowledgments: J.U. Balderas and G.L. Jimenez for their technical support. Especial thanks to G. Righini for his invitation to write this review.

Conflicts of Interest: The authors declare no conflict of interest. The founding sponsors had no role in the design of the study; in the collection, analyses, or interpretation of data; in the writing of the manuscript, and in the decision to publish the results.

References

1. Viguie, J.C.; Spitz, J. Chemical Vapor Deposition at Low Temperatures. *J. Electrochem. Soc.* **1975**, *122*, 585–588. [CrossRef]
2. Kodas, T.T.; Hampden-Smith, M.J. *Aerosol Processing of Materials*; Wiley-VCH: New York, NY, USA, 1999; ISBN 0-471-24669-7.
3. Blandenet, G.; Court, M.; Lagarde, Y. Thin layers deposited by the pyrosol process. *Thin Solid Films* **1981**, *77*, 81–90. [CrossRef]

4. Langlet, M.; Joubert, J.C. *Chemistry of Advanced Materials*; Rao, C.N.R., Ed.; Blackwell Science: Oxford, UK, 1993; Chapter 4; ISBN 0-632-03385-1.

5. Mwakikunga, B.W. Progress in Ultrasonic Spray Pyrolysis for Condensed Matter Sciences Developed From Ultrasonic Nebulization Theories since Michael Faraday. *Crit. Rev. Solid State Mater. Sci.* **2014**, *39*, 46–80. [CrossRef]

6. Patil, P.S. Versatility of chemical spray pyrolysis technique. *Mater. Chem. Phys.* **1999**, *59*, 185–198. [CrossRef]

7. Perednis, D.; Gauckler, L.J. Thin Film Deposition Using Spray Pyrolysis. *J. Electroceram.* **2005**, *14*, 103–111. [CrossRef]

8. Ortiz, A.; Falcony, C.; García, M.; Sánchez, A. Terbium-doped zinc oxide films deposited by spray pyrolysis. *J. Phys. D Appl. Phys.* **1987**, *20*, 670–671. [CrossRef]

9. Aguilar-Frutis, M.; Garcia, M.; Falcony, C. Optical and electrical properties of aluminum oxide films deposited by spray pyrolysis. *Appl. Phys. Lett.* **1998**, *72*, 1700–1702. [CrossRef]

10. Carmona-Téllez, S.; Falcony, C.; Aguilar–Frutis, M.; Alarcón-Flores, G.; García-Hipólito, M.; Martínez-Martínez, R. White light emitting transparent double layer stack of Al_2O_3:Eu^{3+}, Tb^{3+}, and Ce^{3+} films deposited by spray pyrolysis. *ECS J. Solid State Sci. Technol.* **2013**, *2*, R111–R115. [CrossRef]

11. Jergel, M.; García, M.; Conde-Gallardo, A.; Falcony, C.; Canseco, M.A.; Plesch, G. Preliminary studies of thin metal oxide films prepared by deposition of an aerosol generated ultrasonically from aqueous nitrate solutions. *Thin Solid Films* **1997**, *305*, 157–163. [CrossRef]

12. Jergel, M.; Conde-Gallardo, A.; García, M.; Falcony, C.; Jergel, M. Metal oxide Co and Co-Fe-Cr films deposited on glass substrates from a metal-organic aerosol atomised by means of ultrasonic excitations. *Thin Solid Films* **1997**, *305*, 210–218. [CrossRef]

13. García, M.; Jergel, M.; Conde-Gallardo, A.; Falcony, C.; Plesch, G. Optical properties of Co and Co-Fe-Cr thin films deposited from an aerosol on glass substrates. *Mater. Chem. Phys.* **1998**, *56*, 21–26. [CrossRef]

14. García, M.; Alonso, J.C.; Falcony, C.; Ortiz, A. Alternating current electroluminescent devices prepared using low temperature remote plasma enhanced CVD SiO_2 and ZnS:Mn deposited by spray pyrolysis. *J. Phys. D Appl. Phys.* **1995**, *28*, 223–225. [CrossRef]

15. Vázquez-Arreguín, R.; Aguilar-Frutis, M.; Falcony-Guajardo, C.; Castañeda-Galván, A.; Mariscal-Becerra, L.; Gallardo-Hernández, S.; Alarcón-Flores, G.; García-Rocha, M. Electrical, Optical and Structural Properties of SnO_2:Sb:F Thin Films Deposited from $Sn(acac)_2$ by Spray Pyrolysis. *ECS J. Solid State Sci. Technol.* **2016**, *5*, Q101–Q107. [CrossRef]

16. Alarcón-Flores, G.; Aguilar-Frutis, M.; Falcony, C.; García-Hipólito, M.; Araiza-Ibarra, J.J.; Herrera-Suárez, H.J. Low interface states and high dielectric constant Y_2O_3 films on Si substrates. *J. Vac. Sci. Technol. B* **2006**, *24*, 1873–1877. [CrossRef]

17. Reyna-García, G.; García-Hipólito, M.; Guzmán-Mendoza, J.; Aguilar-Frutis, M.; Falcony, C. Electrical, optical and structural characterization of high-k dielectric ZrO_2 thin films deposited by the pyrosol technique. *J. Mater. Sci.: Mater. Electron.* **2004**, *15*, 439–446. [CrossRef]

18. Zaleta-Alejandre, E.; Meza-Rocha, A.N.; Rivera-Alvarez, Z.; Sandoval, I.M.; Araiza, J.J.; Aguilar-Frutis, M.; Falcony, C. Optical Characteristics of Silica Coatings Deposited by Ozone Assisted Spray Pyrolysis Technique. *ECS J. Solid State Sci. Technol.* **2013**, *2*, N145–N148. [CrossRef]

19. Carmona-Téllez, S.; Guzmán-Mendoza, J.; Aguilar-Frutis, M.; Alarcón-Flores, G.; García-Hipólito, M.; Canseco, M.A.; Falcony, C. Electrical, optical, and structural characteristics of Al_2O_3 thin films prepared by pulsed ultrasonic sprayed pyrolysis. *J. Appl. Phys.* **2008**, *103*, 034105. [CrossRef]

20. Aguilar-Frutis, M.; García, M.; Falcony, C.; Plesch, G.; Jimenez-Sandoval, S. A study of the dielectric characteristics of aluminum oxide thin films deposited by spray pyrolysis from Al $(acac)_3$. *Thin Solid Films* **2001**, *389*, 200–206. [CrossRef]

21. Hou, X.; Choy, K.L. Processing and Applications of Aerosol-Assisted CVD. *Chem. Vap. Depos.* **2006**, *12*, 583–596. [CrossRef]

22. Avaru, B.; Patil, M.N.; Gogate, P.R.; Pandit, A.B. Ultrasonic atomization: Effect of liquid phase properties. *Ultrasonics* **2006**, *44*, 146–158. [CrossRef] [PubMed]

23. Ukoba, K.O.; Eloka-Eboka, A.C.; Inambao, F.L. Review of nanostructured NiO thin film deposition using the spray pyrolysis Technique. *Renew. Sustain. Energy Rev.* **2018**, *82*, 2900–2915. [CrossRef]

24. Filipovic, L.; Selberherr, S.; Mutinati, G.C.; Brunet, E.; Steinhauer, S.; Köck, A.; Teva, J.; Kraft, J.; Siegert, J.; Schrank, F. Methods of simulating thin film deposition using spray pyrolysis Techniques. *Microelectron. Eng.* **2014**, *117*, 57–66. [CrossRef]

25. Robertson, J. High dielectric constant gate oxides for metal oxide Si transistors. *J. Rep. Prog. Phys.* **2006**, *69*, 327–396. [CrossRef]

26. Wilk, G.D.; Wallace, R.M.; Anthony, J.M. High-k gate dielectrics: Current status and materials properties considerations. *J. Appl. Phys.* **2001**, *89*, 5243–5275. [CrossRef]

27. Ishida, M.; Katakabe, I.; Nakamura, T.; Ohtake, N. Epitaxial Al_2O_3 films on Si by low-pressure chemical vapor deposition. *Appl. Phys. Lett.* **1988**, *52*, 1326–1328. [CrossRef]

28. Sawada, K.; Ishida, M.; Nakamura, T.; Ohtake, N. Metalorganic molecular beam epitaxy of γ-Al_2O_3 films on Si at low growth temperatures. *Appl. Phys. Lett.* **1988**, *52*, 1672–1674. [CrossRef]

29. Saraie, J.S.; Ngan, S. Photo-CVD of Al_2O_3 Thin Films. *Jpn. J. Appl. Phys.* **1990**, *29*, L1877–L1880. [CrossRef]

30. Dhonge, B.P.; Mathews, T.; Sundari, S.T.; Thinaharan, C.; Kamruddin, M.; Dash, S.; Tyagi, A.K. Spray pyrolytic deposition of transparent aluminum oxide (Al_2O_3) films. *Appl. Surf. Sci.* **2011**, *258*, 1091–1096. [CrossRef]

31. Khatibani, A.B.; Rozati, S.M. Growth and molarity effects on properties of alumina thin films obtained by spray pyrolysis. *Mater. Sci. Semicond. Process.* **2014**, *18*, 80–87. [CrossRef]

32. Sanchez-Alarcón, R.I.; Oropeza-Rosario, G.; Gutiérrez-Villalobos, A.; Muro-López, M.A.; Martínez-Martínez, R.; Zaleta-Alejandre, E.; Falcony, C.; Alarcón-Flores, G.; Fragoso, R.; Hernández-Silva, O.; et al. Ultrasonic spray-pyrolyzed $CuCrO_2$ thin films. *J. Phys. D Appl. Phys.* **2016**, *49*, 175102. [CrossRef]

33. Kim, J.S.; Marzouk, H.A.; Reucroft, P.J.; Robertson, J.D.; Hamrin, C.E. Fabrication of aluminum oxide thin films by low-pressure metalorganic chemical vapor deposition technique. *Appl. Phys. Lett.* **1993**, *62*, 681–683. [CrossRef]

34. Alarcón-Flores, G.; Aguilar-Frutis, M.; García-Hipólito, M.; Guzmán-Mendoza, J.; Canseco, M.A.; Falcony, C. Optical and structural characteristics of Y_2O_3 thin films synthesized from yttrium acetylacetonate. *J. Mater. Sci.* **2008**, *43*, 3582–3588. [CrossRef]

35. Ronda, C.R. (Ed.) *Luminescence: From Theory to Applications*; Wiley-VCH Verlag GmbH & Co. KGaA: Weinheim, Germany, 2008; ISBN 978-3-527-31402-7.

36. Vij, D.R. (Ed.) *Luminescence of Solids*; Plenum Press: New York, NY, USA, 1998; ISBN 0-306-45643-5.

37. Blasse, G.; Grabmaier, B.C. *Luminescent Materials*; Springer: Berlin/Heidelberg, Germany, 1994; ISBN 3-540-58019-0.

38. Yen, W.M.; Shionoya, S.; Yamamoto, H. (Eds.) *Phosphor Handbook*, 2nd ed.; CRC Press: Boca Raton, FL, USA, 2007; ISBN 0-8493-3564-7.

39. Félix-Quintero, H.; Angulo-Rocha, J.; Murrieta, S.H.; Hernández, A.J.; Camarillo, G.E.; Flores, J.M.C.; Alejo-Armenta, C.; García-Hipólito, M.; Ramos-Brito, F. Study on grow process and optical properties of ZnO microrods synthesized by hydrothermal method. *J. Lumin.* **2017**, *182*, 107–113. [CrossRef]

40. Montes, E.; Martínez-Merlín, I.; Guzmán-Olguín, J.C.; Guzmán-Mendoza, J.; Martín, I.R.; García-Hipólito, M.; Falcony, C. Effect of pH on the optical and structural properties of HfO_2:Ln^{3+}, synthesized by hydrothermal route. *J. Lumin.* **2016**, *175*, 243–248. [CrossRef]

41. Jiang, Y.; Chen, J.; Xie, Z.; Zheng, L. Syntheses and optical properties of α- and β-Zn_2SiO_4: Mn nanoparticles by solvothermal method in ethylene glycol–water system. *Mater. Chem. Phys.* **2010**, *120*, 313–318. [CrossRef]

42. Park, K.; Kim, H.; Hakeem, D.A. Photoluminescence properties of Eu^{3+}- and Tb^{3+}-doped $YAlO_3$ phosphors for White LED applications. *Ceram. Int.* **2016**, *42*, 10526–10530. [CrossRef]

43. El-Ghoul, J.; Omri, K.; Alyamani, A.; Barthou, C.; El-Mir, L. Synthesis and luminescence of SiO_2/Zn_2SiO_4 and SiO_2/Zn_2SiO_4:Mn composite with sol-gel methods. *J. Lumin.* **2013**, *138*, 218–222. [CrossRef]

44. Pereyra-Perea, E.; Estrada-Yañez, M.R.; García, M. Preliminary studies on luminescent terbium-doped ZrO_2 thin films prepared by the sol–gel process. *J. Phys. D Appl. Phys.* **1998**, *31*, L7–L10. [CrossRef]

45. Alonso, J.C.; Haro-Poniatowski, E.; Diamant, R.; Fernandez-Guasti, M.; Garcia, M. Photoluminescent thin films of terbium chloride-doped yttrium oxide deposited by the pulsed laser ablation technique. *Thin Solid Films* **1997**, *303*, 76–83. [CrossRef]

46. Aguilar-Castillo, A.; Aguilar-Hernández, J.R.; García-Hipólito, M.; López-Romero, S.; Swarnkar, R.K.; Báez-Rodríguez, A.; Fragoso-Soriano, R.J.; Falcony, C. White light generation from HfO_2 films co-doped with Eu^{3+} + Tb^{3+} ions synthesized by pulsed laser ablation technique. *Ceram. Int.* **2017**, *43*, 355–362. [CrossRef]

47. Liu, L.; Wang, Y.; Su, Y.; Ma, Z.; Xie, Y.; Zhao, H.; Chen, C.; Zhang, Z.; Xie, E. Synthesis and White Light Emission of Rare Earth-Doped HfO$_2$ Nanotubes. *J. Am. Ceram. Soc.* **2011**, *94*, 2141–2145. [CrossRef]

48. Deng, H.; Xue, N.; Hei, Z.; He, M.; Wang, T.; Xie, N.; Yu, R. Close-relationship between the luminescence and structural characteristics in efficient nanophosphor Y$_2$Mo$_4$O$_{15}$:Eu^{3+}. *Opt. Mater. Express* **2015**, *5*, 490–496. [CrossRef]

49. Stojadinović, S.; Tadić, N.; Vasilić, R. Structural and photoluminescent properties of ZrO$_2$:Tb^{3+} coatings formed by plasma electrolytic oxidation. *J. Lumin.* **2018**, *197*, 83–89. [CrossRef]

50. Yasaka, P.; Kaewkhao, J. White emission materials from glass doped with rare earth ions: A review. *AIP Conf. Proc.* **2016**, *1719*, 020002. [CrossRef]

51. Félix-Quintero, H.; Camarillo-Garcia, I.; Hernández-Alcántara, J.; Camarillo-García, E.; Cordero-Borboa, A.; Flores-Jiménez, C.; García-Hipólito, M.; Ramos-Brito, F.; Acosta-Najarro, D.; Murrieta-Sánchez, H. RGB emission of Mn^{2+} doped zinc phosphate glass. *J. Non-Cryst. Solids* **2017**, *466–467*, 58–63. [CrossRef]

52. Rakov, N.; Bispo, L.R.A.; Maciel, G.S. Temperature sensing performance of dysprosium doped aluminum oxide powders. *Opt. Commun.* **2012**, *285*, 1882–1884. [CrossRef]

53. Cedillo Del Rosario, G.; Cruz-Zaragoza, E.; García-Hipólito, M.; Marcazzó, J.; Hernández-A, J.M.; Murrieta-S, H. Synthesis and stimulated luminescence property of Zn(BO$_2$)$_2$:Tb^{3+}. *Appl. Radiat. Isot.* **2017**, *127*, 103–108. [CrossRef] [PubMed]

54. Roman-Lopez, J.; Valverde, M.; García-Hipólito, M.; Lozano, I.B.; Diaz-Góngora, J.A.I. Photoluminescence, thermo- and optically stimulated luminescence properties of Eu^{3+} doped Sr$_2$P$_2$O$_7$ synthesized by the solvent evaporation method. *J. Alloys Compd.* **2018**, *756*, 126–133. [CrossRef]

55. García-Hipólito, M.; Falcony, C.; Aguilar-Frutis, M.A.; Azorín-Nieto, J. Synthesis and characterization of luminescent ZrO$_2$:Mn, Cl powders. *Appl. Phys. Lett.* **2001**, *79*, 4369–4371. [CrossRef]

56. Ramos-Brito, F.; García-Hipólito, M.; Martínez-Martínez, R.; Martínez-Sánchez, E.; Falcony, C. Preparation and characterization of photoluminescent praseodymium-doped ZrO$_2$ nanostructured powders. *J. Phys. D Appl. Phys.* **2004**, *37*, L13–L16. [CrossRef]

57. Ortiz, A.; Falcony, C.; García, M.; López, S. Spray deposition of TbCl$_3$ doped In$_2$O$_3$ photoluminiscent films. *Thin Solid Films* **1988**, *165*, 249–255. [CrossRef]

58. Calderón-Olvera, R.M.; Albanés-Ojeda, E.A.; García-Hipólito, M.; Hernández-Alcántara, J.M.; Álvarez-Perez, M.A.; Falcony, C.; Alvarez-Fregoso, O. Characterization of luminescent SrAl$_2$O$_4$ films doped with terbium and europium ions deposited by ultrasonic spray pyrolysis technique. *Ceram. Int.* **2018**, *44*, 7917–7925. [CrossRef]

59. Azorín, J.; Rivera, T.; Falcony, C.; Martínez, E.; García, M. Ultraviolet thermoluminescent dosimetry using terbium-doped zirconium oxide thin films. *Radiat. Prot. Dosim.* **1999**, *85*, 317–319. [CrossRef]

60. Rivera, T.; Azorín, J.; Falcony, C.; Martínez, E.; García, M. Determination of thermoluminescence kinetic parameters of terbium-doped zirconium oxide. *Radiat. Phys. Chem.* **2001**, *61*, 421–423. [CrossRef]

61. García-Hipólito, M.; Martínez, R.; Alvarez-Fregoso, O.; Martínez, E.; Falcony, C. Cathodoluminescence and photoluminescence properties of terbium doped ZrO$_2$ films prepared by pneumatic spray pyrolysis technique. *J. Lumin.* **2001**, *93*, 9–15. [CrossRef]

62. García-Hipólito, M.; Martínez, E.; Alvarez-Fregoso, O.; Falcony, C.; Aguilar-Frutis, M.A. Preparation and characterization of Eu doped zirconia luminescent films synthesized by pyrosol technique. *J. Mater. Sci. Lett.* **2001**, *20*, 1799–1801. [CrossRef]

63. García-Hipólito, M.; Alvarez-Fregoso, O.; Martínez, E.; Falcony, C.; Aguilar-Frutis, M.A. Characterization of ZrO$_2$:Mn, Cl luminescent coatings synthesized by the pyrosol technique. *Opt. Mater.* **2002**, *20*, 113–118. [CrossRef]

64. Quan, Z.W.; Wang, L.S.; Lin, J. Synthesis and characterization of spherical ZrO$_2$:Eu^{3+} phosphors by spray pyrolysis process. *Mater. Res. Bull.* **2005**, *40*, 810–820. [CrossRef]

65. Ramos-Brito, F.; García-Hipólito, M.; Alejo-Armenta, C.; Alvarez-Fregoso, O.; Falcony, C. Characterization of luminescent praseodymiun-doped ZrO$_2$ coatings deposited by ultrasonic spray pyrolysis technique. *J. Phys. D Appl. Phys.* **2007**, *40*, 6718–6724. [CrossRef]

66. Martínez-Hernández, A.; Guzmán-Mendoza, J.; Rivera-Montalvo, T.; Sánchez-Guzmán, D.; Guzmán-Olguín, J.C.; García-Hipólito, M.; Falcony, C. Synthesis and cathodoluminescence characterization of ZrO$_2$:Er^{3+} films. *J. Lumin.* **2014**, *153*, 140–143. [CrossRef]

67. Báez-Rodríguez, A.; Alvarez-Fregoso, O.; García-Hipólito, M.; Guzmán-Mendoza, J.; Falcony, C. Luminescent properties of ZrO$_2$:Dy^{3+} and ZrO$_2$:Dy^{3+} + Li$^+$ films synthesized by an ultrasonic spray pyrolysis technique. *Ceram. Int.* **2015**, *41*, 7197–7206. [CrossRef]

68. Martínez-Olmos, R.C.; Guzmán-Mendoza, J.; Báez-Rodríguez, A.; Alvarez-Fregoso, O.; García-Hipólito, M.; Falcony, C. Synthesis, characterization and luminescence studies in ZrO$_2$:Dy^{3+} and ZrO$_2$:Dy^{3+}, Gd^{3+} films deposited by the Pyrosol method. *Opt. Mater.* **2015**, *46*, 168–174. [CrossRef]

69. Ramos-Guerra, A.I.; Guzmán-Mendoza, J.; García-Hipólito, M.; Alvarez-Fregoso, O.; Falcony, C. Multicolored photoluminescence and structural properties of zirconium oxide films co-doped with Tb^{3+} and Eu^{3+} ions. *Ceram. Int.* **2015**, *41*, 11279–11286. [CrossRef]

70. Falcony, C.; Ortiz, A.; Dominguez, J.M.; Farías, M.H.; Cota-Araiza, L.; Soto, G. Luminescent Characteristics of Tb Doped Al$_2$O$_3$ Films Deposited by Spray Pyrolysis. *J. Electrochem. Soc.* **1992**, *139*, 267–271. [CrossRef]

71. Falcony, C.; García, M.; Ortiz, A.; Miranda, O.; Gradilla, I.; Soto, G.; Cota-Araiza, L.; Farías, M.H.; Alonso, J.C. Blue Photoluminescence from CeCl$_3$ Doped Al$_2$O$_3$ Films. *J. Electrochem. Soc.* **1994**, *141*, 2860–2863. [CrossRef]

72. Martínez, E.; García, M.; Ramos-Brito, F.; Alvarez-Fregoso, O.; López, S.; Granados, S.; Chavez-Ramírez, J.; Martinez-Martinez, R.; Falcony, C. Characterization of Al$_2$O$_3$:Eu^{3+} luminescent coatings prepared by Spray Pyrolysis Technique. *Phys. Status Solidi B* **2000**, *220*, 677–681. [CrossRef]

73. Esparza-García, A.E.; García-Hipólito, M.; Aguilar-Frutis, M.A.; Falcony, C. Luminescent and morphological characteristics of Al$_2$O$_3$:Tb films deposited by spray pyrolysis using acetylacetonates as precursors. *J. Electrochem. Soc.* **2003**, *150*, H53–H56. [CrossRef]

74. Martinez-Martinez, R.; García-Hipólito, M.; Ramos-Brito, F.; Hernandez-Pozos, J.L.; Caldiño, U.; Falcony, C. Blue and red photoluminescence from Al$_2$O$_3$:Ce:Mn films deposited by spray pyrolysis. *J. Phys. Condens. Matter* **2005**, *17*, 3647–3656. [CrossRef]

75. Martinez-Martinez, R.; Rickards, J.; García-Hipólito, M.; Trejo-Luna, R.; Martínez-Sánchez, E.; Alvarez-Fregoso, O.; Ramos-Brito, F.; Falcony, C. RBS characterization of Al$_2$O$_3$ films doped with Ce and Mn. *Nucl. Instrum. Methods Phys. Res. Sect. B* **2005**, *241*, 450–453. [CrossRef]

76. Martinez-Martinez, R.; García-Hipólito, M.; Huerta, L.; Rickards, J.; Caldiño, U.; Falcony, C. Studies on blue and red photoluminescence from Al$_2$O$_3$:Ce^{3+}:Mn^{2+} coatings synthesized by spray pyrolysis technique. *Thin Solid Films* **2006**, *515*, 607–610. [CrossRef]

77. Martinez-Martinez, R.; Alvarez, E.; Speghini, A.; Falcony, C.; Caldiño, U. White light generation in Al$_2$O$_3$:Ce^{3+}:Tb^{3+}:Mn^{2+} films deposited by ultrasonic spray pyrolysis. *Thin Solid Films* **2010**, *518*, 5724–5730. [CrossRef]

78. Martinez-Martinez, R.; Rivera, S.; Yescas-Mendoza, E.; Alvarez, E.; Falcony, C.; Caldiño, U. Luminescence properties of Ce^{3+}– Dy^{3+} codoped aluminium oxide films. *Opt. Mater.* **2011**, *33*, 1320–1324. [CrossRef]

79. Gonzalez, W.; Alvarez, E.; Martinez-Martinez, R.; Yescas-Mendoza, E.; Camarillo, I.; Caldiño, U. Cold white light generation through the simultaneous emission from Ce^{3+}, Dy^{3+} and Mn^{2+} in 90Al$_2$O$_3$·2CeCl$_3$·3DyCl$_3$·5MnCl$_2$ thin film. *J. Lumin.* **2012**, *132*, 2130–2134. [CrossRef]

80. Huerta, E.F.; Padilla, I.; Martinez-Martinez, R.; Hernandez-Pozos, J.L.; Caldiño, U.; Falcony, C. Extended decay times for the photoluminescence of Eu^{3+} ions in aluminum oxide films through interaction with localized states. *Opt. Mater.* **2012**, *34*, 1137–1142. [CrossRef]

81. Martínez-Martínez, R.; Yescas, E.; Álvarez, E.; Falcony, C.; Caldiño, U. White light generation in rare-earth-doped amorphous films produced by ultrasonic spray pyrolysis. *Adv. Sci. Technol.* **2013**, *82*, 19–24. [CrossRef]

82. Padilla-Rosales, I.; Martinez-Martinez, R.; Cabañas, G.; Falcony, C. The effect of Bi^{3+} and Li$^+$ co-doping on the luminescence characteristics of Eu^{3+}-doped aluminum oxide films. *J. Lumin.* **2015**, *165*, 185–189. [CrossRef]

83. García-Hipólito, M.; Alvarez-Fragoso, O.; Guzmán, J.; Martínez, E.; Falcony, C. Characterization of HfO$_2$:Mn luminescent coatings deposited by spray pyrolysis. *Phys. Status Solidi A* **2004**, *201*, R127–R130. [CrossRef]

84. García-Hipólito, M.; Caldiño, U.; Alvarez-Fragoso, O.; Alvarez-Pérez, M.A.; Martínez-Martínez, R.; Falcony, C. Violet-blue luminescence from hafnium oxide layers doped with CeCl$_3$ prepared by the spray pyrolysis technique. *Phys. Status Solidi A* **2007**, *204*, 2355–2361. [CrossRef]

85. Guzmán–Mendoza, J.; Albarran-Arreguín, D.; Alvarez-Fragoso, O.; Alvarez-Pérez, M.A.; Falcony, C.; García-Hipólito, M. Photoluminescent characteristics of hafnium oxide layers activated with trivalent terbium (HfO$_2$:Tb^{+3}). *Radiat. Eff. Defects Solids* **2007**, *162*, 723–729. [CrossRef]

86. Chacón-Roa, C.; Guzmán-Mendoza, J.; Aguilar-Frutis, M.A.; García-Hipólito, M.; Alvarez-Fragoso, O.; Falcony, C. Characterization of luminescent samarium doped HfO_2 coatings synthesized by spray pyrolysis technique. *J. Phys. D Appl. Phys.* **2008**, *41*, 015104. [CrossRef]

87. Martínez-Martínez, R.; García-Hipólito, M.; Speghini, A.; Bettinelli, M.; Falcony, C.; Caldiño, U. Blue-green-red luminescence from $CeCl_3$- and $MnCl_2$- doped hafnium oxide layers prepared by ultrasonic spray pyrolysis. *J. Phys. Condens. Matter.* **2008**, *20*, 395205. [CrossRef]

88. Martínez-Martínez, R.; Alvarez, E.; Speghini, A.; Falcony, C.; Caldiño, U. Cold white light generation from hafnium oxide films activated with Ce^{3+}, Tb^{3+}, and Mn^{2+} ions. *J. Mater. Res.* **2010**, *25*, 484–490. [CrossRef]

89. Guzmán-Mendoza, J.; Aguilar-Frutis, M.A.; Alarcón-Flores, G.; García-Hipólito, M.; Maciel-Cerda, A.; Azorín-Nieto, J.; Rivera-Montalvo, T.; Falcony, C. Synthesis and characterization of hafnium oxide films for thermo and photoluminescence applications. *Appl. Radiat. Isot.* **2010**, *68*, 696–699. [CrossRef]

90. Martínez-Martínez, R.; Lira, A.C.; Speghini, A.; Falcony, C.; Caldiño, U. Blue-yellow photoluminescence from $Ce^{3+} \rightarrow Dy^{3+}$ energy transfer in $HfO_2:Ce^{3+}:Dy^{3+}$ films deposited by ultrasonic spray pyrolysis. *J. Alloys Compd.* **2011**, *509*, 3160–3165. [CrossRef]

91. Reynoso-Manríquez, R.; Díaz-Góngora, J.A.I.; Guzmán-Mendoza, J.; Rivera-Montalvo, T.; Guzmán-Olguín, J.C.; Cerón-Ramírez, P.V.; García-Hipólito, M.; Falcony, C. Photo, -cathodo- and thermoluminescent properties of dysprosium-doped HfO_2 films deposited by ultrasonic spray pyrolysis. *Appl. Radiat. Isot.* **2014**, *92*, 91–95. [CrossRef] [PubMed]

92. Martínez-Merlín, I.; Guzmán-Mendoza, J.; García-Hipólito, M.; Sánchez-Resendiz, V.M.; Lartundo-Rojas, R.J.; Fragoso, R.J.; Falcony, C. Transparent and low Surface roughness $HfO_2:Tb^{3+}$, Eu^{3+} luminescent thin films deposited by USP technique. *Ceram. Int.* **2016**, *42*, 2446–2455. [CrossRef]

93. Guzman-Olgun, J.C.; Montes, E.; Guzmán-Mendoza, J.; Baez-Rodrıguez, A.; Zamora-Peredo, L.; García-Hipólito, M.; Alvarez-Fregoso, O.; Martínez-Merlín, I.; Falcony, C. Tunable white light emission from hafnium oxide films co-doped with trivalent terbium and europium ions deposited by Pyrosol technique. *Phys. Status Solidi A* **2017**, *214*, 1700269. [CrossRef]

94. Kang, Y.C.; Roh, H.S.; Park, S.B. Preparation of Y_2O_3:Eu phosphor particles of filled morphology at high precursor concentrations by Spray Pyrolysis. *Adv. Mater.* **2000**, *12*, 451–453. [CrossRef]

95. Hao, J.; Studenikin, S.A.; Cocivera, M. Blue, green and red cathodoluminescence of Y_2O_3 phosphor films prepared by spray pyrolysis. *J. Lumin.* **2001**, *93*, 313–319. [CrossRef]

96. Kang, Y.C.; Roh, H.S.; Park, S.B.; Park, H.D. Use of LiCl flux in the preparation of Y_2O_3:Eu phosphor particles by spray pyrolysis. *J. Eur. Ceram. Soc.* **2002**, *22*, 1661–1665. [CrossRef]

97. Kang, Y.C.; Roh, H.S.; Park, S.B.; Park, H.D. High luminescence Y_2O_3:Eu phosphor particles prepared by modified spray pyrolysis. *J. Mater. Sci. Lett.* **2002**, *21*, 1027–1029. [CrossRef]

98. Joffin, N.; Dexpert-Ghys, J.; Verelst, M.; Baret, G.; Garcia, A. The influence of microstructure on luminescent properties of Y_2O_3:Eu prepared by spray pyrolysis. *J. Lumin.* **2005**, *113*, 249–257. [CrossRef]

99. Wang, L.S.; Zhou, Y.H.; Quan, Z.W.; Lin, J. Formation mechanisms and morphology dependent luminescence properties of Y_2O_3:Eu phosphors prepared by spray pyrolysis process. *Mater. Lett.* **2005**, *59*, 1130–1133. [CrossRef]

100. Jungz, K.Y.; Han, K.H. Densification and Photoluminescence Improvement of Y_2O_3 Phosphor Particles Prepared by Spray Pyrolysis. *Electrochem. Solid-State Lett.* **2005**, *8*, H17–H20. [CrossRef]

101. Marinkovic, K.; Mancic, L.; Gomez, L.S.; Rabanal, M.E.; Dramicanin, M.; Milosevic, O. Photoluminescent properties of nanostructured $Y_2O_3:Eu^{3+}$ powders obtained through aerosol synthesis. *Opt. Mater.* **2010**, *32*, 1606–1611. [CrossRef]

102. Lojpur, V.; Nikolic, M.; Mancic, L.; Milosevic, O.; Dramicanin, M.D. Up-conversion luminescence in Ho^{3+} and Tm^{3+} co-doped $Y_2O_3:Yb^{3+}$ fine powders obtained through aerosol decomposition. *Opt. Mater.* **2012**, *35*, 38–44. [CrossRef]

103. Meza-Rocha, A.N.; Huerta, E.F.; Zaleta-Alejandre, E.; Rivera-Álvarez, Z.; Falcony, C. Enhanced photoluminescence of $Y_2O_3:Er^{3+}$ thin films by Li^+ co-doping. *J. Lumin.* **2013**, *141*, 173–176. [CrossRef]

104. Meza-Rocha, A.N.; Canto, C.; Andrade, E.; De Lucio, O.; Huerta, E.F.; González, F.; Rocha, M.F.; Falcony, C. Visible and near infra-red luminescent emission from $Y_2O_3:Er^{3+}$ films co-doped with Li^+ and their elemental composition by ion beam analysis. *Ceram. Int.* **2014**, *40*, 14647–14653. [CrossRef]

105. Meza-Rocha, A.N.; Huerta, E.F.; Caldiño, U.; Zaleta-Alejandre, E.; Murrieta-S, H.; Hernández-A, J.M.; Camarillo, E.; Rivera-Álvarez, Z.; Righini, G.C.; Falcony, C. Li$^+$ co-doping effect on the photoluminescence time decay behavior of Y$_2$O$_3$:Er^{3+} films. *J. Lumin.* **2014**, *154*, 106–110. [CrossRef]

106. Alarcón-Flores, G.; García-Hipólito, M.; Aguilar-Frutis, M.; Carmona-Téllez, S.; Martinez-Martinez, R.; Campos-Arias, M.P.; Jiménez-Estrada, M.; Falcony, C. Luminescent and Structural Characteristics of Y$_2$O$_3$:Tb^{3+} Thin Films as a Function of Substrate Temperature. *ECS J. Solid State Sci. Technol.* **2014**, *3*, R189–R194. [CrossRef]

107. Alarcón-Flores, G.; García-Hipólito, M.; Aguilar-Frutis, M.; Carmona-Téllez, S.; Martinez-Martinez, R.; Campos-Arias, M.P.; Zaleta-Alejandre, E.; Falcony, C. Synthesis and fabrication of Y$_2$O$_3$:Tb^{3+} and Y$_2$O$_3$:Eu^{3+} thin films for electroluminescent applications: Optical and structural characteristics. *Mater. Chem. Phys.* **2015**, *149–150*, 34–42. [CrossRef]

108. Jayaramaiah, J.R.; Nagabhushana, K.R.; Lakshminarasappa, B.N. Effect of lithium incorporation on luminescence properties of nanostructured Y$_2$O$_3$:Sm^{3+} thin films. *J. Anal. Appl. Pyrolysis* **2017**, *123*, 229–236. [CrossRef]

109. Nanto, H.; Minami, T.; Takata, S. Photoluminescence in sputtered ZnO thin films. *Phys. Status Solidi A* **1981**, *65*, K131–K134. [CrossRef]

110. Falcony, C.; Ortiz, A.; García, M.; Helman, J.S. Photoluminiscence characteristics of undoped and terbium chloride doped zinc oxide films deposited by spray pyrolysis. *J. Appl. Phys.* **1988**, *63*, 2378–2381. [CrossRef]

111. Studenikin, S.A.; Golego, N.; Cocivera, M. Fabrication of green and orange photoluminescent, undoped ZnO films using spray pyrolysis. *J. Appl. Phys.* **1998**, *84*, 2287–2294. [CrossRef]

112. Ortiz, A.; García, M.; Falcony, C. Lithium doped zinc oxide photoluminescence films prepared by spray pyrolysis. *Mater. Chem. Phys.* **1990**, *24*, 383–388. [CrossRef]

113. Ortiz, A.; García, M.; Falcony, C. Photoluminescent properties of indium-doped zinc oxide films prepared by spray pyrolysis. *Thin Solid Films* **1992**, *207*, 175–180. [CrossRef]

114. Vanheusden, K.; Seager, C.H.; Warren, W.L.; Tallant, D.R.; Caruso, J.; Hampden-Smith, M.J.; Kodas, T.T. Green photoluminescence efficiency and free-carrier density in ZnO phosphor powders prepared by spray pyrolysis. *J. Lumin.* **1997**, *75*, 11–16. [CrossRef]

115. Ortiz, A.; Falcony, C.; Hernández A, J.; García, M.; Alonso, J.C. Photoluminescent characteristics of lithium doped zinc oxide films deposited by spray pyrolysis. *Thin Solid Films* **1997**, *293*, 103–107. [CrossRef]

116. Yoon, K.H.; Cho, J.Y. Photoluminescence characteristics of zinc oxide thin films prepared by spray pyrolysis technique. *Mater. Res. Bull.* **2000**, *35*, 39–46. [CrossRef]

117. El Hichou, A.; Bougrine, A.; Bubendorff, J.L.; Ebothé, J.; Addou, M.; Troyon, M. Structural, optical and cathodoluminescence characteristics of sprayed undoped and fluorine-doped ZnO thin films. *Semicond. Sci. Technol.* **2002**, *17*, 607–613. [CrossRef]

118. Bougrine, A.; El Hichou, A.; Addou, M.; Ebothé, J.; Kachouane, A.; Troyon, M. Structural, optical and cathodoluminescence characteristics of undoped and tin-doped ZnO thin films prepared by spray pyrolysis. *Mater. Chem. Phys.* **2003**, *80*, 438–445. [CrossRef]

119. Panatarani, C.; Lenggoro, I.W.; Okuyama, K. The crystallinity and the photoluminescent properties of spray pyrolized ZnO phosphor containing Eu^{2+} and Eu^{3+} ions. *J. Phys. Chem. Solids* **2004**, *65*, 1843–1847. [CrossRef]

120. Ratheesh Kumar, P.M.; Vijayakumar, K.P.; Sudha Kartha, C. On the origin of blue-green luminescence in spray pyrolysed ZnO thin films. *J. Mater. Sci.* **2007**, *42*, 2598–2602. [CrossRef]

121. Choi, M.H.; Ma, T.Y. Erbium concentration effects on the structural and photoluminescence properties of ZnO:Er films. *Mater. Lett.* **2008**, *62*, 1835–1838. [CrossRef]

122. Liu, K.; Yang, B.F.; Yan, H.; Fu, Z.; Wen, M.; Chen, Y.; Zuo, J. Strong room-temperature ultraviolet emission from nanocrystalline ZnO and ZnO:Ag films grown by ultrasonic spray pyrolysis. *Appl. Surf. Sci.* **2008**, *255*, 2052–2056. [CrossRef]

123. Bouzid, K.; Djelloul, A.; Bouzid, N.; Bougdira, J. Electrical resistivity and photoluminescence of zinc oxide films prepared by ultrasonic spray pyrolysis. *Phys. Status Solidi A* **2009**, *206*, 106–115. [CrossRef]

124. Behera, D.; Acharya, B.S. Study of the microstructural and photoluminescence properties of Li-doped ZnO thin films prepared by spray pyrolysis. *Ionics* **2010**, *16*, 543–548. [CrossRef]

125. Soumahoro, I.; Schmerber, G.; Douayar, A.; Colis, S.; Abd-Lefdil, M.; Hassanain, N.; Berrada, A.; Muller, D.; Slaoui, A.; Rinnert, H.; et al. Structural, optical, and electrical properties of Yb-doped ZnO thin films prepared by spray pyrolysis method. *J. Appl. Phys.* **2011**, *109*, 033708. [CrossRef]

126. Shinde, S.S.; Shinde, P.S.; Oh, Y.W.; Haranath, D.; Bhosale, C.H.; Rajpure, K.Y. Investigation of structural, optical and luminescent properties of sprayed N-doped zinc oxide thin films. *J. Anal. Appl. Pyrolysis* **2012**, *97*, 181–188. [CrossRef]

127. Shinde, S.S.; Shinde, P.S.; Oh, Y.W.; Haranath, D.; Bhosale, C.H.; Rajpure, K.Y. Structural, optoelectronic, luminescence and thermal properties of Ga-doped zinc oxide thin films. *Appl. Surf. Sci.* **2012**, *258*, 9969–9976. [CrossRef]

128. Tomakin, M. Structural and optical properties of ZnO and Al-doped ZnO microrods obtained by spray pyrolysis method using different solvents. *Superlattices Microstruct.* **2012**, *51*, 372–380. [CrossRef]

129. Bayoud, S.; Addou, M.; Bahedi, K.; El Jouad, M.; Sofiani, Z.; Lamrani, M.A.; Bernéde, J.C.; Ebothé, J. The lithium effect on the blue and red emissions of Er-doped zinc oxide thin films. *Phys. Scr.* **2013**, *T157*, 014045. [CrossRef]

130. Panatarani, C.; Muharam, D.G.; Wibawa, B.M.; Joni, I.M. Blue Luminescence of ZnO:Zn Nanocrystal Prepared by One Step Spray Pyrolysis Method. *Mater. Sci. Forum* **2013**, *737*, 20–27. [CrossRef]

131. Velázquez Lozada, E.; Torchynska, T.V.; Casas Espinola, J.L.; Pérez Millan, B. Emission of ZnO:Ag nanorods obtained by ultrasonic spray pyrolysis. *Phys. B* **2014**, *453*, 111–115. [CrossRef]

132. Vijayalakshmi, K.; Karthick, K. The Role of Substrates on the Structural, Optical, and Morphological Properties of ZnO Nanotubes Prepared by Spray Pyrolysis. *Microsc. Res. Tech.* **2014**, *77*, 211–215. [CrossRef] [PubMed]

133. Kumar, V.; Ntwaeaborwa, O.M.; Soga, T.; Dutta, V.; Swart, H.C. Rare Earth Doped Zinc Oxide Nanophosphor Powder: A Future Material for Solid State Lighting and Solar Cells. *ACS Photonics* **2017**, *4*, 2613–2637. [CrossRef]

134. Angulo-Rocha, J.; Velarde-Escobar, O.; Yee-Rendón, C.; Atondo-Rubio, G.; Millan-Almaraz, R.; Camarillo-García, E.; García-Hipólito, M.; Ramos-Brito, F. Morphological, structural and optical properties of ZnO thin solid films formed by nanoleafs or micron/submicron cauliflowers. *J. Lumin.* **2017**, *185*, 306–315. [CrossRef]

135. Narayanan, N.; Deepak, N.K. Enhancement of visible luminescence and photocatalytic activity of ZnO thin films via Cu doping. *Optik* **2018**, *158*, 1313–1326. [CrossRef]

136. Karakaya, S. Effect of fluorine and boron co-doping on ZnO thin films: Structural, luminescence properties and Hall effect measurements. *J. Mater. Sci.: Mater. Electron.* **2018**, *29*, 4080–4088. [CrossRef]

137. Ortiz, A.; Falcony, C.; García, M.; Sánchez, A. Zinc sulphide films doped with terbium chloride and samarium chloride, prepared by spray pyrolysis. *Semicond. Sci. Technol.* **1988**, *3*, 537–541. [CrossRef]

138. Ortiz, A.; García, M.; Sánchez, A.; Falcony, C. Spray pyrolysis deposition of In, Al and Cu doped ZnS photoluminiscent films. *J. Electrochem. Soc.* **1989**, *136*, 1232–1235. [CrossRef]

139. Falcony, C.; García, M.; Ortiz, A.; Alonso, J.C. Luminescent properties of ZnS:Mn films deposited by spray pyrolysis. *J. Appl. Phys.* **1992**, *72*, 1525–1527. [CrossRef]

140. Liu, S.; Zhang, H.; Swihart, M.T. Spray pyrolysis synthesis of ZnS nanoparticles from a single-source precursor. *Nanotechnology* **2009**, *20*, 235603. [CrossRef] [PubMed]

micromachines

MDPI

Article

About the Implementation of Frequency Conversion Processes in Solar Cell Device Simulations

Alexander Quandt [1,2,*], Tahir Aslan [2], Itumeleng Mokgosi [2], Robert Warmbier [3],
Maurizio Ferrari [1,4] and Giancarlo Righini [1,5]

[1] Historical Museum of Physics and Study & Research Centre "Enrico Fermi", 00184 Roma, Italy;
     maurizio.ferrari@unitn.it (M.F.); giancarlo.righini@centrofermi.it (G.R.)
[2] School of Physics and Materials for Energy Research Group, University of the Witwatersrand,
     2050 Johannesburg, South Africa; 772891@students.wits.ac.za (T.A.); itu.mokgosi@gmail.com (I.M.)
[3] Department of Physics, University of Johannesburg, 2006 Auckland Park, South Africa; rwarmbier@uj.ac.za
[4] Institute of Photonics and Nanotechnologies, National Research Council of Italy, 38123 Povo, Italy
[5] Nello Carrara Institute of Applied Physics, National Research Council of Italy, 50019 Sesto Fiorentino, Italy
*   Correspondence: alex.quandt@wits.ac.za

Received: 16 July 2018; Accepted: 20 August 2018; Published: 30 August 2018

Abstract: Solar cells are electrical devices that can directly convert sunlight into electricity. While solar cells are a mature technology, their efficiencies are still far below the theoretical limit. The major losses in a typical semiconductor solar cell are due to the thermalization of electrons in the UV and visible range of the solar spectrum, the inability of a solar cell to absorb photons with energies below the electronic band gap, and losses due to the recombination of electrons and holes, which mainly occur at the contacts. These prevent the realization of the theoretical efficiency limit of 85% for a generic photovoltaic device. A promising strategy to harness light with minimum thermal losses outside the typical frequency range of a single junction solar cell could be frequency conversion using rare earth ions, as suggested by Trupke. In this work, we discuss the modelling of generic frequency conversion processes in the context of solar cell device simulations, which can be used to supplement experimental studies. In the spirit of a proof-of-concept study, we limit the discussion to up-conversion and restrict ourselves to a simple rare earth model system, together with a basic diode model for a crystalline silicon solar cell. The results of this show that these simulations are very useful for the development of new types of highly efficient solar cells.

Keywords: photovoltaics; frequency conversion; device simulations

---

1. Introduction

The conversion efficiencies of silicon-based solar cells are getting closer and closer to the theoretical limit of 34% for single-junction silicon-based solar cells, as estimated by Shockley and Queisser in 1961 [1]. These impressive numbers must, however, be compared to the theoretical efficiency limit of 85% for a generic photovoltaic device [2], which shows that there is plenty of room for improvements. The main losses in a typical semiconductor solar cell are due to the thermalization of electrons in the UV and visible range of the solar spectrum, the inability of such devices to absorb photons with energies below the electronic band gap, and losses due to the recombination of electrons and holes, which occur particularly at the contacts. A recent survey of novel design strategies to overcome these problems has been given by Polman and Atwater [3], but their main focus was on multi-junction solar cells. Another promising strategy for single junction solar cells to harness light outside their typical frequency ranges is frequency conversion using rare earth ions, as suggested by Trupke [4]. From a technical point of view, the most common way to introduce frequency conversion into a standard solar cell is through additional functional glass layers containing rare-earth ions. In device simulations of solar cells, we may cater for the new light management features through calculating the additional photon fluxes that result from the frequency conversion processes in these additional functional layers.

In this work, we want to demonstrate this approach for a model solar cell. In this proof-of-concept approach, a standard solar cell model is augmented by adding the effects of an additional up-conversion (UC) glass layer. After describing some of the technical background related to up-conversion, we also highlight some of the relevant processes involved in typical UC applications for solar cells by using a rate equation model. We also explain in some detail how a typical device simulation is carried out, and we emphasize the crucial role of the size of the model glass layer for the performance of a model c-Si solar cell device.

2. Up-Conversion in Solar Cells

In the context of solar cells, up-conversion is used to absorb sub-band-gap photons from the solar spectrum and convert them into higher-energy photons in the normal absorption range of the solar cell. The up-converted photons can be utilized by the solar cell to produce electron-hole pairs, which drive the corresponding photocurrents [2]. Note that about 20% of the solar energy reaching the surface of the Earth is not utilized by conventional silicon solar cells, as these photons have energies smaller than the band gap of the semiconductors in the solar cell [5]. Therefore, UC provides an attractive method to harness these lower-energy photons, reducing spectral losses. This increases the photocurrent and improves the overall efficiency of the solar cell. A simplified diagram illustrating the UC process is given in Figure 1.

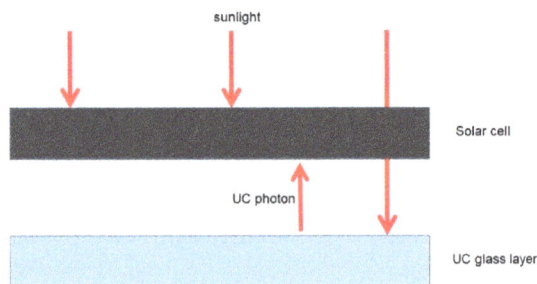

Figure 1. Schematic for up-conversion using glass layers in solar cell applications.

In general, the UC process requires a luminescent material with multiple energy levels that have an appropriate energy spacing. This allows for the emission of photons with frequencies close to the band gap of the solar cell. It is well-known that lanthanide ions, such as Er^{3+}, have these properties, which can easily be embedded into glass layers for the purpose of UC [6,7]. One of the approaches that are currently being studied in Reference [8] is the systematic doping of optical materials with one or more of these so-called activator ions, with the possible inclusion of sensitizer ions, such as Yb^{3+}, as co-dopants.

3. Current Densities and Generation Rate of Cell with Frequency Conversion

In the following, we briefly describe the main contributions to the current density generated by a solar cell. These can be divided into an electronic contribution from the P-N junction, the standard photocurrent, and an additional current from the up-converted photons.

3.1. Dark Current Density

If an external voltage is applied to a P-N junction, a net current density J_{dark} will be obtained. This current density is generated under dark conditions and yields an important reference value. This current is given by Reference [9]:

$$J_{dark} = \left(J_{diff,0} + J_{scr,0} + J_{rad,0} \right) \left(e^{\frac{qV}{kT}} - 1 \right) \tag{1}$$

where $J_{diff,0}$ is the diffusion current density, $J_{scr,0}$ is the recombination current density, and $J_{rad,0}$ is the irradiative recombination current density. k is the Boltzmann's constant, T is the temperature of the device in Kelvin, and q is the electronic charge. The dark current is completely described by electronic semiconductor physics.

3.2. Photo-Current Density

The photo current density J_p stems from the free charge carriers generated through photon absorption and is given by [9,10]:

$$J_p = q\eta_c \int_{\lambda_1}^{\lambda_2} (1 - R(\lambda))\alpha(\lambda) \exp[-\alpha(\lambda)X]\Phi_\lambda(\lambda)d\lambda \tag{2}$$

where η_c is the charge carrier generation quantum efficiency which is equal to 1, $\alpha(\lambda)$ is the absorption coefficient, $R(\lambda)$ is the reflectivity, X is thickness of the absorbing layer, and $\Phi_\lambda(\lambda)$ is the incident spectral photon flux at wavelength λ. In this work, we shall restrict ourselves to the standard AM1.5 spectrum as found in Reference [9]. Reflections and other losses are not included in this model.

3.3. Additional Current Density from UC

In the case of solar cells with UC, an extra current density must be added to the standard photocurrent density in Equation (2). To this end, we estimate the additional photon flux ($\Phi_{ex}(d)$) due to frequency conversion, using:

$$\Phi_{ex,i}(d) = \frac{N_i}{\tau_i}d \tag{3}$$

where N_i denotes the population density of state i from which electrons decay to release upconverted photons. τ_i is the lifetime of this state, and d is the thickness of the frequency conversion layer. Equation (2) needs to be used with caution, as it does not consider extra losses which might be introduced by a thick conversion layer, e.g., extra reflection. It also assumes N_i to be position independent, which also only holds for thin conversion layers.

The generation rate $G_{ex,i}$ accounts for the creation of electron-hole pairs in the solar cell from the extra photon flux $\Phi_{ex,i}$. It can be estimated from:

$$G_{ex,i} = \alpha(\lambda_0)\, \Phi_{ex,i}(\lambda_0) \tag{4}$$

where $\alpha(\lambda_0)$ is the absorption coefficient of the solar cell at the up-converted wavelength. The extra current density produced by a single emission line from the up-conversion is then given by:

$$J_{ex,i} = q\, G_{ex,i}L_D \tag{5}$$

where L_D is the size of the depletion layer in the solar cell.

The total current can then be estimated from the summation of the above current densities, and is given by:

$$J_{tot} = J_{dark} + J_p + J_{ex,i} \tag{6}$$

if more than one UC channel exists, the last term in Equation (5) becomes a sum over the corresponding current density contributions.

The power conversion efficiency of a solar cell is defined by:

$$\eta = \frac{V_m \times J_m}{P_{in}} \qquad (7)$$

here, P_{in} is the input solar power intensity, and V_m and J_m are the voltage and the current densities at the maximum power point of the solar cell. V_m and J_m can be easily determined from the current-voltage curve.

It should be noted that the above model is, of course, highly idealised, and many interactions between the UC layer and the rest of the solar cell are not treated. However, it is a strength of this model to treat the reference cell and the UC layer as separate systems, as this allows us to compute a qualitative estimate of the potential benefit of a UC layer to the solar cell efficiency without the necessity for a more cumbersome simulation.

In the following, we want to focus our discussion of UC on a simple-term scheme for trivalent Erbium (Er^{3+}) and a rate equations-based approach.

4. Er^{3+} Up-Conversion Term Scheme

To describe up-conversion properly, four physical processes need to be considered, which are: ground state absorption (GSA), excited state absorption (ESA), spontaneous emission (SPE), stimulated emission (STE), and energy transfer (ET) [11]. Many lanthanides are suitable for up-conversion applications. Of those, the trivalent Erbium (Er^{3+}) features a rather simple term scheme, which makes Er^{3+} a good study case without adding unnecessary complexity.

A Er^{3+} term scheme is shown in Figure 2, which only considers the processes and states relevant for UC. The processes of GSA and ESA are assumed to be resonant. Three metastable states are shown, from which some further electronic transitions (accompanied by the emission of photons) originate with branching ratios β_{ij}. Here, i denotes the initial state of such a transition, whereas j denotes the final state [12].

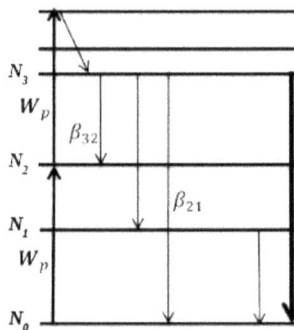

Figure 2. Energy level diagram for Er^{3+} with all the important processes for up-conversion, as described in Reference [12].

The rate equations for this basic system are given by [12]:

$$\frac{dN_3}{dt} = \frac{\sigma_{ESA}}{\sigma_{ESA} + \sigma_{GSA}} W_p N_2 - \frac{N_3}{\tau_3} \qquad (8)$$

$$\frac{dN_2}{dt} = \frac{\sigma_{GSA}}{\sigma_{ESA} + \sigma_{GSA}} W_p N_0 - \frac{N_2}{\tau_2} - \frac{\sigma_{ESA}}{\sigma_{ESA} + \sigma_{GSA}} W_p N_2 + \beta_{32} \frac{N_3}{\tau_3} \tag{9}$$

$$\frac{dN_1}{dt} = \beta_{31} \frac{N_3}{\tau_3} + \beta_{21} \frac{N_2}{\tau_2} - \frac{N_1}{\tau_1} \tag{10}$$

$$N_{tot} = N_0 + N_1 + N_2 + N_3 \tag{11}$$

W_p is the pump rate used to populate the excited states. In the case of solar cells, $W_p = \sigma_p \Phi_p$ is given by the incoming photon flux $\Phi_p \approx 1.2 \times 10^{17}$ cm^{-2}s^{-1} around the up-conversion wavelength, as well as the absorption cross-section $\sigma_p \approx 1.0 \times 10^{19}$ cm^2 [13]. The exact line position and line width depend on the host material of the UC layer, meaning this approximation shall suffice here. N_{tot} is the total population density of the active erbium ions within the up-conversion glass layer. σ_i and τ_i are the cross sections and lifetimes for state i, respectively. β_{ij} denotes the branching ratio from state i to state j. N_0, N_1, N_2, and N_3 are the population densities for various states of the Er^{3+} ion.

Equations (8)–(10) describe the population changes of the meta-stabile excited states, while Equation (11) guarantees conservation of electrons. An extra equation for the ground state population N_0 is not needed, as the system of equations is later solved in the steady-state case, for which the system is already complete. Equation (8) includes the change in N_3 through pumping from N_2, as well the emission. The population of N_2 is fed by pumping from the ground state N_0 and emission from N_3, and loses population through pumping to N_3 and emission. N_1 is populated through emissions from N_2 and N_3, and emits itself back to the ground state.

5. Results and Discussions

A reference 1-D crystalline silicon (c-Si) solar cell is modelled using Equations (1) and (2) [9,10]. For the calculation of J_{dark} in Equation (1), the sum of the different current densities was specified as $J_{diff,0} + J_{scr,0} + J_{rad,0} = 1.95 \times 10^{-9}$ A/m^2. The operating temperature of the device was chosen to be 300 K. J_p is estimated by computing Equation (2), where the absorber thickness X was chosen to be 510×10^{-6} m. The resulting I–V curve for this model solar cell is given by the solid curve in Figure 3. Alternatively, a reference solar cell which can be modelled by various device simulation packages such as GPVDM [12] can be used. In this approach, the resulting photon fluxes from the UC layer are used to augment the photon flux values on the actual spectrum file in the program codes. We stick to using our reference solar cell model for this work.

In the steady state, the population densities N_0, N_1, N_2, and N_3 do not change over time, and can be calculated as a function of the pump power W_p, provided N_{tot} and all necessary constants in Equations (8)–(11) are known. The important parameters needed to solve for the population densities are given in Table 1.

Table 1. Parameters for solving the rate equations. (* obtained from Reference [13]).

Parameter	Symbol	Value
*Lifetime	τ_1	10 ms
*Lifetime	τ_2	4.3 ms
*Lifetime	τ_3	0.37 ms
*GSA cross section	σ_{GSA}	0.25×10^{-20} cm^2
*ESA cross section	σ_{ESA}	1.7×10^{-20} cm^2
*Branching ratio	β_{20}	0.85
*Branching ratio	β_{21}	0.15
*Branching ratio	β_{30}	0.67
*Branching ratio	β_{31}	0.27
*Branching ratio	β_{32}	0.02
Pump wavelength	λ_p	974 nm
Conversion material length	d	0.3 mm
*Total Er^{3+} concentration	N_{tot}	1×10^{19} cm^{-3}
Depletion layer width	L_D	1×10^{-6} m

In the Er^{3+} term scheme, N_3 leads to photoemission with 550 nm wavelength, which is the relevant transition for up-conversion applications. The corresponding lifetime τ_3 for N_3 is 0.37 ms [13]. The values in Table 1 are used to estimate the additional current density from the UC material layer. All the parameters for crystalline silicon used in the subsequent calculations are taken from [9].

The generation rate $G_{ex,3}$ from the up-converted photons is calculated using an absorption coefficient of $\alpha \approx 10^6 \frac{1}{m}$ for c-Si at 550 nm. With this, the photon flux calculated from Equation (3) is $\Phi_{ex,3} \approx 5.35 \times 10^{19} \frac{1}{s \times m^2}$. Using Equation (8), the corresponding generation rate is obtained as $G_{ex,3} \approx 5.35 \times 10^{25} \frac{1}{s \times m^3}$. This should be compared to the generation rate $G_{Si} \approx 4.48 \times 10^{27} \frac{1}{s \times m^3}$ of a c-Si solar cell without UC.

It is important to note that the estimated photon flux strongly depends on the size of the glass layer. A very thick glass layer, like the present one, may strongly exaggerate the additional photon flux due to UC. Also, for very thick glass layer attenuation (Beer's law), the additional photon flux needs to be considered. Therefore, our guideline is to use glass layers of a size that leads to measurable increases in overall efficiency (i.e., in the range of %), but with the final aim of identifying optimum UC systems, where these glass layers may be kept very small.

A certain increase in photon flux does not imply a proportional increase in solar cell efficiency, because the solar cell is a nonlinear device. It is necessary to calculate the corresponding current densities. Using Equation (5), we may determine the additional photocurrent density derived from the UC process. The total current density can then be determined by using Equation (6). The resulting J–V curves are shown in Figure 3, for a c-Si solar cell with (solid black curve) and without (dashed red curve) the UC current density contribution.

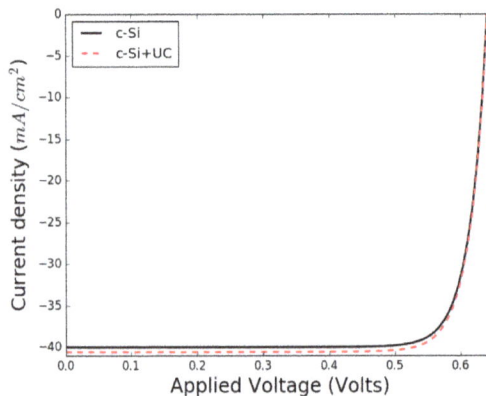

Figure 3. Simulated J–V curve for an ideal c-Si solar cell with and without up-conversion (UC).

Without the frequency conversion layer, the efficiency of the model c-Si solar cell is 21.4%, with a short circuit current of $J_{sc} = 400 \frac{A}{m^2}$ and an open circuit voltage of $V_{oc} = 0.56$ V. Using Equation (5), the additional current density that is derived from the UC glass layer is $J_{ex,3} = 8 \frac{A}{m^2}$. This results in a total short-circuit current of $J_{sc+ex} = 408 \frac{A}{m^2}$. The open circuit voltage is not affected.

The efficiency of the model c-Si solar cell, including contributions from UC, is 22.7%. The overall increase in efficiency for the c-Si solar cell by adding our Er^{3+}-based luminescent glass layer model is 1.3%. The increase in efficiency in this proof-of-concept example is, of course, small. It stems from one specific up-conversion channel only, and the example did not include any sensitizer co-dopants. More importantly though, this modelling approach can be applied to more complex and realistic up-conversion set-ups to find good up-conversion configurations.

Some experimental work has been done studying the impact of frequency conversion, such as UC, on solar cell device performance. In their recent publication, Kumar et al. [14] found a UC improvement of about 0.4% for a dye-sensitized solar cell that uses Er^{3+} and Yb^{3+} co-doped ZnO up-conversion (UC) nanoparticles-based phosphors. Grigoroscuta et al. [15] have also studied the effect of a phosphor film of Yb^{3+}/Er^{3+}-co-doped CeO_2 on the performance of a silicon-based solar cell and have noticed a significant increase in the cells' performance because of the thin UC film. A simplified modelling approach such as this can be useful in aiding the development of more advanced solar cells.

6. Conclusions

In this paper we presented a simple model approach to estimate the impact of frequency conversion on the overall efficiency of solar cell devices. A rate equation model was used to estimate the population densities for a typical UC process in Er^{3+} yielding an extra photon flux in the visible range. The total photocurrent is enhanced by the addition of an up-conversion layer. As the solar cell is a non-linear device, already relatively small increases in the current density can lead to considerable improvements in the overall device efficiency.

Systems other than Er^{3+} can be studied using the same method with Tm^{3+} being a good example, provided the basic parameters for the rate equation model are known. As an alternative to the present approach, the additional photon flux due to UC can also be used as a direct input parameter for existing device simulation codes [16], which seems to compare quite well to the presented approach.

Author Contributions: A.Q., T.A. and I.M. conceived the original idea. The methodology was developed collectively by all authors. The formal analysis and computational framework was carried out by T.A., I.M., R.W. The work presented was validated by A.Q., R.W., G.R. and M.F. The resources used to complete the work presented where made available by A.Q., G.R. and M.F. The original draft was prepared by T.A., I.M., A.Q. and R.W., with subsequent editing being done by all authors. The overall visualization of the work presented was done by T.A., I.M. and R.W. The overall supervision of the work was done by A.Q. and R.W. The authors responsible for the project administration were A.Q., G.R. and M.F. Funding for the work presented was organized by G.R. and A.Q.

Funding: This research was funded in part by the bilateral project Plasmonics for a better efficiency of solar cells between South Africa and Italy (contributo del Ministero degli Affari Esteri e della Cooperazione Internazionale, Direzione Generale per la Promozione del Sistema Paese). This research was also funded in part by the National Research Foundation of South Africa grant number 115457.

Acknowledgments: The authors would like to thank the Materials for Energy Research Group (MERG) and the DST-NRF Centre of Excellence in Strong Materials (CoE-SM) at the University of the Witwatersrand for support.

Conflicts of Interest: The authors declare no conflict of interest.

References

1. Shockley, W.; Queisser, H.J. Detailed balance limit of efficiency of p-n junction solar cells. *J. Appl. Phys.* **1961**, *32*, 510–519. [CrossRef]
2. Würfel, P.; Würfel, U. *Physics of Solar Cells: From Principles to New Concepts*, 3nd ed.; Wiley-VCH: Weinheim, Germany, 2009.
3. Polman, A.; Atwater, H.A. Photonic design principles for ultrahigh-efficiency photovoltaics. *Nat. Mater.* **2012**, *11*, 174–177. [CrossRef] [PubMed]
4. Trupke, T.; Green, M.A.; Würfel, P. Improving solar cell efficiencies by up-conversion of sub-band-gap light. *J. Appl. Phys.* **2002**, *92*, 4117–4122. [CrossRef]
5. Fischer, S.; Goldschmidt, J.C.; Löper, P.; Bauer, G.H.; Brüggemann, R.; Krämer, K.; Biner, D.; Hermle, M.; Glunz, S.W. Enhancement of silicon solar cell efficiency by upconversion: Optical and electrical characterization. *J. Appl. Phys.* **2010**, *108*, 044912. [CrossRef]
6. Lian, H.; Hou, Z.; Shang, M.; Geng, D.; Zhang, Y.; Lin, J. Rare earth ions doped phosphors for improving efficiencies of solar cells. *Energy* **2013**, *57*, 270–283. [CrossRef]
7. Yang, W.; Li, X.; Chi, D.; Zhang, H.; Liu, X. Lanthanide-doped upconversion materials: Emerging applications for photovoltaics and photocatalysis. *Nanotechnology* **2014**, *25*, 482001. [CrossRef] [PubMed]
8. Vega, M.; Alemany, P.; Martin, I.R.; Llanos, J. Structural properties, Judd–Ofelt calculations, and near infrared to visible photon up-conversion in Er^{3+}/Yb^{3+} doped $BaTiO_3$ phosphors under excitation at 1500 nm. *RSC Adv.* **2017**, *7*, 10529–10538. [CrossRef]

9. Smets, A.; Jager, K.; Isabella, O.; Swaaij, R.V.; Zemann, M. *Solar Energy*; UIT Cambridge Ltd.: England, UK, 2016.

10. Nelson, J. *The Physics of Solar Cells*; Imperial College Press: London, UK, 2007.

11. Fischer, S.; Steinkemper, H.; Löper, P.; Hermle, M.; Goldschmidt, J.C. Modeling upconversion of erbium doped microcrystals based on experimentally determined Einstein coefficients. *J. Appl. Phys.* **2012**, *111*, 013109. [CrossRef]

12. MacKenzie, R.C.; Kirchartz, T.; Dibb, G.F.; Nelson, J. Modeling Nongeminate Recombination in P3HT: PCBM solar cells. *J. Phys. Chem. C* **2011**, *115*, 9806–9813. [CrossRef]

13. Finlayson, D.M.; Sinclair, B. *Advances in Lasers and Applications*; CRC Press: Boca Raton, FL, USA, 1998.

14. Kumar, V.; Anurag, P.; Sanjay, K.; Ntwaeaborwa, O.M.; Swart, H.C.; Viresh, D. Synthesis and characterization of Er^{3+}-Yb^{3+} doped ZnO upconversion nanoparticles for solar cell application. *J. Alloy. Compd.* **2018**, *766*, 429–435. [CrossRef]

15. Grigoroscuta, M.; Secu, M.; Trupina, L.; Enculescu, M.; Besleaga, C.; Pintilie, I.; Badica, P. Enhanced near-infrared response of a silicon solar cell by using an up-conversion phosphor film of Yb/Er–co-doped CeO_2. *Sol. Energy* **2018**, *171*, 40–46. [CrossRef]

16. Quandt, A.; Warmbier, R.; Mokgosi, I.; Aslan, T. Solar cell device simulations. In Proceedings of the 19th International Conference on Transparent Optical Networks (ICTON), Girona, Spain, 2–6 July 2017.

micromachines

MDPI

Article

Luminescent Properties of Eu³⁺-Doped Hybrid SiO₂-PMMA Material for Photonic Applications

Pablo Marco Trejo-García [1], Rodolfo Palomino-Merino [1,*], Juan De la Cruz [1], José Eduardo Espinosa [1], Raúl Aceves [2], Eduardo Moreno-Barbosa [1] and Oscar Portillo Moreno [3]

[1] Facultad de Ciencias Físico Matemáticas, Benemérita Universidad Autónoma de Puebla, Avenida San Claudio y 18 Sur, Colonia San Manuel, Ciudad Universitaria, C.P. 72570 Puebla, Mexico; pablo.trejogarcia@alumno.buap.mx (P.M.T.-G.); 217570100@alumnos.fcfm.buap.mx (J.D.l.C.); espinosa@fcfm.buap.mx (J.E.E.); emoreno@fcfm.buap.mx (E.M.-B.)

[2] Departamento de Investigación en Física, Universidad de Sonora, Apartado Postal 5-088, C.P. 83190 Hermosillo, Sonora, Mexico; raceves@cifus.uson.mx

[3] Facultad de Ciencias Químicas, Benemérita Universidad Autónoma de Puebla, P.O. Box 1067, Colonia San Manuel, Ciudad Universitaria, C.P. 72570 Puebla, Mexico; osporti@yahoo.com.mx

* Correspondence: palomino@fcfm.buap.mx; Tel.: +52-222-229-5500 (ext. 2125)

Received: 19 May 2018; Accepted: 19 July 2018; Published: 1 September 2018

Abstract: Hybrid organic-inorganic materials are of great interest for various applications. Here, we report on the synthesis and optical characterization of silica-PMMA samples with different Eu³⁺ molar concentrations. The optical properties of this material make it suitable for photonic applications. The samples were prepared using the sol-gel method, mixing tetraethyl orthosilicate (TEOS) as a silica glass precursor and methyl methacrylate (PMMA) as a polymer component. Europium nitrate pentahydrate was then added in six different molar concentrations (0.0, 0.1, 0.25, 0.5, 0.75, and 1%) to obtain as many different samples of the material. The absorption spectra were obtained applying the Kubelka–Munk formula to the diffuse reflectance spectra of the samples, all in the wavelength range between 240 and 2500 nm. The emission and excitation measurements were made in the visible range. Five bands could be identified in the emission spectra, related to electronic transitions of the ion Eu³⁺ ($^4D_0 \rightarrow {}^7F_i$, i from 0 to 4). In the excitation spectra, the following bands were detected: $^7F_0 \rightarrow {}^5G_3$ (379 nm), $^7F_0 \rightarrow {}^5G_2$ (380 nm), $^7F_0 \rightarrow {}^5L_6$ (392 nm), $^7F_0 \rightarrow {}^5D_3$ (407 nm), $^7F_0 \rightarrow {}^5D_2$ (462 nm), and $^7F_0 \rightarrow {}^5D_1$ (530 nm). The emission decay times were measured for the different samples and showed an inverse dependence with the Eu³⁺ concentration.

Keywords: europium; luminescence; hybrid materials; microdevices

---

1. Introduction

The study and development of hybrid materials has taken off since the end of 20th century and the beginning of the 21st. This is in spite of the fact that hybrid materials were developed for thousands of years when the production of paints was the driving force to try novel mixtures of dyes and/or inorganic pigments. This takeoff has happened due to the availability of novel physico-chemical characterization methods, as well as new perspectives of material creation made possible by nanoscience. Since then, bottom-up strategies going from the molecular level up to material design have led to the creation of materials with very different and applicable physico-chemical properties. Despite all these changes, when the synthetization of hybrid materials is considered, the sol-gel method has continued to be widely used, as it is cost effective and allows for the synthesis of materials of high purity and homogeneity. On the other hand, the search for a better and better matrix able to host a dopant has been a key objective of research for years. An example of this are SiO₂ matrixes, which incorporate lanthanides, are homogenous and transparent, and have controlled porosity; they can be useful in many photonic applications, including the development of doped fiber amplifiers (DFA) for continuous and pulsed lasers, single-mode silica optical fibers that have the ability to provide high bandwidth and long distance communications, or, finally, materials for thermoluminescent applications [1,2].

Methyl methacrylate (PMMA) is an organic polymer, which is transparent, malleable and flexible. This material has recently been employed to make polymeric optical fibers (POFs) [3], as well as highly tunable Bragg gratings [4]. Kuriki and Koike [5] synthesized lanthanide (La)-doped POFs and showed that the fibers were capable of incorporating the inorganic host in high concentrations. They also found that the lanthanides in these systems were pumped more effectively than in conventional crystal systems. This happened because of energy transfer processes in the chelate complexes [5]. Basu and Vasantharajan doped polystyrene, polymethylmethacrylate, and polyurethane matrices with europium for the fabrication of temperature-sensitive coatings (TSP) [6]. The undoped hybrid PMMA-SiO$_2$ presents other characteristics that its components do not present; for example, it has a higher glass transition temperature, a higher optical transparency, a better thermal stability [7], a better adhesion strength than the pure PMMA that allows it to be employed as anticorrosive coating [8], and size-controlled silica particles, which allow it to effectively reduce the gas permeability of the polymer membrane [9].

On the other hand, the Eu^{3+} ion has played an important role in the development of many optical devices, such as lasers, phosphor materials, coatings, luminescent probes, POFs, displays, and so on. This great interest is based on the intense luminescence it emits in the red range, which is generated by the large energy gap between the ground and the first excited state of this element. Moreover, Eu^{3+} is widely used because the ground energy state ($^7F_0$) and the most important emitting excited state ($^5D_0$) are nondegenerate and do not split because of the crystal-field effect. This allows us to easily understand the absorption and emission spectra and their dependence on the environment in which the Eu^{3+} is located. Based on this fact, Eu^{3+} has been widely used as a luminescent probe for host structures and defect studies [10]. The most important transitions in the luminescence spectra of Eu^{3+} are those from the $^5D_0$ excited state to ^7F_J levels, (with a low J value equal to 0, 1, or 2). The interpretation of the spectra turns out to be easy because of the small number of possible crystal-field transitions. Other interesting characteristics refer to the facts that, because the different $^5D_0 \rightarrow {}^7F_J$ lines are well separated, overlapping between the crystal-field levels is difficult [11] and that the luminescence intensity ratio of the magnetic ($^5D_0 \rightarrow {}^7F_1$) and electric bands ($^5D_0 \rightarrow {}^7F_2$), both in the red range of the spectrum, shows a larger symmetry factor with a larger symmetry of local crystal fields [12].

In this study, six hybrid samples, made of inorganic silica and organic PMMA, were synthetized using the sol-gel method. Five samples were doped with different molar concentrations of Eu^{3+}, and one was left undoped. The goal of the study was to characterize the optical properties of these materials because of their potential applications. Due to their transparency, the ease of their fastening to silica and to substrates—because of their vitreous part, increased flexibility conferred by the polymer component, increased mechanical properties, and the fluorescence properties conferred by the dopant—and their capacity to change the refractive index by changing the ratio between the precursors, these materials can be used in the near future as a coating for optical fibers for the development of optical devices, such as sensors, which can act under the excitation of the dopant as a coating layer for scratch resistance or corrosion protection and as an optical filter device. In addition, because the material is made through the sol-gel method, the manufacturing of fibers at room temperature is allowed.

2. Materials and Methods

Material Preparation: Six different SiO$_2$-PMMA samples were prepared with the sol-gel method. During this process, tetraethyl orthosilicate (TEOS), PMMA, and ethanol were mixed. After that, trimethoxysilyl propyl methacrylate (TMSPM) was added and used as a bonding agent between the polymer and the SiO$_2$ molecules. The molar ratios were 1:1:0.22:4.75:4.75

(TEOS/PMMA/TMSPM/H_2O/Ethanol). Sodium hydroxide (NaOH) was added as a catalyst for the hydrolysis to increase the pH of the solution up to a value of 9. Benzoyl peroxide (BPO) was used as a catalyst for the methyl methacrylate (MMA) polymerization in a 1 mass % ratio with respect to the amount of PMMA used [3]. After these steps were completed, europium nitrate pentahydrate was added in molar concentrations of 0, 0.1, 0.25, 0.5, 0.75, and 1 mol %. All this to form one undoped sample and five Eu-doped samples of the hybrid material studied in this work. The gel formation process lasted for a period of 25 days, finishing when the samples were completely dry.

Spectra Characterization: Absorption spectra were obtained applying the Kubelka–Munk equation as follows:

$$(1 - R_\infty)^2/2 \times R_\infty \equiv F(R) \tag{1}$$

where $F(R)$ is the Kubelka–Munk function and R_∞ is the reflectance of a layer so thick as to completely hide the substrate. For this investigation, the diffuse reflectance spectra of the samples were recorded in the range between 240 and 2500 nm using a CARY 5000 spectrophotometer (Agilent Technologies, Santa Clara, CA, USA). The band gap energy (Eg) was calculated by the interpolation of a line in the graph of $(F(R)h\nu)^n$ versus $h\nu$ (Tauc plot); "n" takes the values of 0.5 or 2 (depending on whether the allowed transition is direct or indirect, respectively) and $h\nu$ is the corresponding photon energy [13]. Emission and excitation sample spectra were always recorded at room temperature (RT). For the emission spectra, the setup array was in frontal face mode in a Nanolog Spectrofluorometer (Jobin-Yvon Horiba, Horiba, Ltd., Kyoto, Japan) equipped with double grating in both the excitation and emission monochromators and with a Xenon lamp of 450 W. For the emission spectra, a range of wavelengths greater than the ones used for excitation were used.

Decay time calculations: Because the decay time of Eu^{3+} is in the order of milliseconds, the measurements of the fluorescent decay time were carried out at RT, using a time-correlated single photon-counting (TCSPC) Fluorolog3-TCSPC (Jobin-Yvon Horiba, Horiba, Ltd.). This was a hybrid steady-state (continuous wave (CW)) system with time-correlated fluorescence dynamics. It was equipped with double grating in both the excitation and emission monochromators, a 450 W Xenon lamp for CW measurements, and a Xenon pulsed lamp (3 ms pulse duration). The excitation wavelength used to carry out the experiments was 393 nm, which corresponded to the stronger wavelength that could excite the electronic state $^5D_0$ (614 nm). Finally, a natural logarithm function was applied to the spectra, and a linear fit was calculated; the slope of this line gave us the fluorescence decay time of each sample.

3. Results

In Figure 1, the absorption spectra of the six SiO_2-PMMA samples under investigation are presented. The wavelengths varied in the 240–2500 nm range. In the inset image of Figure 1, an enlargement of the wavelength region between 300 and 750 nm is presented. It can be observed that the absorption edge for all the samples was approximately 350 nm. All the samples presented the same bands with different amplitudes. In the inset image, due to the Eu^{3+} ion presence, three bands were identified that were associated with the electronic transitions $^7F_0 \rightarrow ^5L_6$ (392 nm), $^7F_0 \rightarrow ^5D_2$ (460 nm), and $^7F_0 \rightarrow ^5D_1$ (538 nm). The first band represented the most intense absorption and was observable for all the doped samples. It appeared due to the direct excitation into the $4f^6$ levels of the Eu^{3+} ions [11]. The band associated with $^7F_0 \rightarrow ^5D_2$ was an electric dipole transition and was used to determine the position of the $^5D_2$ level; it was observable for dopant concentrations higher than 0.1 mol %. The magnetic dipole transition ($^7F_0 \rightarrow ^5D_1$) presented small intensities and was observed only for the samples doped at 0.75 and 1 mol %.

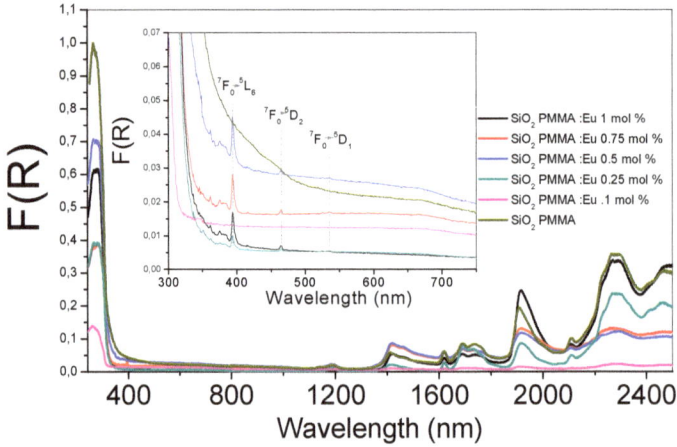

Figure 1. Absorption spectra of the six SiO$_2$- methyl methacrylate (PMMA) samples. The absorption spectra for the undoped and doped hybrid samples are shown in the wavelength range between 240 and 2500 nm. In the inset, a zoom of the range 300–750 nm is presented, where the electronic transitions of the Eu^{3+} ion are identified. The line color of each sample is the same in both graphs.

Figure 2 was obtained drawing the Tauc plot [$(F(R)h\nu)^n$ vs $h\nu$] for the absorption spectra. The main graph corresponds to an n value of 2 and the inset to $n = 0.5$. It can be observed that for $n = 2$ in the absorption edge (3.75−4.25 eV), the samples presented a linear behavior, while in contrast with $n = 0.5$, they presented a polynomial behavior. This property is typical of a direct band gap material. The results reported in the Table 1 show the direct gap value for each dopant concentration; the data were calculated from the extrapolated crossing of the linear segment of the absorption edge with the energy-axis. These values showed that there were no significant changes in the band gap due to dopant presence.

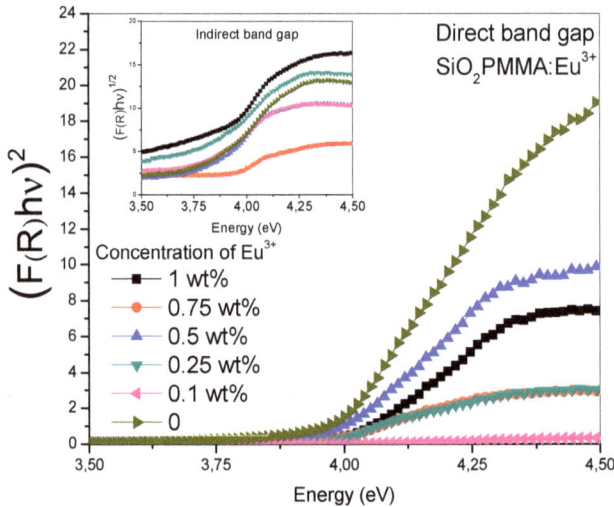

Figure 2. $(F(R)h\nu)^n$ vs photon energy (Tauc plot) in the absorption edge range for the SiO$_2$-PMMA:Eu^{3+} samples. In the inset, the $(F(R)h\nu)^{1/2}$ vs photon energy relation is plotted for $n = 0.5$, while in the main graph $n = 2$.

Table 1. Dopant concentration (mol %) vs direct gap energy (eV) for the six samples.

Concentration of Eu^{3+} (mol %)	Direct Gap (eV)
0	3.97 ± 0.20
0.1	4.01 ± 0.10
0.25	3.97 ± 0.12
0.5	3.98 ± 0.22
0.75	3.96 ± 0.12
1	4.01 ± 0.22

Figure 3 presents the emission spectra using 325 and 393 nm as excitation wavelengths and the excitation spectrum detected at 616 nm of a SiO_2-PMMA:Eu^{3+} sample with 1 mol % Eu^{3+}. The excitation wavelength of 325 nm was used to observe the complete emission band due to the SiO_2-PMMA matrix, which is in the range of 340–500 nm. In contrast, the excitation wavelength of 393 nm was used to excite the Eu^{3+} ions and to induce photoluminescence due to the direct population of the 4f levels. In the same figure, the excitation spectrum recorded at 616 nm displays a set of bands associated with the 4f electronic transitions: $^7F_0 \rightarrow ^5G_3$ (377 nm), $^7F_0 \rightarrow ^5G_2$ (380 nm), $^7F_0 \rightarrow ^5L_6$ (392 nm), $^7F_0 \rightarrow ^5D_3$ (413 nm), $^7F_0 \rightarrow ^5D_2$ (462 nm), and $^7F_0 \rightarrow ^5D_1$ (530nm) [11,14]. Because the emission spectrum of the SiO_2-PMMA matrix (340−500 nm) presents a decrement at around 392 nm and the excitation spectrum at 616 nm has the most intense band associated with the electronic transition $^7F_0 \rightarrow ^5L_6$ (392 nm) in the same wavelength position, it can be argued that the decrement is caused by the radiative energy transfer from the matrix to the Eu^{3+}.

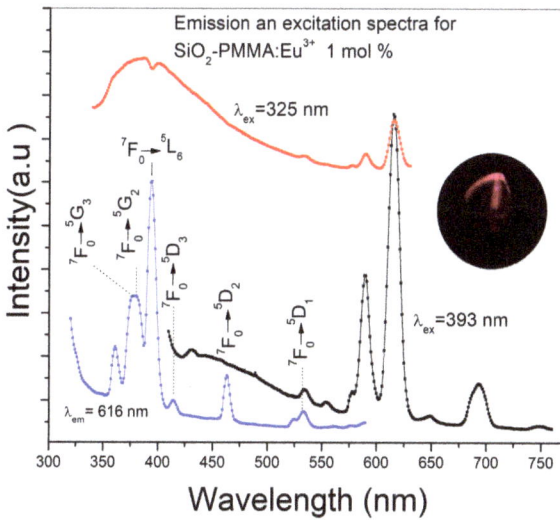

Figure 3. Emission and excitation spectra of a SiO_2-PMMA:Eu^{3+} sample doped at 1 mol % using different excitation wavelengths. The red line corresponds to the excitation at 325 nm and the black line to the excitation at 393 nm. The excitation spectrum was obtained by detection at 616 nm (blue line). Electronic transitions associated with higher energies were identified in the excitation spectrum. In the inset, an image of the emission of the sample using an excitation wavelength of 393 nm can be observed.

Figure 4 presents the emission spectra for all the samples using excitation wavelengths of 325 and 393 nm. The excitation wavelength of 325 nm was used to observe the matrix luminescence, which is composed by a broad band from 350 nm to 430 nm. The optimum excitation wavelength of 393 nm was used to observe the electronic transitions of Eu^{3+}. For this ion, five bands related to the electronic transitions $^5D_0 \rightarrow ^7F_i$ (i from 0 to 4) with maxima in 578, 590, 615, 649, and 693 nm could be identified. It should be noted that for the dopant concentrations used, fluorescence quenching was never observed; also, due to the fact that the dopant concentrations used were small, radiative energy transfer was not observed for the samples with dopant concentration smaller than 1 mol %.

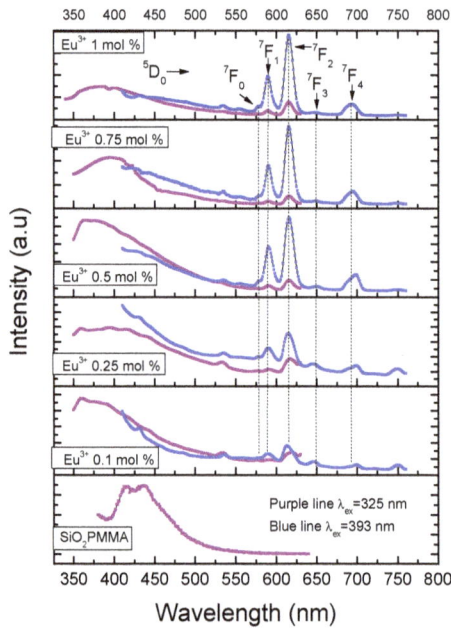

Figure 4. Emission spectra after excitation at 325 nm (purple line) and 393 nm (blue line) for the different hybrid material samples. The intensities of the two emission spectra are not at the same scale; they have been normalized to fit into the same plot frame.

The presence of a single band associated with the $^5D_0 \rightarrow ^7F_0$ transition, which could not be split by the crystal field, indicated that the Eu^{3+} ion occupied a single site with C_{nv}, C_n, or C_s symmetries [11]. The intensity of the $^5D_0 \rightarrow ^7F_1$ transition, which is related to the magnetic dipole nature, grew as did the molar concentration of Eu^{3+}. Even though this electronic transition is observed in all the doped samples, for the slightly doped samples (0.1 and 0.25 mol %) a blue-pink emission was observed due to the fact that the emission of the matrix has a strong component in the blue region. The $^5D_0 \rightarrow ^7F_2$ band has an electric dipole nature, and the increase on the intensity of this band, which was associated with low symmetries of the Eu^{3+}, grew faster than that of the $^5D_0 \rightarrow ^7F_1$. This fact could be confirmed trough the measurement of the ratio $I(^5D_0 \rightarrow ^7F_2)/I(^5D_0 \rightarrow ^7F_1)$. The following values were obtained: 1.185 (0.1 mol %), 1.212 (0.25 mol %), 1.574 (0.5 mol %), 1.907 (0.75 mol %), and 1.957 (1 mol %). These results meant that the samples tended to have a lower symmetry as the dopant increased and confirmed the red shift of the luminescence observed in the samples due to the fast growth in intensity of this band compared with the others. This fact is also observable in the International Commission on Illumination (CIE) chromaticity diagram shown in Figure 5. In this diagram, the shift from the blue region to the pale red region can be observed as the dopant concentration increases.

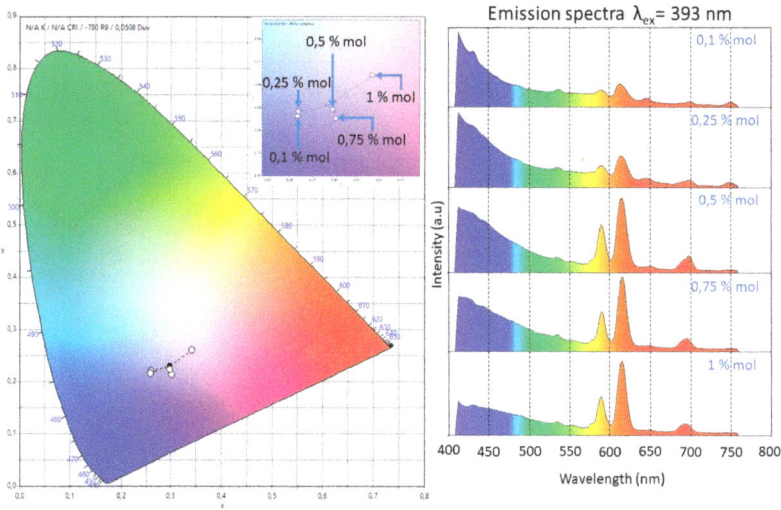

Figure 5. CIE 1931 2° chromaticity diagram. Color chromaticity of the emission in the range of 410–760 nm of the five doped samples under the excitation wavelength at 393 nm. The inset in the diagram represents an amplification of the blue-pale red zone. The red shift in the emission was observed as an increase of the molar relation of Eu^{3+}.

The red shift behaviour had been previously reported and observed in $(Y_{1-x}Eu_x)_2O_2S$, $(Y_{1-x}Eu_x)_2O_3$, and $(Y_{1-x}Eu_x)_2VO_4$ materials and was associated with the favoring of the $^5D_0$ level [15]. The transition $^5D_0\rightarrow^7F_3$ was very weak for all the molar concentrations of the dopants. The intensity ratio between this band and that associated with the $^5D_0\rightarrow^7F_1$ band decreased as the molar concentration grew.

The observation of a weak emission associated with the $^5D_0\rightarrow^7F_4$ transition for all the samples confirmed the low symmetry of the material and supported the idea that the chemical composition of the host matrix influenced the intensity of this band. In other words, when SiO_2 is present in the matrix, the $^5D_0\rightarrow^7F_2$ transition is the most important in the emission spectrum, while when SiO_2 is absent, the main role is played by the $^5D_0\rightarrow^7F_4$ transition. This was first observed by Bortoluzzi et al. [16] for $Eu(Tp)_3$ in PMMA polymer matrix and $Eu(Tp)_3$ (Tp = hydrotris (pyrazol-1-yl) borate). A similar behavior was observed by Blasse and Bril for $GdOCl:Eu^{3+}$, in which the transition $^5D_0\rightarrow^7F_4$ dominated the spectrum, while for $GdOBr:Eu^{3+}$ it was the $^5D_0\rightarrow^7F_2$ transition that was dominant [17].

The decay time curves for all the doped samples first were fitted to a single exponential function with the following expression $I = I_0\,exp(-t/\tau)$, which indicates that the Eu^{3+} ions are located in similar sites suffering the same crystal field [4]. Subsequently, applying the natural logarithm function to these results, we obtain what is presented graphically in Figure 6, where it can be observed that the $^5D_0$ lifetime shortens with the increase of the concentration of the Eu^{3+} ions due to the cross-relaxation process among Eu^{3+} ions. The lifetimes obtained are 275 ± 10 μs (1 mol %), 285 ± 6 μs (0.75 mol %), 308 ± 5 μs (0.5 mol %), 446 ± 20 μs (0.25 mol %), and 617 ± 30 μs (0.1 mol %). The cross-relaxation process leads to fluorescence quenching (i.e., to the decrease of fluorescence intensity when the rare earth concentration is increased). One has to consider, in fact, that, with the increasing concentration of rare earth ions, the ion spacing decreases and may be small enough to allow them to interact and transfer energy [18].

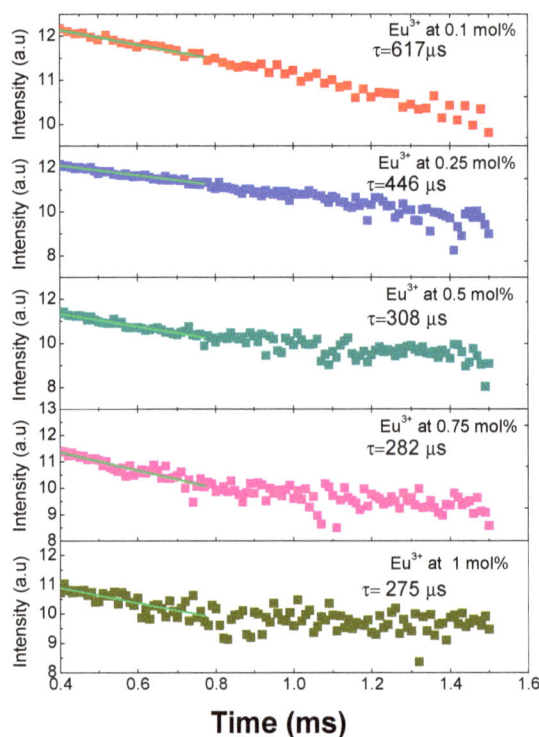

Figure 6. Decay curves and lifetimes of SiO_2-PMMA:Eu^{3+} samples with concentration 0.1 mol % (red points), 0.25 mol % (blue points), 0.5 mol % (cyan points), 0.75 mol % (pink points), and 1 mol % (dark yellow points).

Another aspect that contributes to the lifetime shortening is the existence of hydroxyl groups in all the samples due to the fact that they were made at room temperature and were not submitted to a thermal treatment. Those groups have been observed to reduce the lifetime emission in silica matrix doped with Eu^{3+} when it has not been submitted to a thermal treatment [19]. For this matrix, the reported lifetimes are between 230 and 776 µs and are temperature dependent. To sum up, because the ionic radius of Eu^{3+} (1.07 Å) is larger than that of Si^{4+} (0.4 Å), the amount of Eu^{3+} in the matrix is restricted, and it can give place to the formation of clusters of the ion in the matrix that contributes to the cross relaxations [20].

One can also note that in a work by Basu and Vasantharajan [6], the decay lifetimes of Eu^{3+} in three different host matrices of polystyrene, polymethylmethacrylate, and polyurethane were compared. They observed that the luminescence intensity and the lifetime were strongly dependent on the matrix type, the temperature, and the amount of oxygen present. In particular, they found that the fastest time decay occurred in the PMMA matrixes, and it is around 364.2 µs.

4. Conclusions

Hybrid materials with optical properties are attracting more and more attention, especially in the application areas of sensing, imaging, and energy. Rare-earth-doped hybrid materials, in particular, may be optimized by playing with the interactions between the organic and the inorganic parts. Here, using the sol-gel method, a series of PMMA-silica hybrid materials, namely one undoped and five Eu^{3+}-doped samples, were synthetized and optically characterized. The transparency of the undoped sample as a result of its low absorption in the visible range guarantees that it can have applications in photonic devices, for example, in the fabrication of organic light emitting diodes (OLEDs), microstructured polymer optical fibres (MPOFs), or polymer light emitting diodes (PLEDs). Also, the undoped and doped samples, as a hybrid coating, can work as a chemical detector through the change in the spectral signature, for example, to detect solvents, humidity, or proteins.

An energy transfer from the matrix of SiO_2-PMMA to the Eu^{3+}, doped at 1 mol %, could be observed in the emission spectra of the sample. This energy was absorbed at 393 nm and used to move the electrons from the $^7F_0$ to the $^5L_6$ level, which then relaxed using nonradiative transitions to the $^5D_0$ level and finally emitted at 616 nm. The emission spectra were dopant concentration dependent. No quenching was found for the emission at the dopant concentrations used, but a decrease in the decay lifetime associated with the increment of dopant concentration was observable. It was attributed to a drop in the distance between the ions of Eu^{3+} and associated with the increment of cross relaxations and non-radiative transitions. The possibility of playing with the organic-inorganic structure (even at a nanoscale) on one side and with the rare earth concentration on another side, opens good prospects for the development of efficient luminescent devices in these hybrid materials.

Author Contributions: For the development of this investigation, the individual contributions are listed below. Conceptualization, P.M.T.-G. and R.A.; Data curation, P.M.T.-G., R.P.-M., J.D.l.C., J.E.E., R.A., and E.M.-B.; Formal analysis, P.M.T.-G., R.P.-M., J.D.l.C., J.E.E., R.A., and E.M.-B.; Investigation, P.M.T.-G., R.P.-M., R.A., and O.P.M.; Methodology, P.M.T.-G. and R.P.-M.; Resources, R.P.-M., R.A., and O.P.M.; Original draft writing, P.M.T.-G. and R.P.-M.; Review and editing of manuscript, P.M.T.-G. and R.P.-M.

Acknowledgments: The authors gratefully recognize the financial support from the Benemérita Universidad Autónoma de Puebla (BUAP) and to the Vicerrectoría de Investigación y Estudios de Posgrado (VIEP-BUAP). The authors appreciate the encouragement, the multiple consultancies, and the comments of Giancarlo C. Righini.

Conflicts of Interest: The authors declare that there is no conflict of interest regarding the publication of this paper.

References

1. Klein, L.C. *Sol Gel Optics: Processing And Application*; Springer Science + Business Media: Totowa, NJ, USA, 1994.
2. Pandey, A.; Sahare, P.D; Kanjilal, D. Thermoluminescence and photoluminescence characteristics of sol–gel prepared pure and europium doped silica glasses. *J. Phys. D Appl. Phys.* **2004**, *37*, 842–846. [CrossRef]
3. Thomas, K.J.; Sheeba, M.; Nampoori, V.P.N.; Vallabhan, C.P.G.; Radhakrishnan, P. Raman spectra of polymethyl methacrylate optical fibres excited by 532 nm diode pumped solid state laser. *J. Opt. A Pure Appl. Opt.* **2008**, *10*, 055303. [CrossRef]
4. Xiong, Z.; Peng, G.D.; Wu, B.; Chu, P.L. Highly tunable bragg gratings in single-mode polymer optical fibers. *IEEE Photonics Technol. Lett.* **1999**, *11*, 352–354. [CrossRef]
5. Kuriki, K.; Koike, Y. Plastic optical fiber lasers and amplifiers containing lanthanide complexes. *Chem. Rev.* **2002**, *102*, 2347–2356. [CrossRef] [PubMed]
6. Basu, B.B.J.; Vasantharajan, N. Temperature dependence of the luminescence lifetime of a europium complex inmobilized in different polymer matrices. *J. Lumin.* **2008**, *128*, 1701–1708. [CrossRef]
7. Song, X.; Wang, X.; Wang, H.; Shong, W.; Du, Q. PMMA–silica hybrid thin films with enhanced thermal properties prepared via a non-hydrolytic sol–gel process. *Mater. Chem. Phys.* **2008**, *109*, 143–147. [CrossRef]
8. Yeh, J.-M.; Weng, C.-J.; Liao, W.-J.; Mau, Y.-W. Anticorrosively enhanced PMMA–SiO_2 hybrid coatings prepared from the sol–gel approach with MSMA as the coupling agent. *Surf. Coat. Technol.* **2006**, *201*, 1788–1795. [CrossRef]
9. Yeh, J.-M.; Hsieh, C.-F.; Yeh, C.-W.; Wu, M.-J.; Yang, H.-C. Organic base-catalyzed sol–gel route to prepare PMMA–silica hybrid materials. *Polym. Int.* **2006**, *56*, 343–349. [CrossRef]
10. Smits, K.; Millers, D.; Zolotarjovs, A.; Drunka, R.; Vanks, M. Luminescence of Eu ion in alumina prepared by plasma electrolytic oxidation. *Appl. Surf. Sci.* **2015**, *337*, 166–171. [CrossRef]
11. Binnemans, K. Interpretation of europium(III) spectra. *Coord. Chem.* **2015**, *295*, 1–45. [CrossRef]
12. Medina, D.Y.; Orozco, S.; Hernandez, I.; Hernandez, R.T.; Falcony, C. Characterization of europium doped lanthanum oxide films prepared by spray pyrolysis. *J. Non-Cryst. Solids* **2011**, *357*, 3740–3743. [CrossRef]

13. Tauc, J.; Grigorovici, R.; Vancu, A. Optical properties and electronic structure of amorphous germanium. *Phys. Status Solidi B* **1966**, *15*, 627–637. [CrossRef]

14. Gálico, D.A.; Mazali, I.O.; Sigoli, F.A. Nanothermometer based on intensity variation and emission lifetime of europium (III) benzoylacetonate complex. *J. Lumin.* **2017**, *192*, 224–230. [CrossRef]

15. Ozawa, L.; Jaffe, P.M. The mechanism of the emission color shift with activator concentration in $^{+3}$ activated phosphors. *J. Electrochem. Soc.* **1971**, *118*, 1678–1679. [CrossRef]

16. Bortoluzzi, M.; Paolucci, G.; Gatto, M.; Roppa, S.; Enrichi, F.; Ciorba, S.; Richards, B.S. Preparation of photoluminescent PMMA doped with tris(pyrazol-1-yl)borate lanthanide complexes. *J. Lumin.* **2012**, *132*, 2378–2384. [CrossRef]

17. Blasse, G.; Bril, A. Fluorescence of Eu^{3+}-Activated Lanthanide Oxyhalides LnOX. *J. Chem. Phys.* **1967**, *46*, 2579–2582. [CrossRef]

18. Righini, G.C.; Ferrari, M. Photoluminescence of rare-earth-doped glasses. *Riv. Nuovo Cimento* **2005**, *28*, 1–53.

19. Artizzu, F.; Loche, D.; Mara, D.; Malfatti, L.; Serpe, A.; Van Deun, R.; Casula, M.F. Lighting up Eu^{3+} luminescence through remote sensitization in silica nanoarchitectures. *J. Mater. Chem. C* **2018**, *6*, 7479–7486. [CrossRef]

20. Zareba-Grodz, I.; Pazik, R.; Tylus, W.; Hermanowicz, K.; Strek, W.; Maruszewski, K. Europium-doped silica–titania thin films obtained by the sol–gel method. *Opt. Mater.* **2007**, *29*, 1103–1106. [CrossRef]

Review

Introduction to Photonics: Principles and the Most Recent Applications of Microstructures

Iraj Sadegh Amiri [1], Saaidal Razalli Bin Azzuhri [2], Muhammad Arif Jalil [3],
Haryana Mohd Hairi [4], Jalil Ali [5], Montree Bunruangses [6] and Preecha Yupapin [7,8,]*

[1] Division of Materials Science and Engineering, Boston University, Boston, MA 02215, USA; amiri@bu.edu
[2] Department of Computer System & Technology, Faculty of Computer Science & Information Technology,
    University of Malaya, 50603 Kuala Lumpur, Malaysia; saaidal@um.edu.my
[3] Department of Physics, Faculty of Science, Universiti Teknologi Malaysia, 81300 Johor Bahru, Malaysia;
    arifjalil@utm.my
[4] Faculty of Applied Sciences, Universiti Teknologi Mara, Pasir Gudang Campus, 81750 Johor, Malaysia;
    haryanahairi@gmail.com
[5] Laser Centre, IBNU SINA ISIR, Universiti Teknologi Malaysia, 81310 Johor Bahru, Malaysia;
    djxxx_1@yahoo.com
[6] Faculty of Industrial Education, Rajamangala University of Technology Phranakorn, Bangkok 10300,
    Thailand; montree.b@rmutp.ac.th
[7] Computational Optics Research Group, Advanced Institute of Materials Science, Ton Duc Thang University,
    District 7, Ho Chi Minh City, Vietnam
[8] Faculty of Electrical & Electronics Engineering, Ton Duc Thang University, District 7,
    Ho Chi Minh City, Vietnam
*   Correspondence: preecha.yupapin@tdtu.edu.vn

Received: 1 February 2018; Accepted: 4 April 2018; Published: 11 September 2018

Abstract: Light has found applications in data transmission, such as optical fibers and waveguides and in optoelectronics. It consists of a series of electromagnetic waves, with particle behavior. Photonics involves the proper use of light as a tool for the benefit of humans. It is derived from the root word "photon", which connotes the tiniest entity of light analogous to an electron in electricity. Photonics have a broad range of scientific and technological applications that are practically limitless and include medical diagnostics, organic synthesis, communications, as well as fusion energy. This will enhance the quality of life in many areas such as communications and information technology, advanced manufacturing, defense, health, medicine, and energy. The signal transmission methods used in wireless photonic systems are digital baseband and RoF (Radio-over-Fiber) optical communication. Microwave photonics is considered to be one of the emerging research fields. The mid infrared (mid-IR) spectroscopy offers a principal means for biological structure analysis as well as nonintrusive measurements. There is a lower loss in the propagations involving waveguides. Waveguides have simple structures and are cost-efficient in comparison with optical fibers. These are important components due to their compactness, low profile, and many advantages over conventional metallic waveguides. Among the waveguides, optofluidic waveguides have been found to provide a very powerful foundation for building optofluidic sensors. These can be used to fabricate the biosensors based on fluorescence. In an optical fiber, the evanescent field excitation is employed to sense the environmental refractive index changes. Optical fibers as waveguides can be used as sensors to measure strain, temperature, pressure, displacements, vibrations, and other quantities by modifying a fiber. For some application areas, however, fiber-optic sensors are increasingly recognized as a technology with very interesting possibilities. In this review, we present the most common and recent applications of the optical fiber-based sensors. These kinds of sensors can be fabricated by a modification of the waveguide structures to enhance the evanescent field; therefore, direct interactions of the measurand with electromagnetic waves can be performed. In this research, the most recent applications of photonics components are studied and discussed.

Micromachines **2018**, *9*, 452

Keywords: light; photon; communications; waveguides; fibers; biosensors

---

1. Introduction

The role of light is significant in our lives today. The importance of light cannot be taken for granted because it is vital to most aspects of our contemporary society. It is used everywhere whether it be building, telecommunication, transportation, entertainment, or clothing. Light has applications in data transmission, such as optical fibers and in optoelectronics. It is used in compact disc players where a laser reflecting off of a CD transforms the returning signal into music. It is also used in laser printing and digital photography. Connections between computers and telephone lines are possible with the help of light (fiber-optic cables). It is used in optical fiber lasers, optical fiber interferometers, optical fiber modulators, and sensors. Light is used in the medical field for image production used in hospitals and in lasers that are used for optometric surgery [1]. Light consists of a series of electromagnetic waves, with particle behavior under certain circumstances. Light is the range of wavelengths in the electromagnetic spectrum (Figure 1).

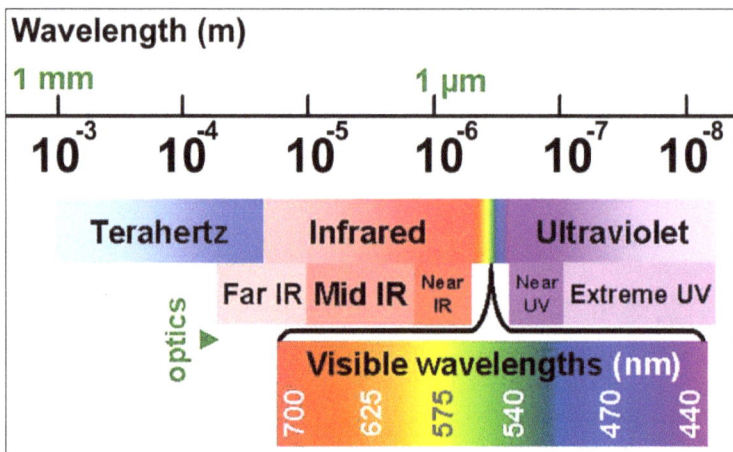

Figure 1. The electromagnetic spectrum.

Photonics is essentially the science that involves generation of a photon (light), its detection, as well as manipulation via transmission, emission, signal processing, modulation, switching, amplification, and sensing. Most importantly, photonics involves the proper use of light as a tool for the benefit of humans [2,3]. Most photonics applications, even though they cover all technical applications over the entire electromagnetic spectrum, range from near-infrared light to visible region. The term "photonics" was derived from the root word "photon", which connotes the tiniest entity of light analogous to an electron in electricity. Just as the electronics revolutionized the 20th century, photonics is doing the same in the 21st century. Photonics is made up of many different technologies including optical fibers, lasers, detectors, quantum electronics, fibers, and materials [4].

The term photonics was first used to designate a field of research area responsible for utilizing light to perform tasks that are conventionally related to the traditional sphere of electronics, like telecommunications, information processing, and so on. Studies in the field of potonics began in 1960 after the discovery of lasers. Other progress followed including optical fibers for transmitting information, the laser diode in the 1970s, as well as erbium fiber amplifiers. These developments made the foundation for the industrial revolution in the telecommunications sector during the late 20th century and supplied the internet infrastructure. Although created before the 1980s, the word photonics was used commonly for the first time in the 1980s as network operators of telecommunications embraced fiber-optic data transmission. Photonics came into being when the "IEEE Lasers and Electro-Optics Society" came up with a journal called "Photonics Technology Letters" towards the end of the 18th century. Through the years, until 2001 with the dot-com crash, research was primarily focused on optical fiber telecommunication. Nevertheless, the field of photonics has a broad range of scientific and technological applications. These include chemical and biological sensors, laser manufacturing, medical therapy and medical diagnosis, optical computing and displaying technology. Advancement of photonics is possible due to the current success recorded concerning the development of silicon photonics. Photonics is related to opto-mechanics, electro-optics, quantum electronics and quantum optics. Nevertheless, these fields mean different things to both the scientific as well as the business community. Quantum optics is often concerned with fundamental theoretical research areas. Photonics, on the other hand, deals with applied research and progress. Optoelectronics is used to refer to the circuits or devices consisting of both electrical and optical components. The word "electro-optics" was utilized in the past to specifically relate to nonlinear interactions between electrical and optical devices. These devices include bulk crystal modulators and later include advanced imaging sensors that are typically employed by both government and private individuals in surveillance activities [5,6].

Photonics is said to be an "All-Pervasive" technology because it allows unlimited light to travel faster than the electrons that are used in electronic computer chips, which means that optical computers will compute thousands of times faster than any electronic computers because of the physical limitations of electronic conduction. More wavelengths can be packed into an optical fiber to allow an increase in the transmission bandwidth that can be in conventional copper wires. There is no electromagnetic interference in light compared to electrons in copper wires [7,8].

2. Applications of Photonics

Photonics have uses in almost every aspect of our life, ranging from daily life to highly innovative science. For instance, information processing, telecommunications, light detection, metrology, lighting, spectroscopy, photonic computing, holography, medical field (surgery, vision correction, health monitoring and endoscopy), fighting machinery, visual art, agriculture, laser material processing, robotics, and biophotonics. Similar to the way electronics have been used extensively since the creation of earlier transistors of 1948, the exceptional use of photonics continuously increases. Economically significant uses of photonic devices include fiber optic telecommunications, optical data storage, displays, optical pumping of high-power lasers and laser printing. Prospective applications of photonics are practically limitless and include medical diagnostics, organic synthesis, information, and communication, as well as fusion energy [9,10]:

- Telecommunication: optical down-converter to microwave, and optical fiber communications.
- Medical applications: laser surgery, poor eyesight correction, tattoo removal and surgical endoscopy.
- Manufacturing processes in industries: involves the use of laser in welding, cutting, drilling, and many surface modification techniques.
- Building and construction: smart structures, laser range finding, and laser leveling.
- Space exploration and aviation: including astronomical telescopes.
- Military operations: command and control, IR sensors, navigation, mine laying, hunt and salvage, and discovery.
- Metrology: range finding, frequency and time measurements.
- Photonic computing: printed circuit boards, and quantum computing.
- Micro-photonics and nanophotonics.

These typically include solid-state devices and photonic crystals [11]. In simple terms, photonics is currently solving and addressing the challenges of a modern world. Photonics enhances the quality of life; it safeguards our health, security, and safety, it drives our economic growth, and it creates jobs as well as global effectiveness. Photonics technology enhances the quality of life in many areas. Specific areas are communications and information technology, advanced manufacturing, defense, health and medicine, and energy [12,13]. Photodetectors are used to detect light. They can be very slow, as in the case of solar cells that are used in harvesting sunlight energy, or very fast like photodiodes that are very fast and are employed in communications in conjunction with digital cameras. Numerous others centered on quantum, thermal, photoelectric and chemical areas also exist. Photonics likewise involves research on photonic systems. The term photonics system has found its application in optical communication systems [14].

3. Advances in Photonics

There has been an exponential growth in the research activities in the field of photonics and optics over the years, as illustrated by the publication and citation trends from the Thomson Reuters web of science database (Figure 2).

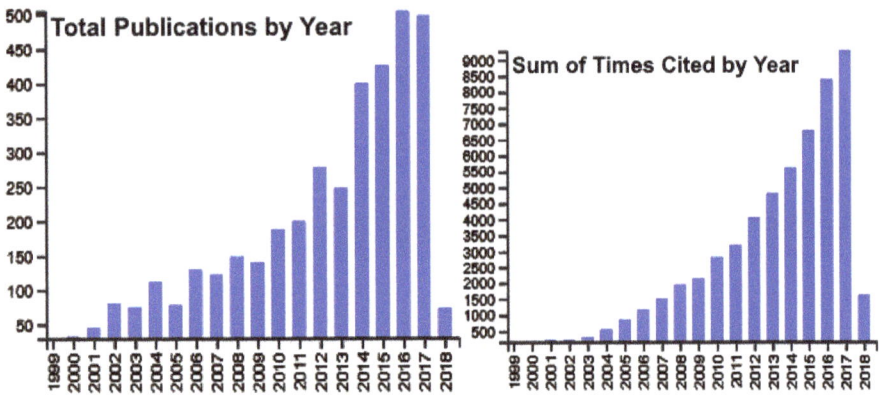

Figure 2. Publications and citation trends in Photonics (Source: Thomson Reuters Web of Science).

Photonic networks are the backbone of data dissemination, specifically in the modern and upcoming wireless communication systems. Photonic networks continue to gain interest for distribution of data from, say, central location to a remote antenna unit at base stations. While the demand for wireless photonic systems continues to rise, there is a need for implementation of low-cost systems [15]. Two of the most popular data transmission methods in wireless photonic systems are digital baseband and RoF (Radio-over-Fiber) optical communication. In addition, further emerging fields are opto-atomics, in which there is an integration of both atomic and photonic devices. Opto-atomics applications include precise time-keeping. Opto-mechanics, metrology, and navigation, as well as polaritonics, are different from photonics due to the presence of polarization as the primary carrier of information. Microwave photonics is considered to be an emerging research field. Microwave photonics is an enabling technology for the generation, control, distribution, measurement, and detection of microwave signals. It also deals with the operation of new systems and devices [16–19]. Part of the various functionalities facilitated by photonics, microwave measurements centered on photonics can offer greater performance regarding broad frequency coverage, significant direct bandwidth, high immunity to electromagnetic interference (EMI) and low frequency-dependent loss. Photonic microwave measurements therefore have been widely investigated in recent times. Moreover, several new methodologies have been offered to address the challenges confronting electronic solutions [20]. Plasmon lasers are among the categories of optical frequency amplifiers that send strong, penetrating, and guiding superficial plasmons underneath the diffraction walls. The interactions between light energy and matter can be intensely improved by the tightly held electric fields in plasmon lasers. This will also bring substantial innovative possibilities to data storage, bio-sensing, optical communications and photolithography [21]. Because they can generate high-intensity nano-scale electromagnetic radiation in a fraction of a second, the modern development of plasmon lasers today has sparked the investigation of nanoscience and technology. This would enable more feature sizes than the conventional lasers [22,23]. They could also be used to package additional information onto storage media such as hard disks or DVDs [24,25]. The mid-IR spectroscopy offers a principal means for biological structure analysis as well as nonintrusive measurements. For instance, the broad cross-section for absorption allows for the detection of traces of vapors at the order of parts-per-trillion (ppt) as well as parts per- billion (ppb).

4. Structure, Types, and Applications of Optical Fibers

Optical fibers are flexible filaments made of very clear glass and can carry information in the form of light from one point to another. They are hair-thin structures formed through the formation of preforms, which are glass rods made into fine threads of glass and secured by plastic coatings. Various vapor deposition processes are employed by fiber manufacturers to draw the preform. The thread drawn from this preform is then usually wrapped into a cable configuration, which is then placed into an operative situation for years of dependable performance [26].

The two most important components of optical fibers are the core and the cladding. The "core", which is the axial part of the optical fiber, is made up of silica glass. The optical fiber core is that area of the fiber where light is transmitted. Sometimes, doping elements are used to modify the fiber refractive index, thereby changing the light velocity through the fiber. The "cladding", on the other hand, is the layer that surrounds the core completely. The cladding refractive index is less than that of the core. This enables the light inside the core to strike the core-cladding interface at a "bouncing angle", is confined inside the core by the total internal reflection, and keeps moving in the appropriate direction along the fiber length to a certain point. The cladding is usually surrounded by another layer known as "coating," which normally is comprised of protective polymer films coated during the process of fiber drawing, before being in contact with any surface. Additional protective layers of "Buffers" are further applied on top of the polymer coatings as shown in Figure 3 [27].

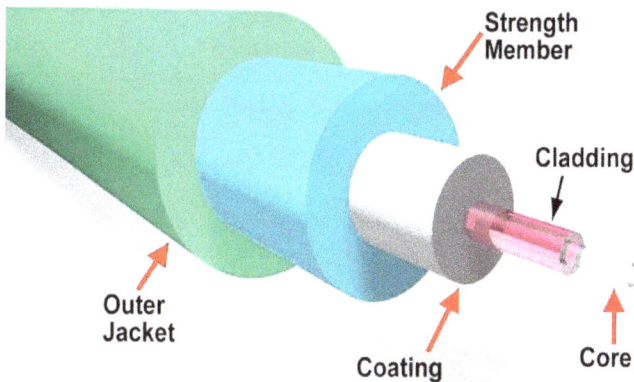

Figure 3. Structure of optical fiber.

The mechanism of the modifications on the fiber surface can be characterized through the transmission spectrum measurement of the fibers. There are so many different possible configurations of fibers corresponding to different application purposes. The most important classification considers fibers as either single-mode fibers and multimode fibers. The concept of application-specific fibers was invented at Bell Laboratories in the mid-1990s, and this is followed by an introduction of fibers designed for network applications. These next designs that are used mainly for signal transmission in communications consist of 10-Gbps laser-optimized multimode fibers (OM3), Zero Water Peak Fiber (ZWPF), Non-Zero Dispersion Fibers (NZDF), and fibers that are specially designed for the marine application. Specially designed fibers, like erbium-doped fibers, and dispersion compensating fibers perform tasks that supplement the transmission fibers. The differences between the different transmission fibers are responsible for variations in the number and range of different wavelengths or pathways via which the light is received or transmitted; this is the distance at which a signal can travel without being amplified or regenerated, and the speed at which this signal can travel.

The silica fibers are the common type of fibers that can transmit light with wavelengths below the mid-infrared range [28]. The silica as an optical waveguide is a strongly absorbing material for wavelengths above 2 µm [29]. This is due to multiphoton absorption that causes vibrational resonance; however, there are different glass materials that can be used to fabricate the optical fibers in which these materials can transmit light at a longer wavelength [30]. The crystalline materials and hollow fiber waveguides are good candidates to perform these kinds of transmissions [31]. For instance, glasses such as the chalcogenides, which may have different compositions of sulfides, selenides or tellurides, have substantially lower vibration frequencies and therefore lower photon energy compared to silica [32]. This is due to the higher mass of chalcogenide ions compared to oxygen ions. Examples of these materials can be such as arsenic (As) or germanium (Ge), where the infrared absorption of the materials starts at longer wavelengths. Hollow waveguide fibers, however, can be used for single-mode transmission, although fibers transmitting light of wavelengths larger than 2 µm can be manufactured using either glass or crystalline materials [33]. The mid-infrared optical fibers have disadvantages of high fabrication cost, less mechanical robustness, and higher propagation loss in the optical communication wavelength range at 1.5 µm compared to silica fibers [34]. These are available as bare fibers and fiber patch cable and are presenting additional protection and fiber connectors at the end of their length. These fibers are mostly multimode waveguides and can be used for particular applications; however, there are many challenges in the fabrication of mid-infrared optical fibers in single-mode construction [35]. Recently, scientists are facing many technical challenges with fabricating the kind of fibers with air holes. For instance, omniguide fibers [36], hollow IR transmitting fibers [37] and holely fibers [38] can provide additional functionalities that are not available in other conventional fibers such as solid core fibers. These have unusual guiding structures and can support new light propagation features applicable to novel photonic devices such as lasers and transmitters. Infrared fibers such as Chalcogenide (CIR) [39] and Polycrystalline (PIR) [40] can be made of two different core materials. In CIR fibers, a high transmittance can be achieved in the wavelength range between 2 to 6 µm, where these fibers exhibit very low optical loss and high flexibility. The PIR fibers show a high transmittance in the range between 4 to 18 µm. In these two types of fibers, the light leakage is eliminated by implementing a special design of the core and cladding, which allows for a high damage tolerance to withstand damages from other even more intense sources such as continuous-wave CO_2 lasers. Infrared fibers have many applications in imaging devices, thermal imaging, evanescent wave sensors and chemical species analyzers [41,42].

Some critical parameters affect the performance of optical fibers transmission systems. These parameters and their specifications vary by fiber type and depend upon the intended use. Two of the more significant parameters of fibers are fiber dispersion and attenuation. Attenuation is the decrease in optical power when it propagates from one place to another. High attenuations affect the distance at which signals can be transmitted. Figure 4 shows the variation in attenuation with wavelengths for a wide range of fiber optic cables.

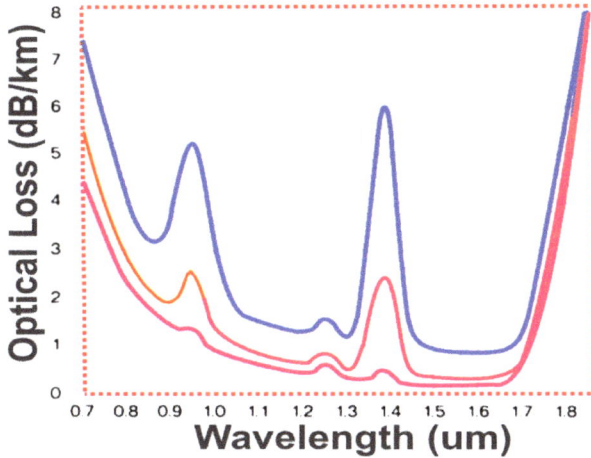

Figure 4. Attenuation against wavelength transmission windows.

Dispersion, on the other hand, is inversely related to the bandwidth and refers to the fiber to carry information. Single-mode fibers are associated with a chromatic dispersion that causes pulse spreading due to the various colors of light passing through the fiber at different speeds. Similarly, multimode fibers are related to the modal dispersion that causes pulse spreading due to the geometry of a multimode fiber core, which allows for the multiple modes lasers to simultaneously separate and propagate at the fiber interface.

Multimode fibers are the first fibers to be produced on a commercial scale. They are called multimode fibers just because they allow several modes or rays of light to propagate through the waveguide simultaneously. These types of fibers have a much wider core diameter, when compared to the single-mode fibers, and allow for the higher number of modes. Multimode fibers are easier to couple than single-mode fibers. Multimode fibers can be classified into graded-index and step-index fibers. Graded-index multimode fibers make use of the differences in compositions of the glass inside the fiber core and recompense the different path lengths of the modes. They offer more bandwidth than step index fibers. Step-index multimode fibers were the first cords designed but are too slow regarding most applications because of the dispersion caused by the different path lengths of the various modes. Step-index fibers are barely used in modern telecommunications. Multimode fibers that are employed in communications possess the core size of 50 or 62.5 microns. The big core sizes allow the fibers to support many diagonal electromagnetic modes for a given polarization and frequency.

Single-mode fibers enjoy lower fiber attenuation than multimode fibers and retain better reliability of each light pulse because they have no dispersion associated with multiple mode fibers. Hence, data can be transferred over a longer distance. Similar to multimode fibers, the earlier single-mode fibers were commonly characterized as step-index fibers (shown in Figure 5), which means the refractive index of the fiber cladding is a step below that of the core rather than graduated as in the case of graded-index fibers. Current single-mode fibers have grown into a more sophisticated design like depressed clad, matched clad, or other mysterious structures.

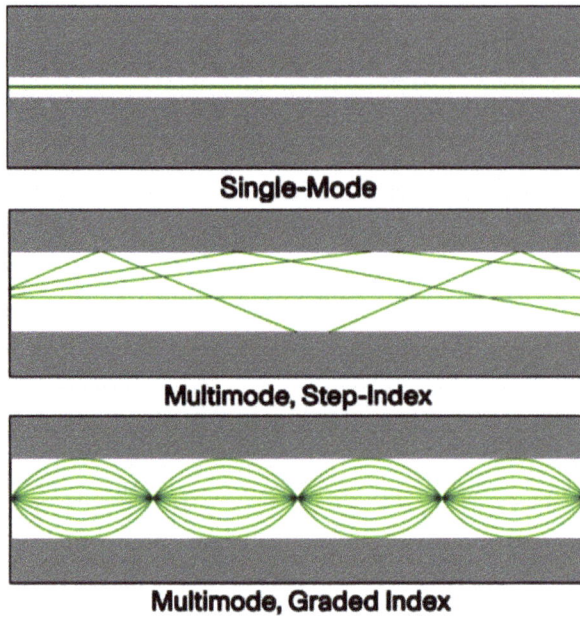

Figure 5. Multimode and single-mode fibers.

The core size of single-mode fibers usually is nine microns. Because only one mode can propagate down the fiber length, the total internal reflection process does not occur; hence, the concept of numerical aperture becomes similar to those of multimode fibers. The numerical aperture of multimode fibers is usually larger than those of single-mode fibers. The most common lasers appropriate for applications over single-mode fiber include distributed feedback (DFB) and Fabry–Perot lasers. The attenuation of single-mode fibers is about 0.2 dB per km [43]. Optical fibers operate based on the principle of total internal reflection. Imagine rays of light striking a distinct boundary separating an optically less dense medium. A less dense medium is the one with a lower reflection index. At an appropriate incidence angle, these rays rather than passing through will be reflected fully. This phenomenon is referred to as the total internal reflection [44,45]. Prisms in binoculars and camera viewfinders make use of total internal reflection. If the incidence angle is represented by the symbol (α), and the angle of refraction as β (see Figure 6 at this boundary to the less-dense medium, ($n_G > n_A$) assuming air and glass are being considered), the condition $\alpha < \beta$ holds. However, the angle β cannot be greater than 90°. This is evident considering the Snell's law of refraction (Equation (1)):

$$\frac{\sin \alpha}{\sin \beta} = \frac{n_A}{n_G} \tag{1}$$

Bearing in mind that ($\sin \beta$) cannot be greater than one,

$$\sin \alpha_{critical} = \frac{n_A}{n_G} < 1 \tag{2}$$

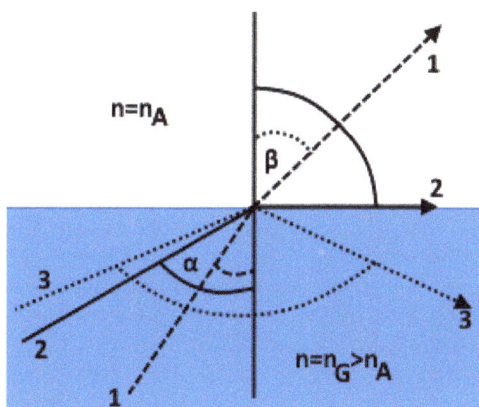

Figure 6. Total internal reflection phenomena

For even greater angles of incidence, the rays of light are entirely reflected back into the denser medium almost without encountering any loss. It is the same principle that guides light around bends as well as inside optical fibers [46].

Optical fibers have applications for assisting us in various aspects of our lives—for example, in amplifiers [47], in telecommunications [48,49], in medicine [50], in aerospace and aviation technology [51,52], in engineering [53,54], nanotechnology [55,56], and in sensing applications. Optical fiber sensors have been studied for over 40 years. Several concepts have been suggested, and many methods have been established for various parameters as well as for various uses. Commercialization of optical fiber sensors has been carried out successfully. However, out of the many methods investigated, only a small amount of applications and methodologies have been commercialized successfully [57]. The optical fiber-based sensors possess many advantages over copper cables for their high sensitivity, small size, large bandwidth, lightweight quality, as well as immunity towards electromagnetic interferences [58–60]. Pressure, temperature, and strain are the extensively investigated parameters, and, for the optical fiber sensors, the Bragg fiber grating sensors are the most widely studied technologies. However, in various applications, optical fiber-based sensors are expected to compete with other existing technologies like electronic-based systems. To get attention, since customers are already familiar with the current technologies, there is a need to demonstrate the superior qualities of optical fiber-based sensors over other contemporary methods. Usually, customers are not interested in the procedures involved in the detection. However, these clients only desire sensors with excellent performance at reasonable costs. Therefore, optical fiber-based sensors should be obtainable in the form of a system that includes signal detection and signal processing.

5. Classification of Optical Fiber Sensors

There have been some approaches to the classification of optical fiber sensors. The increasing complexity of several types of optical fiber sensors is what prompted the development of adequate and appropriate classification systems. Factors such as physical quantity transduced by the sensors, detection systems, as well as sensor type have been considered in so many classifications. To develop the most suitable classification scheme for optical fiber sensors, an emphasis is given to the most important aspects and, hence, a classification method is adopted. Previous work that attempted to offer classification methods that cover the majority of the essential optical fiber sensors is cited in [61]. With the continuous increase in the development of optical fiber sensors, so many classification systems that were adopted previously became unsuitable. Other classification systems were given based on the modulation type chosen [62,63]. Hence, factors like wavelength, intensity, phase, and polarization were regarded as the primary classification standards. The disadvantage of this type of classification, however, is that the technique used is given emphasis rather than the sensor itself. This may be insignificant in applications where the most suitable technology is targeted for measuring a parameter of interest like pressure or temperature. This second method that considered variables like temperature, pressure, magnetic field, electric field vibration and flows in classifying sensors has also been adopted [64]. However, this approach is also associated with some disadvantages when applied in a similar way to the other methods of measuring various parameters like displacement. Other factors such as novelty and geographical location were also considered in the classification of sensors [65]. In the most extensively used system of classification, optical fiber sensors are classified as intrinsic or extrinsic sensors [66,67]. Extrinsic sensors are those in which the fiber guides the light wave, and the interaction between the magnitude of the parameter measured and light occurs outside the fiber. These types of sensors have been used successfully for some applications. For intrinsic sensors, on the other hand, interactions between light and the measured parameter occur inside the fiber. Figure 7 shows a comparison between the intrinsic and the extrinsic optical fiber sensors [61].

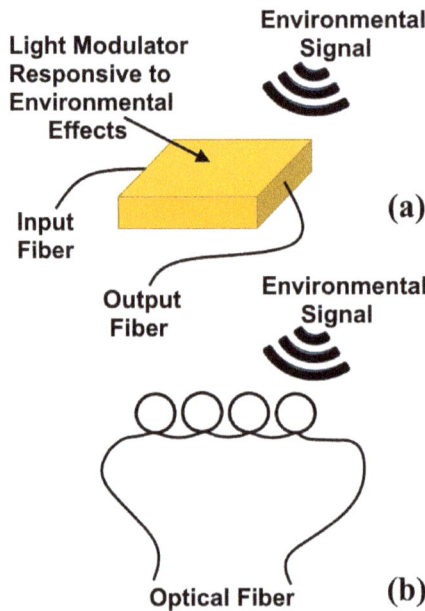

Figure 7. Schematic comparison between (**a**) extrinsic and (**b**) intrinsic sensors

An important parameter to be considered in intrinsic sensors is the nature of the optical guidance of the fiber—that is, whether it is multimode, single, or otherwise. Another important sub-class of the intrinsic sensors is interferometric sensors [61].

5.1. Intrinsic Optical Fiber Sensors

Optical fibers can be applied as sensors in measuring temperature, strain, pressure or other parameters through fiber modification in such a way that the parameter of interest controls the polarization, intensity, wavelength, phase, and the time in light passes through the cord. The simplest sensors are those that vary the light intensity because they require only a simple detector and source of light. Intrinsic sensors can offer distributed detection for comprehensive coverage. This broad sensing ability associated with intrinsic sensors is very useful [68]. An optical fiber that has a temporary loss, which depends on temperature, can be used to measure temperature. This measurement can be possible by analyzing the Raman scattering of the optical fiber. Nonlinear optical effects that can change the light polarization, which depends on electric field or voltage, can be used in sensing electrical voltage. Other types of fibers are specially designed for special applications such as direction recognition [69–71]. Other optical fibers have applications in sonar and seismic detection. Examples of these types of fibers are hydrophones. Oil industries, as well as the navy in some countries, make use of the hydrophones systems. Microphone systems that involve the use of optical fibers have been developed by Sennheiser (Germany). In applications where high electric or magnetic fields are required, optical fiber based headphones and microphones are very useful. These applications include team communication among medics working on a patient in an MRI (Magnetic resonance imaging) system during surgeries that are MRI-guided [72]. In oil industries, optical fibers are used to measure temperature and pressure in oil wells [73,74]. These types of applications very much require optical fiber sensors since they can withstand very high temperatures compared to the semiconductor sensors. Optical fiber sensors can be used for interferometric sensings such as fiber optic gyroscopes, which are utilized for navigation in some cars and the Boeing 767 aircraft (USA). Optical fibers are used in making hydrogen sensors. Some optical fiber sensors have been designed for simultaneous measurement of collocated temperature and strain with high precision using Fiber Bragg gratings [75]. This approach is predominantly beneficial when obtaining data related to complex or small configurations [76]. Sensors based on Fiber Bragg grating are also very suitable for remote sensing. Detection of temperature and strain over considerable distances of up to 120 kilometers is also possible using "Brillouin scattering effects" [77].

Fiber-optic sensors have also found applications in electrical changeover gear for transmission of light between an electrical arc-flash to a digitally protecting relay in order to allow fast falling off a breaker to decrease the arc blast energy [78]. Fiber optic sensors that are based on Fiber Bragg grating improve performance, productivity, and protection in some manufacturing processes. Integration of Fiber Bragg grating technology enables sensors to offer full investigation and complete information on insights with precise resolution. These types of sensors are normally used in various industries such as aerospace, automotive, telecommunication, and energy. Fiber Bragg gratings are sensitive to mechanical tension, static pressure, and compression and changes in fiber temperatures. Central wavelength adjustment of light emitting source provides the effectiveness of Fiber Bragg grating optical fiber sensors [79,80]. The structure of the side-polished fibers (SPFs) has a cladding section that is partially removed on one side; therefore, by modification of the cladding, the evanescent field of the propagating light within the core can interact with surrounding materials that present different refractive indices.

Researchers have investigated many applications of the SPFs, especially in nonlinear optics photonics technologies [81,82]. We can illustrate the setup of the fiber modification as shown in Figure 3. In this case, a single mode fiber type that is known as SMF-28 is used for the fabrication, where it should be tightly suspended above the polishing wheel when the polishing process starts. The polishing section is only a few centimeters. Therefore, the SMF-28, which is striped, is suspended over the polishing wheel as illustrated in Figure 8. The SMF-28 should be adjusted in such a way that the center of the stripping section should be placed at $L_0/2$.

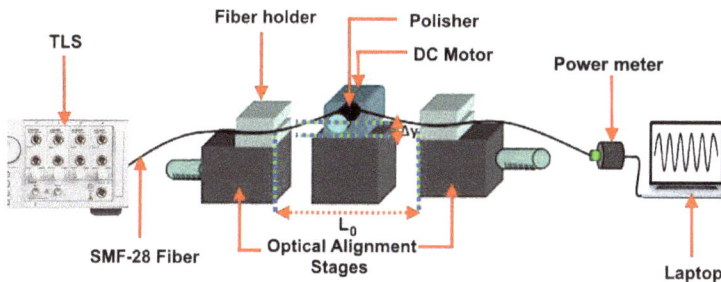

Figure 8. Polisher design setup.

The double-sided scotch tape is wrapped around a shaft of the DC motor. Therefore, the silicon carbide paper sticks to the double-sided scotch tape to create the uniform polishing wheel in such a way that it is perpendicular to the suspended SMF-28 (Figure 9). The position of the polisher should be adjusted to create the contact between the fiber and the wheel. Figure 9 shows the experimental fabrication of the SPF.

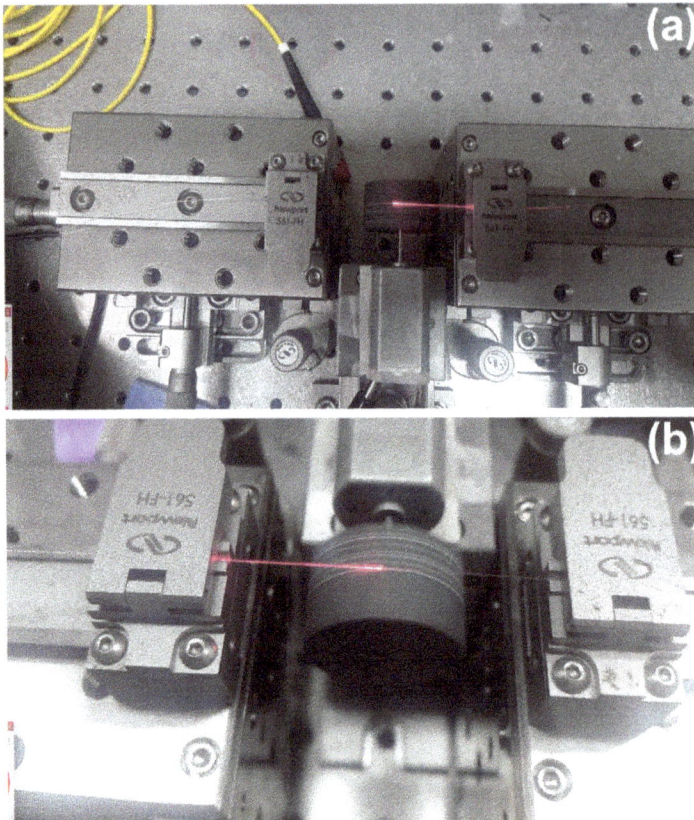

Figure 9. Polisher design setup; (**a**) the stage used to hold the fiber; and (**b**) the polishing process, where the light is figuring out from the fiber due to a removal of the cladding.

5.2. Extrinsic Optical Fiber Sensors

This type of fiber optic sensor makes use of optical fiber cables, usually the multimode type, to pass controlled light from either an electronic sensor linked to an optical transmitter or a non-fiber optical sensor. The advantage of extrinsic sensors is that they extend to places that cannot be otherwise accessible—for example, measuring the inside of aircraft engines using fibers to pass radiation to a radiation pyrometer that is situated on the exterior part of the machines. Similarly, extrinsic fiber optic sensors can be utilized in measuring the internal temperature of electrical transformers, in which the presence of a high electromagnetic field makes it impossible to measure using other measurement techniques. Extrinsic fiber optic sensors offer an outstanding shield of the frequency signal from being corrupted by noise. Regrettably, several traditional sensors release electrical outputs that must be changed to optical signals for fiber use. Extrinsic sensors found application in measuring temperature, rotation, acceleration, vibration, velocity, as well as displacement [83].

6. Fiber Bragg Grating and Applications

Even though the development of fiber gratings has been reported on since 1978 [84], it was only in 1989 that serious research on fiber gratings begun. This serious research activity followed the discovery of the regulated and operational methodology for their fabrication [85]. Fiber gratings have been widely used in amplifier gain flattening filters, fiber laser, and dispersion compensators for optical communication purposes. Rigorous studies have also been carried out on fiber grating sensors, and thus many have been commercialized. Different types of fiber gratings are shown in Figure 10.

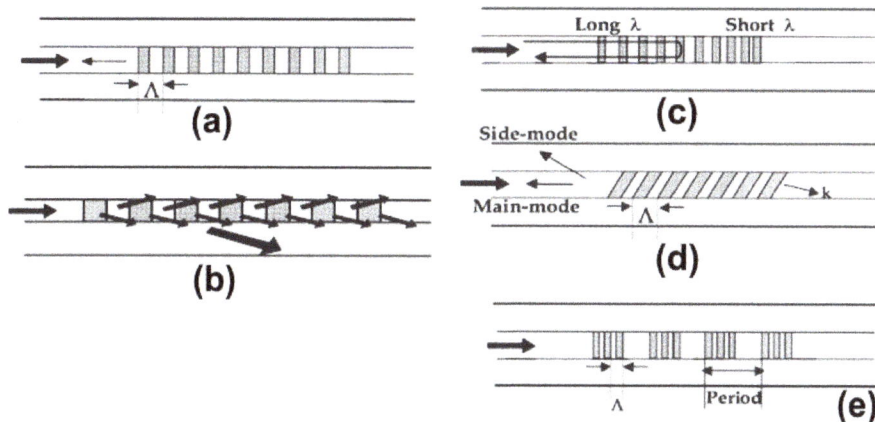

Figure 10. Types of fiber gratings. (**a**) Fiber Bragg grating; (**b**) long-period fiber grating; (**c**) chirped fiber grating; (**d**) tilted fiber grating; (**e**) sampled fiber grating

A Fiber Bragg Grating (FBG) is a periodic perturbation of the refractive index alongside some meters on the fiber length. In the design of FBG, the core is exposed to ultraviolet light [86]. The perturbation index inside the single-mode fiber core serves as a filter that reflects incident optical fields. The reflection of the incident optical field is maximally achieved when the perturbation index and the wavelengths of the incident fields match by [57]:

$$\lambda_B = 2n_{eff} A \tag{3}$$

where λ is the grating period and n_{eff} is the effective index of refraction of the fiber (see Figure 11). If the grating period changes, or if the effective index of refraction changes, due to changes in temperature or applied strain, the grating period and the effective refractive index will also change, thereby shifting the wavelength of the mean reflectance. These characteristics can be utilized for the purpose of sensors [86].

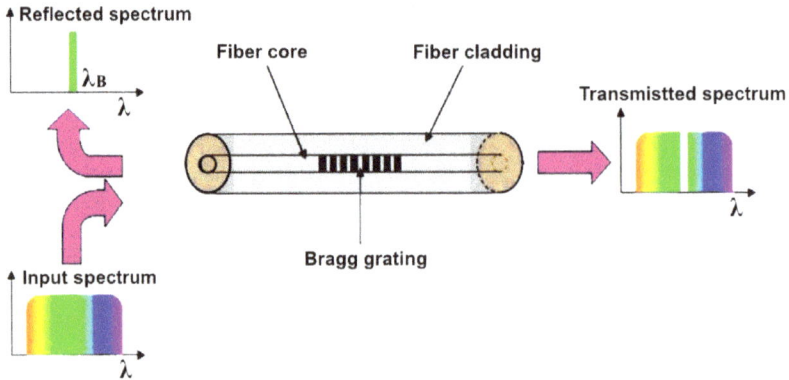

Figure 11. Schematic representation of the principle of Fiber Bragg grating.

There has been an increasing demand for sensors in almost all spheres of modern technology. The use of sensors that are based on Fiber Bragg grating technology has the potential to provide a lasting solution [87]. The strengths of distributed sensors can be further harnessed either through changing the sensitivity parameters of the FBG sensors or through coupling of both pressure and temperature sensors on one fiber. These types of sensing principles that are multi-parameter-based have been illustrated in [88,89]. Moreover, a remarkable range of operational temperature between 37 and 573 K has been demonstrated in [90]. The oil and gas industries processes in severe situations or space are potential marketplaces that nowadays assent to and value the sensors that are based on Fiber Bragg grating technologies. Individual markets try to use these technologies in diverse ways and most are very successful. An example of current advances in the oil and gas industries is Fiber Bragg gratings-based flowmeters that can be utilized in the downhole and harsh surface conditions, where temperatures can be above 573 K and have pressure of 99 atmospheres [89,90]. A substantial additional market where Fiber Bragg grating technologies have been widely recognized over the years is in structural health monitoring. In building constructions, bridges and many other types of large structures, Fiber Bragg grating sensors are employed to monitor continuously and verify the structural quality of these structures [91]. Optical fiber sensors like Fiber Bragg grating sensors are appropriate for composite material process monitoring because of their low invasiveness. The advantage of Fiber Bragg grating over other sensors is that they allow access to some physical parameters in the material. Hence, they can be used to examine the thick laminates and provide access to the manifestation of exothermic phenomena or residual strains [92].

7. Waveguides and Applications

A waveguide consists of a hollow, metal tube that is a unique form of transmission line. The technology of applying hollow pipes to streamline the movement of electromagnetic waves first appeared in 1897. The 1930s, following the development of the first microwave-producing equipment, necessitated the creation of a hollow waveguide for them. The success of these hollow waveguides motivated scientists to invent waveguides in the infrared region of the electromagnetic spectrum. These waveguides were initially used for medical purposes, but other areas of applications followed (Figure 12). [93].

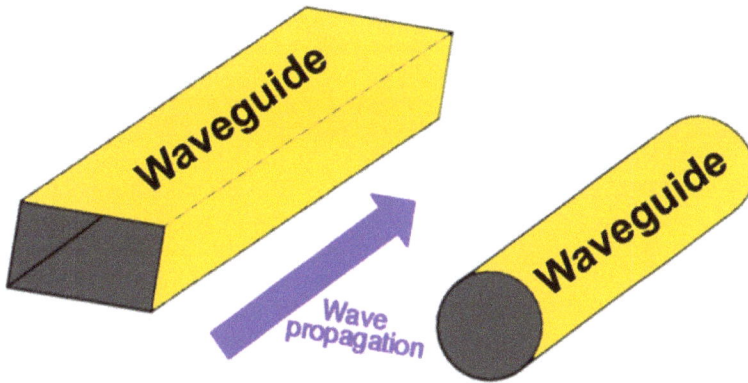

Figure 12. Rectangular and circular waveguides.

It directs the waves in a similar way river banks head a tidal wave [93]. Nevertheless, since waveguides are regarded as single-conductor materials, there is a difference in the way electrical energy is propagated down a waveguide as compared to the way in which it is propagated through a two-conductor transmission system. Figure 13 shows the propagation of the TEM mode in the waveguide.

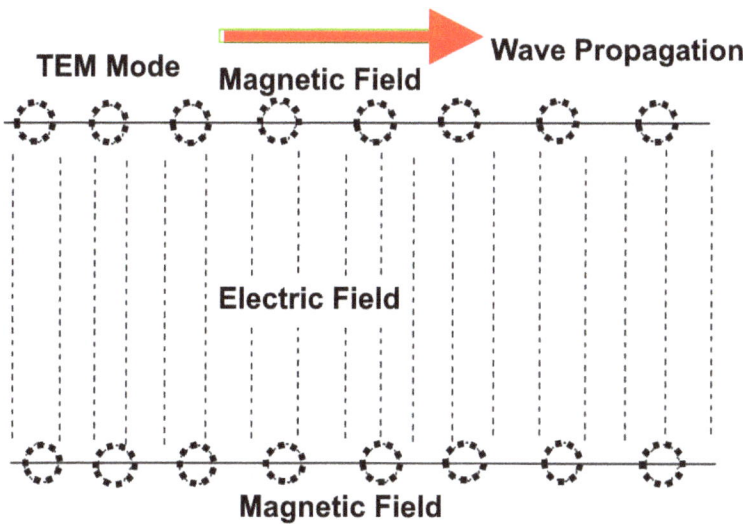

Figure 13. Transverse electromagnetic (TEM) mode propagation of a waveguide.

From Table 1, it could be concluded that a major improvement in waveguides is the decrease in propagation losses. The waveguide dimensions become impossibly large for lower frequencies, while, when the rate is higher, the dimensions become impracticably smaller. Waveguides present numerous advantages when compared to their optical fiber counterparts. Waveguides can transmit wavelengths above 20 μm. Their air core enables them to deliver high power lasers. Waveguides have relatively simple structures and they are cheap compared to the existing optical fibers. These qualities allow their use in applications where transmission of electromagnetic radiation requires a material of high mechanical, optical and thermal properties. Even though waveguides are associated with some setbacks such as losses upon bending as well as small numerical aperture, they appear to be the best option for use in sensors. Waveguides can be classified into two classes based on the principles on which they operate. These are attenuated total internal reflection and leaky-type waveguides. In attenuated total internal reflection-based waveguides, hollow core substances are present, which are enclosed by a wall that has a refractive index lower than that of the wavelength of the transmitted light. In the design of metallic waveguides, the inner wall is formed using a smooth metallic surface by depositing a metal film on the inner surface of plastic or a glass tube. Dielectric waveguides, on the other hand, are achieved by the formation of alternating high-low refractive index structure formed by the addition of multiple dielectric layers onto the metal surface. There has been a dramatic increase in the use of substrate integrated waveguides over the last decade due to their compact, low profile, and many other advantages over the conventional metallic waveguides. They resemble the conventional waveguides in their performances and can be made with printed circuit boards [94].

Table 1. Rectangular waveguide sizes (source: https://www.everythingrf.com/tech-resources/ waveguides-sizes).

Waveguide Name			Recommended Frequency (GHz)	Cutoff Frequency Lowest Order Mode (GHz)	Cutoff Frequency Next Mode (GHz)	Inner Dimensions of Waveguide Opening	
EIA	RCSC	IEC				A Inch	B Inch
-	WG9	-	2.20 to 3.30	1.686	3.372	3.5	1.75
WR340	WG9A	R26	2.20 to 3.30	1.736	3.471	3.4	1.7
WR284	WG10	R32	2.60 to 3.95	2.078	4.156	2.84	1.34
-	WG11	-	3.30 to 4.90	2.488	4.976	2.372	1.122
WR229	WG11A	R40	3.30 to 4.90	2.577	5.154	2.29	1.145
WR187	WG12	R48	3.95 to 5.85	3.153	6.305	1.872	0.872
WR159	WG13	R58	4.90 to 7.05	3.712	7.423	1.59	0.795
WR137	WG14	R70	5.85 to 8.20	4.301	8.603	1.372	0.622
WR112	WG15	R84	7.05 to 10	5.26	10.52	1.122	0.497
WR102	-	-	7.00 to 11	5.786	11.571	1.02	0.51
WR90	WG16	R100	8.20 to 12.40	6.557	13.114	0.9	0.4
WR75	WG17	R120	10.00 to 15	7.869	15.737	0.75	0.375
WR62	WG18	R140	12.40 to 18	9.488	18.976	0.622	0.311
WR51	WG19	R180	15.00 to 22	11.572	23.143	0.51	0.255
WR42	WG20	R220	18.00 to 26.50	14.051	28.102	0.42	0.17
WR34	WG21	R260	22.00 to 33	17.357	34.715	0.34	0.17
WR28	WG22	R320	26.50 to 40	21.077	42.154	0.28	0.14
WR22	WG23	R400	33.00 to 50	26.346	52.692	0.224	0.112
WR19	WG24	R500	40.00 to 60	31.391	62.782	0.188	0.094
WR15	WG25	R620	50.00 to 75	39.875	79.75	0.148	0.074
WR12	WG26	R740	60 to 90	48.373	96.746	0.122	0.061
WR10	WG27	R900	75 to 110	59.015	118.03	0.1	0.05
WR8	WG28	R1200	90 to 140	73.768	147.536	0.08	0.04
WR6	WG29	R1400	110 170	90.791	181.583	0.065	0.0325
WR7	WG29	R1400	110 to 170	90.791	181.583	0.065	0.0325
WR5	WG30	R1800	140 to 220	115.714	231.429	0.051	0.0255
WR4	WG32	R2200	172 to 260	137.243	274.485	0.043	0.0215
WR3	WG32	R2600	220 to 330	173.571	347.143	0.034	0.017

The need to explore waves at a millimeter scale for the subsequent generation of mobile communications has contributed significantly to the advancement of modern telecommunications components based on microwaves [95]. Substrate-integrated waveguides have found many potential applications outside the telecommunication industry. They are widely used in the automotive industry, as well as in biomedical devices for sensing applications. In terms of their dispersion and propagation characteristics, substrate integrated waveguides are similar to rectangular waveguides.

Another important emerging area is opted fluidics, which combines the benefits of optics and microfluidics in order to achieve highly compact and highly functional materials. Specifically, fluidic elements are integrated into the photonics structure [96,97]. A lot of optofluidic sensors are produced for healthcare and pharmaceutical researchers. Moreover, sensors are also available for biochemical analyses, environmental monitoring as well as biomedical researchers [98]. Integration of optical and fluidic structures can simply be achieved via optofluidic waveguides. Optofluidic waveguides are found to provide a very powerful foundation for building optofluidic sensors. The liquid core serves as the medium through which the light is guided with the help of highly reflective mirrors that are achieved by the sidewalls of the core. High sensitivity is usually achieved by taking advantage of flow or guidance of the light and the fluid through the same medium or channel. This direct interaction provides high sensitivity due to the small volume of the liquid as well as strong optical connections. Sensitivity can be achieved as low as the molecular level and this is the ultimate desire for any given analytical procedure. Just like other waveguide systems such as slot waveguides and photonic crystals, they have a high attractive capacity for optofluidic integrations with planar systems [99,100]. These types of waveguides allow simultaneous confinement and propagation of light because of the interference of light that occurs at the claddings that are made up of alternating thin layers of low- and high-refractive index. The operating principle of these types of waveguides can be understood easier when one considers the 1D structure as presented below in Figure 14. Assuming that two layers of the cladding are considered, and their refractive indexes are respectively n_1 and n_2 (with $n_1 > n_2$), propagation of light through the core is achieved by Fresnel reflections at the cladding interfaces. Repeated reflections at the interfaces cause the entrapped light to interfere. Interference cladding is designed to certify certain conditions to strengthen the intensity of the reflected light [101].

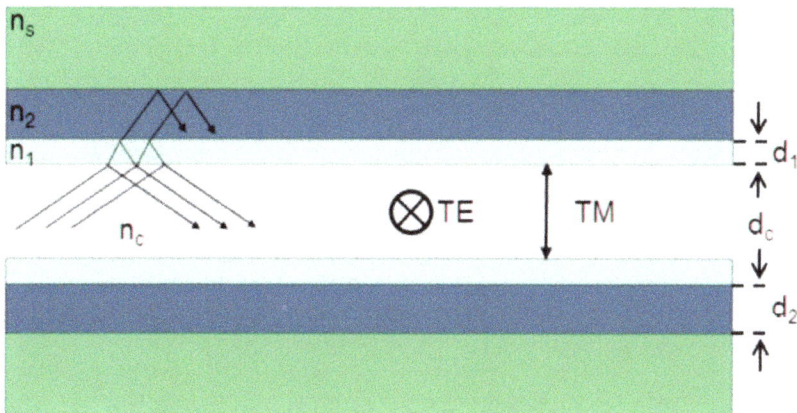

Figure 14. 1D structure of narrow waveguide.

Presently, optical biosensors have emerged as the favorite choice to replace bulky laboratory instruments for such applications where a bulky quantity of samples is required to be simultaneously analyzed, like microplate array systems. Optical biosensors possess some desirable characteristics like small size, low-cost and being easier to use. These pleasant characteristics allow them to be used in online monitoring and sensing for the analysis of samples that are complex, and for real-time or online monitoring of experimental procedures [102]. Therefore, optical bio-sensing is now an active research field and commercialization of a number of platforms has already been realized. The application areas include environmental monitoring, clinical application, and food safety and control [103]. The most powerful and reliable biosensors are undoubtedly those based on fluorescence. Fluorescence intensity, decay time, and emission anisotropy are among the parameters that could be measured and used for sensing [104]. There is, therefore, a variety of options towards improving the performance of biosensors. Optical waveguides are dielectric configurations having two extreme wavelengths in the infrared and the ultraviolet regions of the electromagnetic spectrum and can be used to transport energy between these two extreme wavelengths. Based on their geometrical shape, they can be categorized into two main classes: planar and cylindrical. Optical waveguides are contained within the first group and are comprised of a cylindrical central dielectric core-cladding by an element characterized by a low refractive index (by z 1%). The planar waveguides are prepared from a dielectric slab core enclosed between two layers of the cladding with slightly lower refractive indices [105]. In both layers, propagation of light and streamlining along the direction of core depend on the famous total internal reflection phenomenon. When light is propagated within a planar or cylindrical waveguide, the light is reflected totally at the interface between two mediums in which one is optically denser provided that the angle of refraction is greater than the critical angle. In fluorescent biosensors that are based on evanescent field excitation, the geometry of the sensing region must be properly designed in order to optimize the excitation ability along the probe length, and also to prevent the coupling of the emitted fluorescence into guided modes that do not propagate in the cladding. This is known as "V-number mismatch" [106,107].

Many industrial operations such as welding and cutting are achieved using carbon dioxide lasers. However, bringing the laser beam from the source of flame to the desired area is a challenge because other equipment may block the path. The only method to successfully deliver the beam is using articulated arms; however, these colossal systems also require large spaces and mirrors requiring regular maintenance and alignment. Attempts have been made to use solid core waveguides to deliver high power carbon dioxide beams, but, due to thermal damage, particularly at the high interface, they have not been successful. Short lifespan associated with some of the promising solid core is also a disadvantage. Transmission of up to 3000 Watts of laser power is required at the initial attempt using circular and rectangular metal-coated waveguides. However, large core radii are needed to achieve the needed power levels and, when subjected to bends and other movements, these large radii waveguides exhibit poor outputs. Moreover, most of the industrial operations require low order outputs to achieve sharp and clean cuts. Therefore, to attain this type of quality with hollow waveguides, smaller size core is needed to filter and transmit the higher order mode in the desired fashion. It is a known fact that, as the size of the core decreases, the loss of the waveguide increases and this, in turn, decreases the power capacity. These setbacks reduce the use of hollow waveguides to industrial uses such as marking cutting of plastic or paper, which requires lower power [108].

Controlling and streamlining the movement of light has been one of the main focus areas of research in the last few decades. Materials such as optical fibers, which operate based on the principle of total internal reflection, have significantly transformed the communication industry [109]. The concept of photonic crystals was proposed independently by John [110] and Yablonovitch [111] in 1987. They use the electronic band concept analogous to semiconductor crystals. Photonic crystals are dielectric materials that are periodic in 1D, 2D or 3D orthogonal directions. They can be fabricated in a simple way and have unusual optical properties. Two-dimensional photonic crystals are the most interesting and can be categorized into two: dielectric materials in air, or air in the dielectric material. The former is fabricated easily via periodic inscription of holes in materials of high dielectric properties such as GaAs, Si, and Ge. Photonic bandgaps are characterized by photonic crystals because of the intermittent disparity in the refractive index. Photonic band gaps have a range of frequencies that cannot allow propagating inside the crystal. Because of this peculiar property, waveguides are formed through inducing of line imperfections in photonic crystal structures [112,113]. These line faults are used to guide light from one place to another. The defects are guided inside the photonic band gap through the streamline via the total internal reflection principle. Because of the asymmetrical boundary, this streamlining creates backscattering, which causes slow light phenomena. Slow light causes optical signals to compress in space, and this enables interaction between light and matter and allows miniaturization. Presentation of photonic crystal waveguides is given in Figure 15a. A variety of dielectric slab materials of the high refractive index can be used to produce photonic crystal slabs. A typical example of these photonic crystals is based on polymethyl methacrylate (PMMA) prepared by [114]. A photonic crystal slab based on Si₃N₄ has been demonstrated to work in the visible region of the electromagnetic spectrum. Moreover, photonic crystals based on InP/InGaAsP structure have been prepared with a slight loss [115].

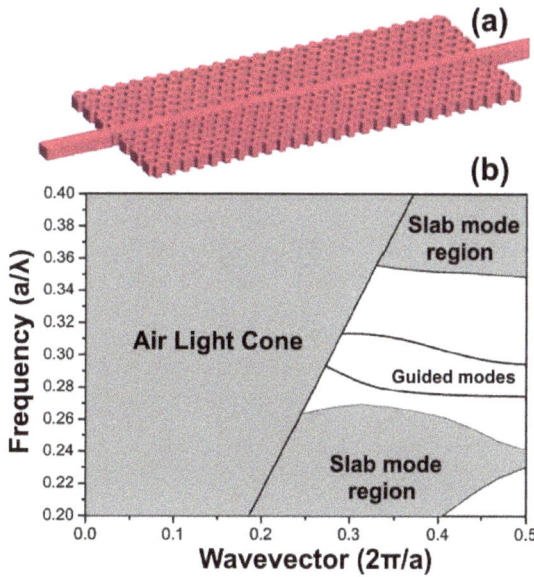

Figure 15. (a) Photonic crystal waveguide slab; (b) photonic band diagram

Several types of research are on-going for the potential applications of photonic crystals. Most frequent among them are related to the photonic integrated circuits. The introduction of defects can be achieved through the photonic band gap. Instead of guiding light through total internal reflection, it could be conducted using line defects in photonic crystals. The use of a photonic bandgap to guide light allows for small bending loss even when the bending angles are large. In the area of sensors, photonic crystals have been widely used in the field of sensors. A photonic crystal slab provides sensitivity to the photonic band gap. Some of these sensors can be designed to detect pressure using a GaAs/AIGaAs slab [116].

8. Conclusions

Waveguides and optical fibers have applications for assisting us in various aspects of our lives. As anticipated, optical fiber-based sensors can be appropriate instruments for monitoring physical parameters such as strain and temperature. A review of some of the recent advances related to the design and application of optical fiber sensors has been given. It has been established that optical fiber grating sensors and side-polished fibers continue to play a significant role in the development of various sensors with the combination of new fiber materials and structures. These new classes of Fiber Bragg grating sensors have the potential for many industrial uses. Each market and application has its separate advantages derived from Fiber Bragg grating based sensor applications. Nevertheless, optimization of the sensor systems is not restricted to the sensors only, but the entire system must be considered. To obtain an optimal Fiber Bragg grating sensors system for application, optimization of the interrogator with regard to wavelengths, resolutions, sweep frequencies as well as costs, among other factors, must be considered.

Fiber optics sensors have been developing for many years but have not achieved great commercial success yet due to the difficulties of introducing modern technologies that could replace current

Micromachines **2018**, *9*, 452

well-established technologies. However, for applications such as sensing in high-voltage and high-power machinery, or in microwave ovens, the fiber optics sensors are well recognized for presenting many advantages. Fiber Bragg grating sensors have developed significantly, and these can now be used to monitor conditions within the wings of airplanes, in wind turbines, bridges, large dams, oil wells and pipelines. In smart structures, which are the main drivers for the further development of fiber-optic sensors, the fiber sensors can monitor and obtain essential information about the strain, vibrations, and other phenomena. Since the year 2000, fiber optics has provided a significant contribution in applications such as optical communications, transmission fibers used underwater, in terrestrial areas, metro and local area networks (LAN). Other special fibers have been used in amplifiers, lasers, sensors and photonics devices. Further improvements of the fiber optics can be done by providing higher bandwidth, transmissions capacities for longer distances, and introducing devices with at a lower cost. For instance, in the LAN fiber world, the use of new wideband multimode fibers is recommended to improve the overall system efficiency. The wideband multimode fibers can be used in wider frequency ranges from visible to infrared such as the short wavelength-division multiplexing ranges 850 to 950 nm. Another rapidly growing technology is free-space communication, where the optical signals can be used for satellite–satellite communications. Recently, optical fibers have been used for transmission from light emitting sources such as high-power lasers, where the sudden changes in wavelength can be controlled easily in these devices.

Author Contributions: I.S.A. conceived and designed the experiments; S.R.B.A. performed the experiments; M.A.J. and J.A. analyzed the data; H.M.H. contributed reagents/materials/analysis tools; I.S.A. wrote the paper; M.B. and P.Y. have performed the revisions.

Acknowledgments: All sources of funding of the study should be disclosed. Please clearly indicate grants that you have received in support of your research work. Clearly state if you received funds for covering the costs to publish in open access.

Conflicts of Interest: The authors declare no conflict of interest.

References

1. Cavin, S.; Wang, X.; Zellweger, M.; Gonzalez, M.; Bensimon, M.; Wagnières, G.; Krueger, T.; Ris, H.; Gronchi, F.; Perentes, J.Y. Interstitial fluid pressure: A novel biomarker to monitor photo-induced drug uptake in tumor and normal tissues. *Lasers Surg. Med.* **2017**, *49*, 773–780. [CrossRef] [PubMed]
2. Mignon, C.; Tobin, D.J.; Zeitouny, M.; Uzunbajakava, N.E. Shedding light on the variability of optical skin properties: Finding a path towards more accurate prediction of light propagation in human cutaneous compartments. *Biomed. Opt. Express* **2018**, *9*, 852–872. [CrossRef] [PubMed]
3. Sun, Q.; He, Y.; Liu, K.; Fan, S.; Parrott, E.P.J.; Pickwell-MacPherson, E. Recent advances in terahertz technology for biomedical applications. *Quant. Imaging Med. Surg.* **2017**, *7*, 345–355. [CrossRef] [PubMed]
4. Gorin, A.; Jaouad, A.; Grondin, E.; Aimez, V.; Charette, P. Fabrication of silicon nitride waveguides for visible-light using PECVD: A study of the effect of plasma frequency on optical properties. *Opt. Express* **2008**, *16*, 13509–13516. [CrossRef] [PubMed]
5. Zheludev, N. The life and times of the LED—A 100-year history. *Nat. Photonics* **2007**, *1*, 189–192. [CrossRef]
6. Gordon, R. Sensors: Single-ion detection. *Nat. Photonics* **2016**, *10*, 697–698. [CrossRef]
7. Hochberg, M.; Baehr-Jones, T. Towards fabless silicon photonics. *Nat. Photonics* **2010**, *4*, 492–494. [CrossRef]
8. Malka, D.; Peled, A. Power splitting of 1×16 in multicore photonic crystal fibers. *Appl. Surf. Sci.* **2017**, *417*, 34–39. [CrossRef]
9. Dersch, R.; Steinhart, M.; Boudriot, U.; Greiner, A.; Wendorff, J.H. Nanoprocessing of polymers: Applications in medicine, sensors, catalysis, photonics. *Polym. Adv. Technol.* **2005**, *16*, 276–282. [CrossRef]
10. Weil, T.; Vosch, T.; Hofkens, J.; Peneva, K.; Müllen, K. The rylene colorant family—tailored nanoemitters for photonics research and applications. *Angew. Chem. Int. Ed.* **2010**, *49*, 9068–9093. [CrossRef] [PubMed]
11. Jariwala, D.; Marks, T.J.; Hersam, M.C. Mixed-dimensional van der Waals heterostructures. *Nat. Mater.* **2017**, *16*, 170–181. [CrossRef] [PubMed]
12. Hedberg, T.; Feeney, A.B.; Helu, M.; Camelio, J.A. Toward a Lifecycle Information Framework and Technology in Manufacturing. *J. Comput. Inf. Sci. Eng.* **2017**, *17*, 021010. [CrossRef] [PubMed]

13. Sundaravadivel, P.; Kougianos, E.; Mohanty, S.P.; Ganapathiraju, M.K. Everything You Wanted to Know about Smart Health Care: Evaluating the Different Technologies and Components of the Internet of Things for Better Health. *IEEE Consum. Electron. Mag.* **2018**, *7*, 18–28. [CrossRef]

14. Xu, T.; Shevchenko, N.A.; Lavery, D.; Semrau, D.; Liga, G.; Alvarado, A.; Killey, R.I.; Bayvel, P. Modulation format dependence of digital nonlinearity compensation performance in optical fibre communication systems. *Opt. Express* **2017**, *25*, 3311–3326. [CrossRef] [PubMed]

15. Farsaei, A.; Wang, Y.; Molavi, R.; Jayatilleka, H.; Caverley, M.; Beikahmadi, M.; Shirazi, A.H.M.; Jaeger, N.; Chrostowski, L.; Mirabbasi, S. A review of wireless-photonic systems: Design methodologies and topologies, constraints, challenges, and innovations in electronics and photonics. *Opt. Commun.* **2016**, *373*, 16–34. [CrossRef]

16. Zou, X.; Lu, B.; Pan, W.; Yan, L.; Stöhr, A.; Yao, J. Photonics for microwave measurements. *Laser Photonics Rev.* **2016**, *10*, 711–734. [CrossRef]

17. Marpaung, D.; Roeloffzen, C.; Heideman, R.; Leinse, A.; Sales, S.; Capmany, J. Integrated microwave photonics. *Laser Photonics Rev.* **2013**, *7*, 506–538. [CrossRef]

18. Capmany, J.; Mora, J.; Pastor, D.; Ortega, B.; Sales, S. Microwave photonic signal processing. *J. Lightwave Technol.* **2013**, *31*, 571–586. [CrossRef]

19. Aulakh, S.K. Application of microwave photonics in electronic warfare. *IJCST* **2013**, *4*, 53–58.

20. Liu, L.; Feng, F.; Hu, Q.; Paau, M.C.; Liu, Y.; Chen, Z.; Bai, Y.; Guo, F.; Choi, M.M.F. Capillary electrophoretic study of green fluorescent hollow carbon nanoparticles. *Electrophoresis* **2015**, *36*, 2110–2119. [CrossRef] [PubMed]

21. Ma, R.; Oulton, R.F.; Sorger, V.J.; Zhang, X. Plasmon lasers: Coherent light source at molecular scales. *Laser Photonics Rev.* **2013**, *7*, 1–21. [CrossRef]

22. Chen, B.; Yang, J.; Hu, C.; Wang, S.; Wen, Q.; Zhang, J. Plasmonic polarization nano-splitter based on asymmetric optical slot antenna pairs. *Opt. Lett.* **2016**, *41*, 4931–4934. [CrossRef] [PubMed]

23. Park, W.; Rhie, J.; Kim, N.Y.; Hong, S.; Kim, D. Sub-10 nm feature chromium photomasks for contact lithography patterning of square metal ring arrays. *Sci. Rep.* **2016**, *6*, 23823. [CrossRef] [PubMed]

24. El-Rabiaey, M.A.; Areed, N.F.F.; Obayya, S.S.A. Novel plasmonic data storage based on nematic liquid crystal layers. *J. Lightwave Technol.* **2016**, *34*, 3726–3732. [CrossRef]

25. Stipe, B.C.; Strand, T.C.; Poon, C.C.; Balamane, H.; Boone, T.D.; Katine, J.A.; Li, J.; Rawat, V.; Nemoto, H.; Hirotsune, A.; et al. Magnetic recording at 1.5 Pb m-2 using an integrated plasmonic antenna. *Nat. Photonics* **2010**, *4*, 484–488. [CrossRef]

26. Petersen, M.R.; Yossef, M.; Chen, A. Gap between Code Requirements and Current State of Research on Safety Performance of Fiber-Reinforced Polymer for Nonstructural Building Components. *Pract. Period. Struct. Des. Constr.* **2017**, *22*, 04017005. [CrossRef]

27. Knight, J.C.; Birks, T.A.; Russell, P.S.J.; Atkin, D.M. All-silica single-mode optical fiber with photonic crystal cladding. *Opt. Lett.* **1996**, *21*, 1547–1549. [CrossRef] [PubMed]

28. Seddon, A.B.; Tang, Z.; Furniss, D.; Sujecki, S.; Benson, T.M. Progress in rare-earth-doped mid-infrared fiber lasers. *Opt. Express* **2010**, *18*, 26704–26719. [CrossRef] [PubMed]

29. Borshchevskaia, N.A.; Katamadze, K.G.; Kulik, S.P.; Klyamkin, S.N.; Chuvikov, S.V.; Sysolyatin, A.A.; Tsvetkov, S.V.; Fedorov, M.V. Luminescence in germania–silica fibers in a 1–2 μm region. *Opt. Lett.* **2017**, *42*, 2874–2877. [CrossRef] [PubMed]

30. Hegedűs, G.; Sarkadi, T.; Czigány, T. Analysis of the Light Transmission Ability of Reinforcing Glass Fibers Used in Polymer Composites. *Materials* **2017**, *10*, 637. [CrossRef] [PubMed]

31. Ertman, S.; Lesiak, P.; Woliński, T.R. Optofluidic photonic crystal fiber-based sensors. *J. Lightwave Technol.* **2017**, *35*, 3399–3405. [CrossRef]

32. Singh, E.; Kim, K.S.; Yeom, G.Y.; Nalwa, H.S. Atomically thin-layered molybdenum disulfide (MoS_2) for bulk-heterojunction solar cells. *ACS Appl. Mater. Interfaces* **2017**, *9*, 3223–3245. [CrossRef] [PubMed]

33. Qiao, P.; Yang, W.; Chang-Hasnain, C.J. Recent advances in high-contrast metastructures, metasurfaces, and photonic crystals. *Adv. Opt. Photonics* **2018**, *10*, 180–245. [CrossRef]

34. Tao, G.; Ebendorff-Heidepriem, H.; Stolyarov, A.M.; Danto, S.; Badding, J.V.; Fink, Y.; Ballato, J.; Abouraddy, A.F. Infrared fibers. *Adv. Opt. Photonics* **2015**, *7*, 379–458. [CrossRef]

35. Peacock, A.C.; Sparks, J.R.; Healy, N. Semiconductor optical fibres: Progress and opportunities. *Laser Photonics Rev.* **2014**, *8*, 53–72. [CrossRef]

36. Wang, Z.; Guo, C.; Jiang, W. Large mode area OmniGuide fiber with superconductor-dielectric periodic multilayers cladding. *Opt. Int. J. Light Electron Opt.* **2014**, *125*, 6789–6792. [CrossRef]

37. Hasan, M.I.; Akhmediev, N.; Chang, W. Mid-infrared supercontinuum generation in supercritical xenon-filled hollow-core negative curvature fibers. *Opt. Lett.* **2016**, *41*, 5122–5125. [CrossRef] [PubMed]

38. Bellanca, G.; Riesen, N.; Argyros, A.; Leon-Saval, S.G.; Lwin, R.; Parini, A.; Love, J.D.; Bassi, P. Holey fiber mode-selective couplers. *Opt. Express* **2015**, *23*, 18888–18896. [CrossRef] [PubMed]

39. Hu, K.; Kabakova, I.V.; Büttner, T.F.; Lefrancois, S.; Hudson, D.D.; He, S.; Eggleton, B.J. Low-threshold Brillouin laser at 2 μm based on suspended-core chalcogenide fiber. *Opt. Lett.* **2014**, *39*, 4651–4654. [CrossRef] [PubMed]

40. Kim, R.; Park, C.H.; Lee, A.; Moon, J.H. Development of the noncontact temperature sensor using the infrared optical fiber coated with antifog solution. *Sci. Technol. Nucl. Install.* **2015**, *2015*, 718592. [CrossRef]

41. Mishra, V. 16 Medical Applications of Fiber-Optic Sensors. *Opt. Fiber Sens. Adv. Tech. Appl.* **2015**, *36*, 455.

42. Pospíšilová, M.; Kuncová, G.; Trögl, J. Fiber-optic chemical sensors and fiber-optic bio-sensors. *Sensors* **2015**, *15*, 25208–25259. [CrossRef] [PubMed]

43. Liu, Y.; Zhang, M.; Zhang, J.; Wang, Y. Single-longitudinal-mode triple-ring Brillouin fiber laser with a saturable absorber ring resonator. *J. Lightwave Technol.* **2017**, *35*, 1744–1749. [CrossRef]

44. Axelrod, D.; Burghardt, T.P.; Thompson, N.L. Total internal reflection fluorescence. *Annual review of biophysics and bioengineering. Ann. Rev. Biophys. Bioeng.* **1984**, *13*, 247–268. [CrossRef] [PubMed]

45. Sönnichsen, C.; Geier, S.; Hecker, N.E.; von Plessen, G.; Feldmann, J.; Ditlbacher, H.; Lamprecht, B.; Krenn, J.R.; Aussenegg, F.R.; Chan, V.Z.; et al. Spectroscopy of single metallic nanoparticles using total internal reflection microscopy. *Appl. Phys. Lett.* **2000**, *77*, 2949–2951. [CrossRef]

46. Mitschke, F. *Fiber Optics: Physics and Technology*; Springer Science & Business Media: New York, NY, USA, 2016.

47. Abedin, K.S.; Yan, M.F.; Taunay, T.F.; Zhu, B.; Monberg, E.M.; DiGiovanni, D.J. State-of-the-art multicore fiber amplifiers for space division multiplexing. *Opt. Fiber Technol.* **2017**, *35*, 64–71. [CrossRef]

48. Predehl, K.; Grosche, G.; Raupach, S.M.F.; Droste, S.; Terra, O.; Alnis, J.; Legero, T.; Hansch, T.W.; Udem, T.; Holzwarth, R.; et al. A 920-kilometer optical fiber link for frequency metrology at the 19th decimal place. *Science* **2012**, *336*, 441–444. [CrossRef] [PubMed]

49. Li, Z.; Heidt, A.M.; Daniel, J.M.O.; Jung, Y.; Alam, S.U.; Richardson, D.J. Thulium-doped fiber amplifier for optical communications at 2 μm. *Opt. Express* **2013**, *21*, 9289–9297. [CrossRef] [PubMed]

50. Marhic, M.E.; Andrekson, P.A.; Petropoulos, P.; Radic, S.; Peucheret, C.; Jazayerifar, M. Fiber optical parametric amplifiers in optical communication systems. *Laser Photonics Rev.* **2015**, *9*, 50–74. [CrossRef] [PubMed]

51. Taffoni, F.; Formica, D.; Saccomandi, P.; Pino, G.; Schena, E. Optical fiber-based MR-compatible sensors for medical applications: An overview. *Sensors* **2013**, *13*, 14105–14120. [CrossRef] [PubMed]

52. Choi, Y.; Yoon, C.; Kim, M.; Yang, T.D.; Fang-Yen, C.; Dasari, R.R.; Lee, K.J.; Choi, W. Scanner-free and wide-field endoscopic imaging by using a single multimode optical fiber. *Phys. Rev. Lett.* **2012**, *109*, 203901. [CrossRef] [PubMed]

53. Woyessa, G.; Fasano, A.; Markos, C.; Stefani, A.; Rasmussen, H.K.; Bang, O. Zeonex microstructured polymer optical fiber: Fabrication friendly fibers for high temperature and humidity insensitive Bragg grating sensing. *Opt. Mater. Express* **2017**, *7*, 286–295. [CrossRef]

54. Gu, F.; Xie, F.; Lin, X.; Linghu, S.; Fang, W.; Zeng, H.; Tong, L.; Zhuang, S. Single whispering-gallery mode lasing in polymer bottle microresonators via spatial pump engineering. *Light Sci. Appl.* **2017**, *6*, e17061. [CrossRef]

55. Kostovski, G.; Stoddart, P.R.; Mitchell, A. The Optical Fiber Tip: An Inherently Light-Coupled Microscopic Platform for Micro-and Nanotechnologies. *Adv. Mater.* **2014**, *26*, 3798–3820. [CrossRef] [PubMed]

56. Lepinay, S.; Staff, A.; Ianoul, A.; Albert, J. Improved detection limits of protein optical fiber biosensors coated with gold nanoparticles. *Biosens. Bioelectron.* **2014**, *52*, 337–344. [CrossRef] [PubMed]

57. Lee, B. Review of the present status of optical fiber sensors. *Opt. Fiber Technol.* **2003**, *9*, 57–79. [CrossRef]

58. Perrotton, C.; Westerwaal, R.J.; Javahiraly, N.; Slaman, M.; Schreuders, H.; Dam, B.; Meyrueis, P. A reliable, sensitive and fast optical fiber hydrogen sensor based on surface plasmon resonance. *Opt. Express* **2013**, *21*, 382–390. [CrossRef] [PubMed]

59. Zeng, W.; Shu, L.; Li, Q.; Chen, S.; Wang, F.; Tao, X. Fiber-based wearable electronics: A review of materials, fabrication, devices, and applications. *Adv. Mater.* **2014**, *26*, 5310–5336. [CrossRef] [PubMed]

60. Zhang, H.; Healy, N.; Shen, L.; Huang, C.C.; Aspiotis, N.; Hewak, D.W.; Peacock, A.C. Graphene-based fiber polarizer with PVB-enhanced light interaction. *J. Lightwave Technol.* **2016**, *34*, 3563–3567. [CrossRef]

61. Grattan, K.; Ning, Y. *Optical Fiber Sensor Technology*; Springer: Berlin/Heidelberg, Germany, 1998; pp. 1–35.

62. Medlock, R.S. Fibre optic intensity modulation sensors. *Appl. Sci.* **1987**, *132*, 123–124.

63. Spooncer, R. Fibre optics in instrumentation. *Handb. Meas. Sci.* **1992**, *3*, 1691–1720.

64. Grattan, K.T.V. Fibre optic sensors—the way forward, measurement. *J. Int. Meas. Confed.* **1987**, *5*, 122. [CrossRef]

65. Grattan, K.T.V. New Developments in Sensor Technology—Fibre and Electro-Optics. *Meas. Control* **1989**, *22*, 165–175. [CrossRef]

66. Barozzi, M.; Manicardi, A.; Vannucci, A.; Candiani, A.; Sozzi, M.; Konstantaki, M.; Pissadakis, S.; Corradini, R.; Selleri, S.; Cucinotta, A. Optical fiber sensors for label-free DNA detection. *J. Lightwave Technol.* **2017**, *35*, 3461–3472. [CrossRef]

67. Ferreira, M.F.S.; Castro-Camus, E.; Ottaway, D.J.; López-Higuera, J.M.; Feng, X.; Jin, W.; Jeong, Y.; Picqué, N.; Tong, L.; Reinhard, B.M.; et al. Roadmap on optical sensors. *J. Opt.* **2017**, *19*, 083001. [CrossRef] [PubMed]

68. Guan, B.; Jin, L.; Cheng, L.; Liang, Y. Acoustic and ultrasonic detection with radio-frequency encoded fiber laser sensors. *IEEE J. Sel. Top. Quantum Electron.* **2017**, *23*, 302–313. [CrossRef]

69. Zhao, D.; Chen, X.; Zhou, K.; Zhang, L.; Bennion, I.; MacPherson, W.N.; Barton, J.S.; Jones, J.D.C. Bend sensors with direction recognition based on long-period gratings written in D-shaped fiber. *Appl. Opt.* **2004**, *43*, 5425–5428. [CrossRef] [PubMed]

70. Zhao, D.; Zhou, K.; Chen, X.; Zhang, L.; Bennion, I.; Flockhart, G.; MacPherson, W.N.; Barton, J.S.; Jones, J.D.C. Implementation of vectorial bend sensors using long-period gratings UV-inscribed in special shape fibres. *Meas. Sci. Technol.* **2004**, *15*, 1647–1650. [CrossRef]

71. Shu, X.; Zhao, D.; Zhang, L.; Bennion, I. Use of dual-grating sensors formed by different types of fiber Bragg gratings for simultaneous temperature and strain measurements. *Appl. Opt.* **2004**, *43*, 2006–2012. [CrossRef] [PubMed]

72. Rapp, M.; Ley, C.J.; Hansson, K.; Sjöström, L. Postoperative computed tomography and low-field magnetic resonance imaging findings in dogs with degenerative lumbosacral stenosis treated by dorsal laminectomy. *Vet. Comp. Orthop. Traumatol.* **2017**, *30*, 143–152. [CrossRef] [PubMed]

73. Yoon, S.; Ye, W.; Heidemann, J.; Littlefield, B.; Shahabi, C. SWATS: Wireless sensor networks for steamflood and waterflood pipeline monitoring. *IEEE Netw.* **2011**, *25*, 50–56. [CrossRef]

74. Mishra, C.; Palai, G. Temperature and pressure effect on GaN waveguide at 428.71 terahertz frequency for sensing application. *Opt. Int. J. Light Electron Opt.* **2015**, *126*, 4685–4687. [CrossRef]

75. Trpkovski, S.; Wade, S.A.; Baxter, G.W.; Collins, S.F. Dual temperature and strain sensor using a combined fiber Bragg grating and fluorescence intensity ratio technique in Er^{3+}-doped fiber. *Rev. Sci. Instrum.* **2003**, *74*, 2880–2885. [CrossRef]

76. Schultz, J.H. Protection of superconducting magnets. *IEEE Trans. Appl. Supercond.* **2002**, *12*, 1390–1395. [CrossRef]

77. Soto, M.A.; Angulo-Vinuesa, X.; Martin-Lopez, S.; Chin, S.; Ania-Castanon, J.D.; Corredera, P.; Rochat, E.; Gonzalez-Herraez, M.; Thevenaz, L. Extending the real remoteness of long-range Brillouin optical time-domain fiber analyzers. *J. Lightwave Technol.* **2014**, *32*, 152–162. [CrossRef]

78. Kramer, A.; Over, D.; Stoller, P.; Paul, T.A. Fiber-coupled LED gas sensor and its application to online monitoring of ecoefficient dielectric insulation gases in high-voltage circuit breakers. *Appl. Opt.* **2017**, *56*, 4505–4512. [CrossRef] [PubMed]

79. Oromiehie, E.; Prusty, B.G.; Compston, P.; Rajan, G. In-situ simultaneous measurement of strain and temperature in automated fiber placement (AFP) using optical fiber Bragg grating (FBG) sensors. *Adv. Manuf. Polym. Compos. Sci.* **2017**, *3*, 52–61. [CrossRef]

80. Yang, S.; Homa, D.; Pickrell, G.; Wang, A. Fiber Bragg grating fabricated in micro-single-crystal sapphire fiber. *Opt. Lett.* **2018**, *43*, 62–65. [CrossRef] [PubMed]

81. Zhao, J.; Cao, S.; Liao, C.; Wang, Y.; Wang, G.; Xu, X.; Fu, C.; Xu, G.; Lian, J.; Wang, Y. Surface plasmon resonance refractive sensor based on silver-coated side-polished fiber. *Sens. Actuators B Chem.* **2016**, *230*, 206–211. [CrossRef]

82. Wang, G. Wavelength-switchable passively mode-locked fiber laser with mechanically exfoliated molybdenum ditelluride on side-polished fiber. *Opt. Laser Technol.* **2017**, *96*, 307–312. [CrossRef]

83. Roland, U.; Renschen, C.P.; Lippik, D.; Stallmach, F.; Holzer, F. A new fiber optical thermometer and its application for process control in strong electric, magnetic, and electromagnetic fields. *Sens. Lett.* **2003**, *1*, 93–98. [CrossRef]

84. Hill, K.O.; Malo, B.; Bilodeau, F.; Johnson, D.C.; Albert, J. Bragg gratings fabricated in monomode photosensitive optical fiber by UV exposure through a phase mask. *Appl. Phys. Lett.* **1993**, *62*, 1035–1037. [CrossRef]

85. Meltz, G.; Morey, W.W.; Glenn, W.H. Formation of Bragg gratings in optical fibers by a transverse holographic method. *Opt. Lett.* **1989**, *14*, 823–825. [CrossRef] [PubMed]

86. Pospori, A.; Marques, C.A.F.; Bang, O.; Webb, D.J.; André, P. Polymer optical fiber Bragg grating inscription with a single UV laser pulse. *Opt. Express* **2017**, *25*, 9028–9038. [CrossRef] [PubMed]

87. Zhang, A.P.; Gao, S.; Yan, G.; Bai, Y. Advances in optical fiber Bragg grating sensor technologies. *Photonic Sens.* **2012**, *2*, 1–13. [CrossRef]

88. Tanaka, Y.; Miyazawa, H. Multipoint Fiber Bragg Grating Sensing Using Two-Photon Absorption Process in Silicon Avalanche Photodiode. *J. Lightwave Technol.* **2018**, *36*, 1032–1038. [CrossRef]

89. Li, T.; Shi, C.; Tan, Y.; Zhou, Z. Fiber Bragg grating sensing-based online torque detection on coupled bending and torsional vibration of rotating shaft. *IEEE Sens. J.* **2017**, *17*, 1999–2007. [CrossRef]

90. Cheng, L.K.; Schiferli, W.; Nieuwland, R.A.; Franzen, A.; den Boer, J.J.; Jansen, T.H. Development of a FBG Vortex Flow Sensor for High-Temperature Applications. In Proceedings of the 21st International Conference on Optical Fibre Sensors (OFS21), Ottawa, ON, Canada, 15–19 May 2011.

91. Zhang, F.; Zhou, Z.; Liu, Q.; Xu, W. An intelligent service matching method for mechanical equipment condition monitoring using the fibre Bragg grating sensor network. *Enterp. Inf. Syst.* **2017**, *11*, 284–309. [CrossRef]

92. Molimard, J.; Vacher, S.; Vautrin, A. Monitoring LCM process by FBG sensor under birefringence. *Strain* **2011**, *47*, 364–373. [CrossRef]

93. Caloz, C.; Itoh, T. *Electromagnetic Metamaterials: Transmission Line Theory and Microwave Applications*; John Wiley & Sons: New York, NY, USA, 2005.

94. Lin, P.T.; Singh, V.; Kimerling, L.; Agarwal, A.M. Planar silicon nitride mid-infrared devices. *Appl. Phys. Lett.* **2013**, *102*, 251121. [CrossRef]

95. Henry, M.; Free, C.E.; Izquierdo, B.S.; Batchelor, J.; Young, P. Millimeter wave substrate integrated waveguide antennas: Design and fabrication analysis. *IEEE Trans. Adv. Packag.* **2009**, *32*, 93–100. [CrossRef]

96. Monat, C.; Domachuk, P.; Eggleton, B.J. Integrated optofluidics: A new river of light. *Nat. Photonics* **2007**, *1*, 106–114. [CrossRef]

97. Psaltis, D.; Quake, S.R.; Yang, C. Developing optofluidic technology through the fusion of microfluidics and optics. *Nature* **2006**, *442*, 381–386. [CrossRef] [PubMed]

98. Fan, X.; White, I.M. Optofluidic microsystems for chemical and biological analysis. *Nat. Photonics* **2011**, *5*, 591–597. [CrossRef] [PubMed]

99. Schmidt, H.; Hawkins, A.R. The photonic integration of non-solid media using optofluidics. *Nat. Photonics* **2011**, *5*, 598. [CrossRef]

100. Malik, A.; Muneeb, M.; Pathak, S.; Shimura, Y.; van Campenhout, J.; Loo, R.; Roelkens, G. Germanium-on-silicon mid-infrared arrayed waveguide grating multiplexers. *IEEE Photonics Technol. Lett.* **2013**, *25*, 1805–1808. [CrossRef]

101. Bhardwaj, S.; Mittholiya, K.; Bhatnagar, A.; Bernard, R.; Dharmadhikari, J.A.; Mathur, D.; Dharmadhikari, A.K. Inscription of type I and depressed cladding waveguides in lithium niobate using a femtosecond laser. *Appl. Opt.* **2017**, *56*, 5692–5697. [CrossRef] [PubMed]

102. Taitt, C.; Anderson, G.P.; Ligler, F.S. Evanescent wave fluorescence biosensors: Advances of the last decade. *Biosens. Bioelectron.* **2016**, *76*, 103–112. [CrossRef] [PubMed]

103. Tompkin, R. Control of Listeria monocytogenes in the food-processing environment. *J. Food Prot.* **2002**, *65*, 709–725. [CrossRef] [PubMed]

104. Peveler, W.J.; Algar, W.R. More Than a Light Switch: Engineering Unconventional Fluorescent Configurations for Biological Sensing. *ACS Chem. Biol.* **2018**. [CrossRef] [PubMed]

105. Liu, X.; Osgood, R.M.; Vlasov, Y.A.; Green, W.M.J. Mid-infrared optical parametric amplifier using silicon nanophotonic waveguides. *Nat. Photonics* **2010**, *4*, 557–560. [CrossRef]

106. Rijal, K.; Leung, A.; Shankar, P.M.; Mutharasan, R. Detection of pathogen Escherichia coli O157: H7 AT 70cells/mL using antibody-immobilized biconical tapered fiber sensors. *Biosens. Bioelectron.* **2005**, *21*, 871–880. [CrossRef] [PubMed]

107. Tao, S.; Gong, S.; Fanguy, J.C.; Hu, X. The application of a light guiding flexible tubular waveguide in evanescent wave absorption optical sensing. *Sens. Actuators B Chem.* **2007**, *120*, 724–731. [CrossRef]

108. Harrington, J.A. A review of IR transmitting, hollow waveguides. *Fiber Integr. Opt.* **2000**, *19*, 211–227. [CrossRef]

109. Dutta, H.S.; Goyal, A.K.; Srivastava, V.; Pal, S. Coupling light in photonic crystal waveguides: A review. *Photonics Nan. Fundam. Appl.* **2016**, *20*, 41–58. [CrossRef]

110. John, S. Strong localization of photons in certain disordered dielectric superlattices. *Phys. Rev. Lett.* **1987**, *58*, 2486–2489. [CrossRef] [PubMed]

111. Yablonovitch, E. Inhibited spontaneous emission in solid-state physics and electronics. *Phys. Rev. Lett.* **1987**, *58*, 2059. [CrossRef] [PubMed]

112. Xiao, T.; Zhao, Z.; Zhou, W.; Takenaka, M.; Tsang, H.K.; Cheng, Z.; Goda, K. Mid-infrared germanium photonic crystal cavity. *Opt. Lett.* **2017**, *42*, 2882–2885. [CrossRef] [PubMed]

113. Shankar, R.; Leijssen, R.; Bulu, I.; Lončar, M. Mid-infrared photonic crystal cavities in silicon. *Opt. Express* **2011**, *19*, 5579–5586. [CrossRef] [PubMed]

114. Senn, T.; Bischoff, J.; Nüsse, N.; Schoengen, M.; Löchel, B. Fabrication of photonic crystals for applications in the visible range by Nanoimprint Lithography. *Photonics Nan. Fundam. Appl.* **2011**, *9*, 248–254. [CrossRef]

115. Kappeler, R.; Kaspar, P.; Jäckel, H.; Hafner, C. Record-low propagation losses of 154 dB/cm for substrate-type W1 photonic crystal waveguides by means of hole shape engineering. *Appl. Phys. Lett.* **2012**, *101*, 131108. [CrossRef]

116. Stomeo, T.; Grande, M.; Qualtieri, A.; Passaseo, A.; Salhi, A.; de Vittorio, M.; Biallo, D.; D'orazio, A.; de Sario, M.; Marrocco, V.; et al. Fabrication of force sensors based on two-dimensional photonic crystal technology. *Microelectron. Eng.* **2007**, *84*, 1450–1453. [CrossRef]

micromachines

MDPI

Communication

Multiple Light Coupling and Routing via a Microspherical Resonator Integrated in a T-Shaped Optical Fiber Configuration System

Georgia Konstantinou [1,2], **Karolina Milenko** [3], **Kyriaki Kosma** [1] **and Stavros Pissadakis** [1,*]

[1] Foundation for Research and Technology-Hellas (FORTH), Institute of Electronic Structure and Laser (IESL), GR-711 10 Heraklion, Greece; georgia.konstantinou@epfl.ch (G.K.); kosma@iesl.forth.gr (K.K.)
[2] EPFL, École polytechnique fédérale de Lausanne, CH-1015 Lausanne, Switzerland
[3] Department of Electronic Systems, Norwegian University of Science and Technology, NO-7491 Trondheim, Norway; karolina.milenko@ntnu.no
* Correspondence: pissas@iesl.forth.gr; Tel.: +30-2810-391348

Received: 18 September 2018; Accepted: 9 October 2018; Published: 15 October 2018

Abstract: We demonstrate a three-port, light guiding and routing T-shaped configuration based on the combination of whispering gallery modes (WGMs) and micro-structured optical fibers (MOFs). This system includes a single mode optical fiber taper (SOFT), a slightly tapered MOF and a $BaTiO_3$ microsphere for efficient light coupling and routing between these two optical fibers. The $BaTiO_3$ glass microsphere is semi-immersed into one of the hollow capillaries of the MOF taper, while the single mode optical fiber taper is placed perpendicularly to the latter and in contact with the equatorial region of the microsphere. Experimental results are presented for different excitation and reading conditions through the WGM microspherical resonator, namely, through single mode optical fiber taper or the MOF. The experimental results indicate that light coupling between the MOF and the single mode optical fiber taper is facilitated at specific wavelengths, supported by the light localization characteristics of the $BaTiO_3$ glass microsphere, with spectral Q-factors varying between 4.5×10^3 and 6.1×10^3, depending on the port and parity excitation.

Keywords: microstructured optical fibers; whispering gallery modes; light localization

1. Introduction

High quality light trapping in dielectric, microspherical cavities can be readily achieved using whispering gallery modes (WGMs) resonation; attracting constant academic and potential industrial interest [1,2]. In the WGM resonation process light is confined at the inner interface of a high refractive index dielectric microcavity, illustrated as closed-loop trajectory where rays circulate through total internal reflection (TIR). Several types of optical geometries and corresponding materials have been employed for demonstrating and implementing WGM resonation in photonic devices. The confinement of light propagation within a closed, spherical symmetry dielectric cavity, leads to a three-fold modal quantization of the permitted localization states of light within the cavity volume [3], denoted with three complementary quantum numbers, per k-vector, being directly dependent upon the optogeometric characteristics of the cavity. The interface mechanism of TIR on the boundary of a curved surface is inherently a low loss light dispersion process, leading to high quality factors Q denoting light localization up to $\times 10^8$, upon materials, excitation protocols and resonator geometries used [4]. A great number of photonic devices based on WGM resonation have been presented including chemical and biological sensors [2,5], lasers [6], wavelength routers and metrological devices [7].

The efficiency of WGMs excitation and related Q-factors of the supported resonances critically depends upon the material properties of the resonator (surface roughness, Rayleigh scattering etc.), its geometry, real and imaginary refractive index, launching conditions and far field or near field signal collection method [8]. Prisms [9], gratings [10], angle cleaved [11], and tapered optical fibers [12] were used for the WGM evanescent field, light coupling into microspheres, leading to different Q-factors and practical consideration including stability of coupling, simplicity of implementation, integration, and interaction with the ambient environment [13].

In addition to the above, MOFs have also been used for reading or exciting WGM resonation inside microspherical cavities, starting with the work of Francois, et al., where a microsphere was wedged onto the end face of a collapsed core MOF [14]; others have used hollow core optical fibers [15,16]. An alternative and high integration WGM excitation approach was presented by Kosma et al. by placing a microspherical resonator inside the air capillary of a MOF taper, for providing a compact and robust operation [17]; similar implementations were also presented by other groups [18,19]. Generally, the use of the end face of standard, tapered and MOFs has been evolved into an efficient configuration for exciting WGMs in microspheres, offering advantages such as single port operation, good yield in fluorescing signal collection, robustness, and versatility in functionalizing and trapping the micro-spherical cavity.

Herein, we present light coupling and routing between the MOF taper and a standard optical fiber taper (SOFT), through a $BaTiO_3$ microsphere semi-immersed inside the capillary of the end face of the MOF taper; where the SOFT is placed perpendicularly to the MOF. This T-shaped light coupling system provides interesting WGMs excitation and collection with all three created fiber ports along with light paths parities. We anticipate that the proposed T-shaped excitation system can lead to numerous devices and applications of WGM resonators for sensing, filtering and frequency stabilization, and can be used in integrated optical devices as an efficient add-drop filter. The use of $BaTiO_3$ microspheres constitutes a base for straightforwardly attaining permanent and/or transient photorefractive tuning of the spectral characteristics of the WGMs using pulsed or continuous wave (CW) laser beams at low pump powers. This type of spectral trimming has already been demonstrated for a variety of materials and WGM resonating geometries [20–22]. Potentially, the photorefractive tuning of the WGM spectral characteristics can also happen for light excitation (for example using 405 nm or 532 nm CW lasers) through an all silica MOF, where the photorefractivity of the MOF itself will be minimum. An additional reason for using $BaTiO_3$ microspheres is the possibility of permanent encapsulation and fixation of the system MOF-$BaTiO_3$-SOFT using standard ultraviolet (UV) glues (see Norland UV adhesives, Norland Products, Inc., Cranbury, NJ, USA) with a refractive index lower than silica, without great compromising of the Q-factors of the excited WGMs [23]. Our investigations include the steps followed for realizing the device presented, and its spectral characterization and discussion for different portal excitations and read-outs.

2. Experimental Section

A grapefruit shaped MOF (drawn by ACREO, Stockholm, Sweden), consisting of two concentric cores, a 3.5% germanium doped silica inner core with diameter 8.5 μm and an outer silica core of diameter 16.1 μm, 20.8 μm diameter 5-air capillaries and 125 μm cladding diameter was used (see Figure 1a). For facilitating efficient wedging of the microspheres available, as well as increasing the evanescence field tail extending outside the microstructured core, this MOF was adiabatically tapered down to ~55% of its initial size and cleaved at the transition region, resulting in a tip diameter of ~68 μm and air capillaries scaling down to a ~11 μm diameter (see Figure 1b,c). For these tapering conditions the fundamental guiding mode is still confined at the Ge doped core of the grapefruit MOF, however, with a more extended modal profile covering the microstructured area. In Figure 2, we show the fundamental guiding mode confinement by using a commercial modal solver.

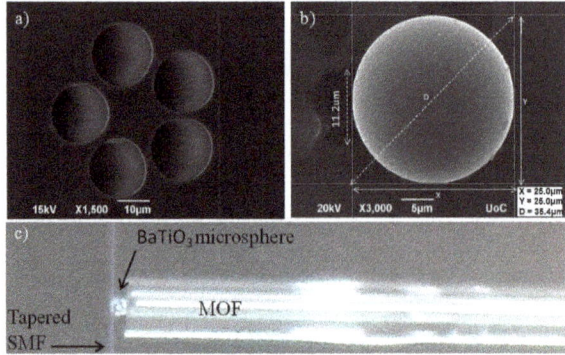

Figure 1. (**a**) Scanning electron microscope (SEM) picture of the un-tapered, grapefruit shaped micro-structured optical fiber (MOF) used, with five air capillaries and germanium doped core, (**b**) SEM image of the attached BaTiO$_3$, microsphere with 25 μm diameter, top fitting in one of the capillaries (diameter: 11.2 μm) of the tapered MOF. The other empty capillaries can be seen in the background, (**c**) optical microscope picture of a side view on the T-shaped light coupling system (single mode optical fiber taper (SOFT) placed perpendicular to MOF in a contact with the microsphere).

Additionally, since the adiabaticity criterion was followed, coupling from the central mode to modes supported at the extended microstructured core area is rather limited; the last is confirmed by the absence of significant beating features in the transmission spectra of the tapered MOF.

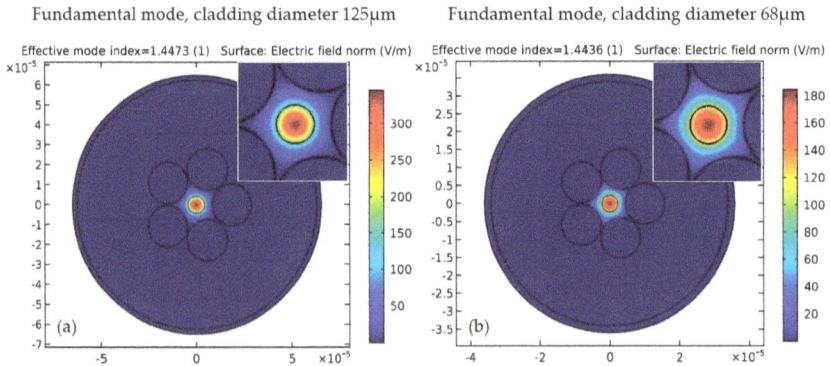

Figure 2. (**a**) Fundamental mode for the MOF in a 2D cross-section image by adopting COMSOL (Multiphysics 3.5, COMSOL Inc., Stockholm, Sweden) simulation. (**b**) Fundamental mode in a 2D cross-section image for a smaller cladding size after the tapering process (68 μm diameter), and accordingly smaller core and air capillaries. Increased spreading of the mode is observed due to reduced spatial confinement.

Barium titanate (BaTiO$_3$) glass microspheres of poly-disperse sizes and typical eccentricities of the order of 10% (Mo-Sci Corporation) were used. The refractive index of these microspheres varies between 1.9 and 2.1, depending on the stoichiometric ratio between Ba and Ti. A BaTiO$_3$ microsphere with a diameter 25 μm was attached to the end face of the MOF taper tip while being semi-immersed inside one of its air capillaries (Figure 1b). Air suction was used to ensure stable positioning and attachment of the microsphere. Figure 1 presents scanning electron microscopy (SEM) images of the employed MOF and the BaTiO$_3$ microsphere. Additionally, a single mode optical fiber (SMF-28,

Corning Inc., Corning, NY, USA) was tapered adiabatically down to 0.9 μm final waist diameter to achieve a broad evanescent field. All optical fibers used were thermally tapered with the use of a Vytran GPX-3000 (Thorlabs, Inc., Newton, NY, USA) optical fiber processing equipment.

The MOF taper with the attached microsphere was aligned perpendicular to the waist of the tapered fiber, so that the SOFT waist was in a contact with the equatorial region of the $BaTiO_3$ microsphere (Figure 3). In this way a three channel device was implemented, with routing point at the $BaTiO_3$ microsphere semi-immersed into the end face of the MOF; namely CH1 and CH2 represent the two portal ends of the SOFT and CH3 the MOF port. Moreover, this three-port optical fiber configuration was characterized with respect to the orientation of the line defined by the MOF core and the $BaTiO_3$ microsphere with the axis of the SOFT (Figure 3). In the parallel orientation (State A) the MOF central core and the center of the $BaTiO_3$ microsphere are in line with the axis of the SOFT, and the k-vectors of light propagating into the MOF core and the circumference of the $BaTiO_3$ along the plane of the SOFT excitation rest along the same axis of propagation. In the perpendicular orientation (State B) the SOFT excites the $BaTiO_3$ microsphere, however the axis between the MOF central core and the $BaTiO_3$ microsphere is positioned at 90° with respect to the axis of the SOFT, thus the k-vectors of the circumferential WGMs without polar components excited do not lay on the same axis with the MOF core. In the B State orientation, the position of the MOF taper core was below and far from the SOFT horizontal plane, eliminating any potential coupling due to lensing and total internal reflection cancellation. Transmission spectra were obtained for both State A and B of this three port, optical T-junction.

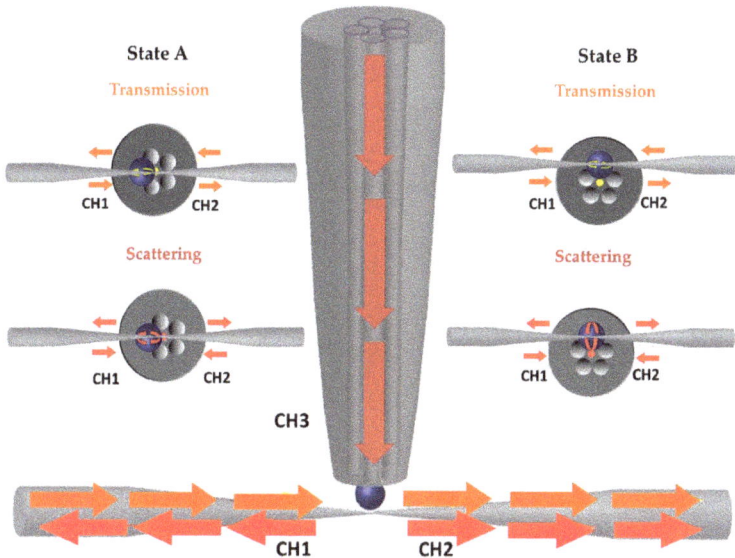

Figure 3. Schematic representation of the experimental setup showing the T-shape excitation system. CH1, CH2 and CH3 represent ports used to excite and measure WGMs. Two possible WGMs excitation configurations are demonstrated in the cross section of MOF taper tip (yellow and red arrows). The yellow arrows are referred to the transmission spectrum (CH1 to CH2 and vice versa) whereas the red arrows are chosen for the scattering spectrum (CH1/CH2 to CH3 and vice versa). Two different states, in terms of orientation between the axis of the SOFT with the axis formed by the MOF taper core and the semi-immersed microsphere, were studied and are presented as State A and State B. In State A the aforementioned two axes are parallel to each other whereas in the State B they are perpendicular.

All spectral measurements were made using the amplified spontaneous emission (ASE) of an erbium doped fiber amplifier with a spectral range 1518–1580 nm, whereas an optical spectrum analyzer (OSA) with a maximum spectral resolution of 0.01 nm was used as a detector. Polarization resolved measurements were obtained using a 5 m length of polarizing optical fiber (providing a polarization resolving ration of ~30 dB) attached to the end of the corresponding reading port (only CH1 or CH2) and connected directly to the OSA, while being supported in a rotating v-groove for being aligned with the coordination system of the microsphere.

3. Results and Discussion

Transmission spectra of the T-junction for different polarization states, port excitation and reading, for State A and State B are presented in Figure 4.

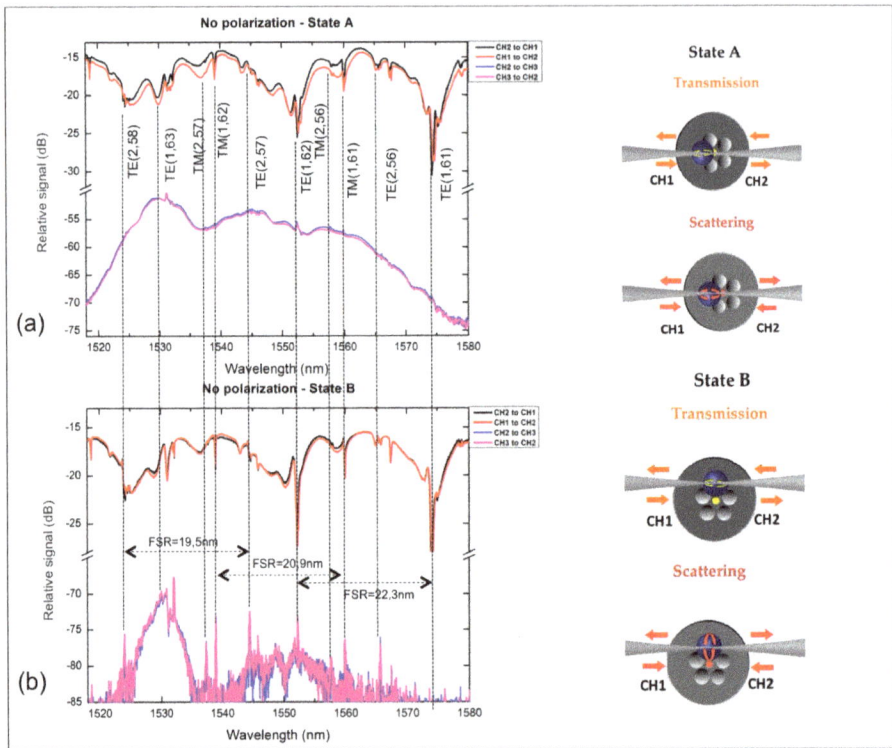

Figure 4. Both states A and B for transmission and scattering are measured and plotted at first with an SMF28 without using polarizing optical fiber. The coupling between the SOFT and the core of the MOF taper in State A (**a**) is obvious by observing the elimination of clear and sharp resonances whereas the geometry of State B (**b**) allows us to obtain more significant resonances in the scattering spectra.

The transmission spectra of the SOFT in contact with the $BaTiO_3$ microsphere show WGMs behaviour (Figure 4, CH1 to CH2 and vice versa) and are typically obtained for microspherical cavities. The azimuthal l and radial q modal orders (TE, TM(q, l)) have been estimated assuming a nominal $BaTiO_3$ microsphere diameter 25 μm (as measured using SEM scans) and a refractive index of 1.9 [3]. A particular feature of the WGM spectra of the State A is a modal distortion by means of broadening observed for particular WGM orders; this is related to the perturbation resulted from the wedging of the $BaTiO_3$ miscosphere into the MOF capillary end face. Typical Q-factors for the spectra of State A of

Figure 4 are ~700, well below values reported for BaTiO$_3$ microspheres of similar diameter and optical fiber taper excitation [23]; this is related to the great mismatch between the effective indices of the WGMs and both the MOF and SOFT [24]. Accordingly, the resonant wavelengths of WGMs excited with the SOFT and scattered from the microsphere were collected and measured with the MOF (CH2 to CH3, State A), proving the possibility of using the T-shaped system for the light routing. Yet, this spectra (CH3 to CH2 and vice versa-State A) are characterized by a strong background continuous signal emerging from the light coupling from the BaTiO$_3$ microsphere to the two waveguide systems, however not through the WGM resonation, but from a standard total internal reflection cancellation and a standard lensing process. This strong and broad background can also shadow WGM mediated coupling from the MOF to the SOFT.

The spectral data for State B excitation presented in Figure 4 exhibit specific similarities with those corresponding to State A. Several of the WGM notches for both polarizations spectrally coincide with those of State A, with relative spectral shifts between the two States that are mostly attributed to possible eccentricity of the capillary wedged BaTiO$_3$ microsphere. A point of particular interest is related to the Q-factor of the WGM excited for State B, which appears much greater compared to the one obtained for the same microsphere, while being positioned and excited at State A. For State B the Q factor was calculated to be 1.63×10^3 for the scattering spectra whereas for State A it was calculated to be ~500 [4,15]. This may be attributed to the fact that for State B the positioning of the wedged microsphere with respect to the MOF symmetry is different, with the MOF core not being aligned with the SOFT axis. This affects the excitation conditions of the systems both through the SOFT and the MOF [25,26]. The WGM peaks measured for light routing between the MOF and the SOFT (CH2 to CH3 and vice versa), exhibit a higher signal to noise ratio with respect to those of State A, since the background signal is lowered by almost 30 dB (at wavelength 1530 nm). This significant reduction of the background, broadband signal denotes that light coupling is minimized through lensing effects and total internal reflection cancellation, but rather takes place through the WGMs resonation facilitated at the BaTiO$_3$ microsphere.

An interesting point for the State B light routing scheme is that while the MOF core and the SOFT axis do not coincide/align in space, light coupling between the two vertically placed waveguide structures still exist through WGM resonations with polar modal numbers $|m| \neq l$, where l refers to the azimuthal WGM order [27]. In Figure 4 we denote resonances for the two first radial orders along with the azimuthal number for TE and TM modes [4,10]. The free spectral range (FSR) measured in the graphs is in good agreement with the theoretically expected values based on the FSR formula and the corresponding resonant wavelength [1]. In such a light coupling scheme WGMs with $|m| \neq l$ will be delocalized from the azimuthal circumference of the BaTiO$_3$ microsphere defined by the excitation plane of the SOFT [28], possibly allowing modal crossing between modes of high difference between m and l modal order, facilitated by the microsphere wedging perturbation and eccentricity of the microsphere.

In another set of measurements, we used the polarization maintaining optical fiber for both States A and B for resolving the two polarization states. In State A because of the position of the sphere with respect to the SOFT, the fiber s-polarization excites mostly TM modes for the microspherical resonator, whereas in State B the fiber s-polarization corresponds mostly to TE modes for the resonator ('slapping' and 'piercing' polarization). The majority of polarization resolved data obtained appeared quite noisy, hindering WGM resonation features especially in the SOFT to MOF coupling. In general, due to the geometry of the specific WGM resonation scheme, we expect polarization cross-coupling to take place [29], decrementing polarization resolved modal measurements as shown in Figure 5.

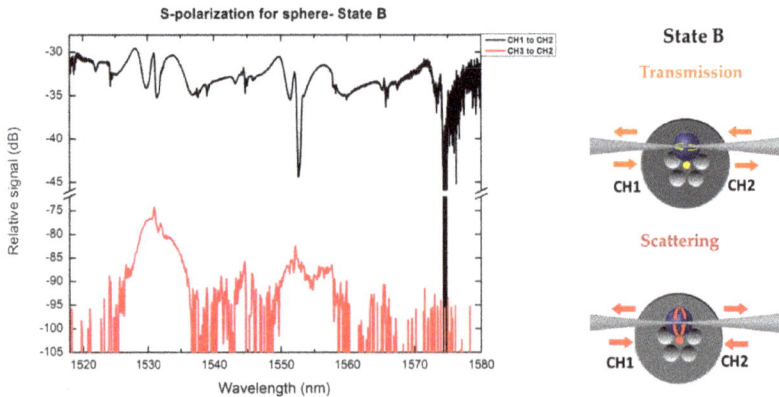

Figure 5. State B: Transmission and corresponding scattering spectra of the T-shape system obtained with the polarizing optical fiber.

4. Conclusions

We communicate results on the investigation of light coupling and routing via a microspherical resonator integrated in a T-shaped optical fiber configuration system, constituted of a $BaTiO_3$ microsphere wedged in the capillary of a grapefruit shaped optical fiber while being excited for different orientations using a single mode optical fiber taper. The specific optical configuration was spectrally characterized in a multi-portal arrangement and the light routing results were discussed in conjunction with basic WGMs formulation, showing the possibility to rout the light in a 90° angle system from the SOFT to the MOF and vice versa. We intend to continue our investigations of the specific light coupling and routing system for potential use in self-feedback systems for active glass microspherical cavities, where the MOF end will be spliced to one end of the SOFT. Another possible application refers to optical routers and microspherical lasers arranged in three dimensions [6], while being pumped through a single optical fiber taper with specific spectral signatures per MOF port [30]. Preliminary results in tuning the coupling between the SOFT and the MOF through the WGM spectral comb by exploiting the photorefractivity of $BaTiO_3$ microsphere, have resulted in typical spectral shifts of the WGMs by ~0.3 nm for a exposure dose of 10.8 J, using a 405 nm CW solid state laser.

Funding: This work was partially supported by the projects of ACTPHAST (Grant Agreement No. 619205) and H2020 Laserlab Europe (ECGA 654148). We acknowledge partial support of this work by the project "HELLAS-CH" (MIS 5002735) which is implemented under the "Action for Strengthening Research and Innovation Infrastructures", funded by the Operational Programme "Competitiveness, Entrepreneurship and Innovation" (NSRF 2014–2020).

Acknowledgments: All authors would like to gratefully thank Maria Konstantaki for experimental assistance with tapered optical fibers fabrication, and, Mo-Sci Corporation (http://www.mo-sci.com/) for providing the $BaTiO_3$ microspheres. Warm acknowledgments are also directed to Walter Margulis (ACREO AB, Lund, Sweden) for kindly providing the grape-fruit shape MOF used.

Conflicts of Interest: The authors declare no conflicts of interest.

References

1. Chiasera, A.; Dumeige, Y.; Féron, P.; Ferrari, M.; Jestin, Y.; Nunzi Conti, G.; Pelli, S.; Soria, S.; Righini, G.C. Spherical whispering-gallery-mode microresonators. *Laser Photonics Rev.* **2010**, *4*, 457–482. [CrossRef]
2. Foreman, M.R.; Swaim, J.D.; Vollmer, F. Whispering gallery mode sensors. *Adv. Opt. Photonics* **2015**, *7*, 168–240. [CrossRef] [PubMed]
3. Matsko, A.B.; Ilchenko, V.S. Optical resonators with whispering-gallery modes—Part I: Basics. *IEEE J. Sel. Top. Quantum Electron.* **2006**, *12*, 3–14. [CrossRef]

4. Ilchenko, V.S.; Bennett, A.M.; Santini, P.; Savchenkov, A.A.; Matsko, A.B.; Maleki, L. Whispering gallery mode diamond resonator. *Opt. Lett.* **2013**, *38*, 4320–4323. [CrossRef] [PubMed]
5. Vollmer, F.; Arnold, S.; Keng, D. Single virus detection from the reactive shift of a whispering-gallery mode. *Proc. Natl. Acad. Sci. USA* **2008**, *105*, 20701–20704. [CrossRef] [PubMed]
6. Cai, M.; Painter, O.; Vahala, K.J.; Sercel, P.C. Fiber-coupled microsphere laser. *Opt. Lett.* **2000**, *25*, 1430–1432. [CrossRef] [PubMed]
7. Kosma, K.; Schuster, K.; Kobelke, J.; Pissadakis, S. An "in-fiber" whispering-gallery-mode bi-sphere resonator, sensitive to nanometric displacements. *Appl. Phys. B* **2018**, *124*, 1. [CrossRef]
8. Riesen, N.; Reynolds, T.; François, A.; Henderson, M.R.; Monro, T.M. Q-factor limits for far-field detection of whispering gallery modes in active microspheres. *Opt. Express* **2015**, *23*, 28896–28904. [CrossRef] [PubMed]
9. Von Klitzing, W.; Long, R.; Ilchenko, V.S.; Hare, J.; Lefèvre-Seguin, V. Frequency tuning of the whispering-gallery modes of silica microspheres for cavity quantum electrodynamics and spectroscopy. *Opt. Lett.* **2001**, *26*, 166–168. [CrossRef] [PubMed]
10. Farnesi, D.; Chiavaioli, F.; Righini, G.C.; Soria, S.; Trono, C.; Jorge, P.; Nunzi Conti, G. Long period grating-based fiber coupler to whispering gallery mode resonators. *Opt. Lett.* **2014**, *39*, 6525–6528. [CrossRef] [PubMed]
11. Hanumegowda, N.M.; Stica, C.J.; Patel, B.C.; White, I.; Fan, X. Refractometric sensors based on microsphere resonators. *Appl. Phys. Lett.* **2005**, *87*, 201107. [CrossRef]
12. Knight, J.C.; Cheung, G.; Jacques, F.; Birks, T.A. Phase-matched excitation of whispering-gallery-mode resonances by a fiber taper. *Opt. Lett.* **1997**, *22*, 1129–1131. [CrossRef] [PubMed]
13. Milenko, K.; Konidakis, I.; Pissadakis, S. Silver iodide phosphate glass microsphere resonator integrated on an optical fiber taper. *Opt. Lett.* **2016**, *41*, 2185–2188. [CrossRef] [PubMed]
14. Francois, A.; Rowland, K.J.; Monro, T.M. Highly efficient excitation and detection of whispering gallery modes in a dye-doped microsphere using a microstructured optical fiber. *Appl. Phys. Lett.* **2011**, *99*, 141111–141113. [CrossRef]
15. Zeltner, R.; Pennetta, R.; Xie, S.; Russell, P.S.J. Flying particle microlaser and temperature sensor in hollow-core photonic crystal fiber. *Opt. Lett.* **2018**, *43*, 1479–1482. [CrossRef] [PubMed]
16. Maslov, A.V. Resonant optical propulsion of a particle inside a hollow-core photonic crystal fiber. *Opt. Lett.* **2016**, *41*, 3062–3065. [CrossRef] [PubMed]
17. Kosma, K.; Zito, G.; Schuster, K.; Pissadakis, S. Whispering gallery mode microsphere resonator integrated inside a microstructured optical fiber. *Opt. Lett.* **2013**, *38*, 1301–1303. [CrossRef] [PubMed]
18. Wang, J.; Yin, Y.; Hao, Q.; Zhang, Y.; Ma, L.; Schmidt, O.G. Strong coupling in a photonic molecule formed by trapping a microsphere in a microtube cavity. *Adv. Opt. Mater.* **2017**, *6*, 1700842. [CrossRef]
19. Zhang, M.; Yang, W.; Tian, K.; Yu, J.; Li, A.; Wang, S.; Lewis, E.; Farrell, G.; Yuan, L.; Wang, P. In-fiber whispering-gallery mode microsphere resonator-based integrated device. *Opt. Lett.* **2018**, *43*, 3961–3964. [CrossRef] [PubMed]
20. Kosma, K.; Konidakis, I.; Pissadakis, S. Photorefractive tuning of whispering gallery modes of a spherical resonator integrated inside a microstructured optical fibre. *Eur. Phys. J. Spec. Top.* **2014**, *223*, 2035–2040. [CrossRef]
21. Savchenkov, A.A.; Matsko, A.B.; Strekalov, D.; Ilchenko, V.S.; Maleki, L. Enhancement of photorefraction in whispering gallery mode resonators. *Phys. Rev. B* **2006**, *74*, 245119. [CrossRef]
22. Canciamilla, A.; Grillanda, S.; Morichetti, F.; Ferrari, C.; Hu, J.; Musgraves, J.D.; Richardson, K.; Agarwal, A.; Kimerling, L.C.; Melloni, A. Photo-induced trimming of coupled ring-resonator filters and delay lines in as2s3 chalcogenide glass. *Opt. Lett.* **2011**, *36*, 4002–4004. [CrossRef] [PubMed]
23. Svitelskiy, O.; Li, Y.; Darafsheh, A.; Sumetsky, M.; Carnegie, D.; Rafailov, E.; Astratov, V.N. Fiber coupling to batio3 glass microspheres in an aqueous environment. *Opt. Lett.* **2011**, *36*, 2862–2864. [CrossRef] [PubMed]
24. Humphrey, M.J.; Dale, E.; Rosenberger, A.T.; Bandy, D.K. Calculation of optimal fiber radius and whispering-gallery mode spectra for a fiber-coupled microsphere. *Opt. Commun.* **2007**, *271*, 124–131. [CrossRef]
25. Cai, M.; Painter, O.; Vahala, K.J. Observation of critical coupling in a fiber taper to a silica-microsphere whispering-gallery mode system. *Phys. Rev. Lett.* **2000**, *85*, 74–77. [CrossRef] [PubMed]
26. Mohd Nasir, M.N.; Senthil Murugan, G.; Zervas, M.N. Spectral cleaning and output modal transformations in whispering-gallery-mode microresonators. *J. Opt. Soc. Am. B* **2016**, *33*, 1963–1970. [CrossRef]

Micromachines **2018**, *9*, 521

27. Senthil Murugan, G.; Panitchob, Y.; Tull, E.J.; Bartlett, P.N.; Hewak, D.W.; Zervas, M.N.; Wilkinson, J.S. Position-dependent coupling between a channel waveguide and a distorted microsphere resonator. *J. Appl. Phys.* **2010**, *107*, 053105. [CrossRef]

28. Attar, S.T.; Shuvayev, V.; Deych, L.; Martin, L.L.; Carmon, T. Level-crossing and modal structure in microdroplet resonators. *Opt. Express* **2016**, *24*, 13134–13141. [CrossRef] [PubMed]

29. Bianucci, P.; Fietz, C.R.; Robertson, J.W.; Shvets, G.; Shih, C.-K. Polarization conversion in a silica microsphere. *Opt. Express* **2007**, *15*, 7000–7005. [CrossRef] [PubMed]

30. Ming, C.; Hunziker, G.; Vahala, K. Fiber-optic add-drop device based on a silica microsphere-whispering gallery mode system. *IEEE Photonic Technol. Lett.* **1999**, *11*, 686–687. [CrossRef]

micromachines

MDPI

Article

Laser-Inscribed Glass Microfluidic Device for Non-Mixing Flow of Miscible Solvents

Valeria Italia [1], Argyro N. Giakoumaki [2,3], Silvio Bonfadini [1,2], Vibhav Bharadwaj [3], Thien Le Phu [2,3], Shane M. Eaton [3,*], Roberta Ramponi [2,3], Giacomo Bergamini [4], Guglielmo Lanzani [1] and Luigino Criante [1,*]

[1] Center for Nano Science and Technology, Istituto Italiano di Tecnologia, 20133 Milano, Italy; valeriaitalia.vi@gmail.com (V.I.); silvio.bonfadini@iit.it (S.B.); guglielmo.lanzani@iit.it (G.L.)
[2] Dipartimento di Fisica, Politecnico di Milano, Piazza Leonardo da Vinci 32, 20133 Milano, Italy; argyrogiak@gmail.com (A.N.G.); thien.lephu@polimi.it (T.L.P.); roberta.ramponi@polimi.it (R.R.)
[3] Istituto di Fotonica e Nanotecnologie-Consiglio Nazionale delle Ricerche (IFN-CNR), Piazza Leonardo da Vinci 32, 20133 Milano, Italy; vibhavbharadwaj@gmail.com
[4] Department of Chemistry Giacomo Ciamician University of Bologna Via Selmi 2, I-40126 Bologna, Italy; giacomo.bergamini@unibo.it
* Correspondence: shane.eaton@gmail.com (S.M.E.); luigino.criante@iit.it (L.C.); Tel.: +39-320-092-1952 (S.M.E.); +39-022-399-9812 (L.C.)

Received: 24 October 2018; Accepted: 24 December 2018; Published: 29 December 2018

Abstract: In recent years, there has been significant research on integrated microfluidic devices. Microfluidics offer an advantageous platform for the parallel laminar flow of adjacent solvents of potential use in modern chemistry and biology. To reach that aim, we worked towards the realization of a buried microfluidic Lab-on-a-Chip which enables the separation of the two components by exploiting the non-mixing properties of laminar flow. To fabricate the aforementioned chip, we employed a femtosecond laser irradiation technique followed by chemical etching. To optimize the configuration of the chip, several geometrical and structural parameters were taken into account. The diffusive mass transfer between the two fluids was estimated and the optimal chip configuration for low diffusion rate of the components was defined.

Keywords: optofluidics; lab-on-a-chip; femtosecond laser; laser micromachining; diffusion

---

1. Introduction

The recent introduction of microfluidics in chemistry and biology has led to a paradigm shift in both fields. Lab-on-a-Chip is now a commonly known concept and significant efforts have been made for the realization of multifunctional integrated systems for chemical analysis [1], cell culture, and biochemical systems investigation [2], but most importantly for multiphase chemical reactions even for miscible solutions. Microfluidic reactors have already been proven valuable due to their high surface-to-volume ratio, the scale-out capabilities for industrial applications, the higher yield over batch reactors, and the versatility of the microfluidic chip set-ups [3]. By manufacturing a microfluidic chip with a suitable geometry, it is possible to manage simultaneously two or more fluids and create dynamic interfaces between them while avoiding active mixing due to laminar flow [4,5]. The present approach exploits (mimics) liquid–liquid interfaces which could not be accessed in batch situations. Diffusive mixing between two interfacing laminar flows is a theoretically, and in some cases experimentally, well-defined process [6,7]. It depends mainly on the area of interaction, the laminarity of the flow, the time of interaction, and the concentration gradient between two or more streams.

A wide variety of materials and techniques have been employed for the fabrication of microfluidic systems [8]. The most popular process is the soft lithographic fabrication of 2D chips on polydimethylsiloxane (PDMS) [9] and other elastomers. The technique is easy, fast, and has a low cost, but although PDMS' porosity is a virtue for long-term cell cultures, it becomes a drawback when it comes to chemical analysis or organic synthesis because it cannot be defined as chemically inert. PDMS can undergo swelling due to solvent adsorption in the pores, creating deformations on the microfluidic channels [10]. Extensive deformation can create leakages, thus compromising the chip. Combined with the incompatibility with many organic solvents, PDMS has a limited range of applications in synthetic chemistry.

Even though a plethora of water-based biological applications have been demonstrated for PDMS microfluidic chips, the use of other solvents is prohibited due to the incompatibility of PDMS. It has been demonstrated that water-based chemical reactions [4] are able to be performed on the interface of two interacting laminar flows in a PDMS microfluidic chip. This high yield and recyclable approach could also be applied in organic chemistry, especially for reactions that involve toxic or expensive reagents. Lee et al. [10] performed an extensive study on the compatibility of PDMS with a variety of organic solvents, and it is evident that, for a broader application of microfluidics in organic chemistry, PDMS is not the optimal material. One of the most promising materials that can overcome the above mentioned challenges is fused silica.

Fused silica microfluidic devices have been well established, but the exploitation of a liquid–liquid interface that is present in a Y-shaped fluidic system requires an extensive study on the behavior of pressure-driven flows. Glass chips differ from PDMS as they are rigid, and they do not exhibit deformation upon high-pressure-driven flow [11]. Fused silica is compatible with a wide variety of organic solvents as well as water, and since it is not gas-permeable, it can be used as a material for the fabrication of microreactors for a wide variety of reactions, including water splitting for hydrogen production.

In this paper, we report the fabrication and characterization of a double Y-branch fused silica microfluidic device for the introduction, interaction, and separation of two miscible solutions characterized by laminar flow, taking advantage of the femtosecond laser irradiation followed by the chemical etching (FLICE) method [12–16]. There are many advantages attributed to this microfabrication technique in comparison with traditional photolithography, including the ability to quickly realize 3D monolithic structures completely buried in the substrate without the requirement of masks or a clean room. With FLICE, there is no need to create complex microfluidic chips in two halves to be subsequently welded as it is a process which often leads to sealing problems. As a preliminary investigation, the laser-fabricated, fused silica microfluidic device was used for the study of the diffusion of Rhodamine 6G (R6G) in the ethanol–ethanol interface. The angle between the two inlets and the height of the chamber were varied, and the diffusion was qualitatively determined for different flow rates. R6G was used as a colorant for one of the two streams due to its optical properties and its well-established diffusivity in ethanol.

2. Materials and Methods

2.1. Methods

Fused silica glass is ideally suited for this application as it possesses several important characteristics: it is chemically inert to a variety of solvents, hydrophilic, thermally and mechanically stable, and optically transparent in a wide range of wavelengths [12,17]. To fabricate the optofluidic device in the bulk of glass, we exploit the FLICE method [12,13], which requires two steps: (1) tightly focused, femtosecond, laser pulses drive a permanent and localized periodic redistribution of material density, which defines the desired structure on the surface or in the bulk of fused silica [18]; (2) chemical etching of the laser-modified volume by a strong acid or a strong base (typically HF or KOH, respectively) to remove the irradiated zone, producing the hollowed-out, microfluidic device [15].

The femtosecond laser used for device fabrication in fused silica was a generatively amplified Yb:KGW system (Pharos, Light Conversion, Vilnius, Lithuania) with 230-fs pulse duration, 515-nm wavelength (frequency doubled), and 500-kHz repetition rate focused with a 0.42-NA microscope objective (M Plan Apo SL50X Ultra-Long Working Distance Plan-Apochromat, Mitutoyo, Kawasaki, Japan). Computer-controlled, 3-axis motion stages (ABL-1000, Aerotech, Pittsburgh, PA, USA) interfaced by CAD-based software (ScaBase, Altechna, Vilnius, Lithuania) with an integrated acousto-optic modulator were used to translate the sample relative to the laser irradiation desiderate patch. An average power (pulse energy) of 200 mW (400 nJ) and a scan speed of 5 mm/s were used to laser-pattern the microfluidic device shown in Figure 1a. A multiscan writing procedure with 7 μm spacing between transverse scans was adopted to form the microfluidic device. The thickness of the fused silica windows was 1 mm, and the buried microfluidic chips were laser-inscribed at a depth of 0.5 mm. The overall fabrication time of a single chip varied between 57 min and 69 min for chamber heights between 100 μm and 500 μm. The sample was etched in a sonication bath of HF (20% vol in water), with a feedback-controlled temperature of 37 °C. The etching rate of the laser-exposed area of the fused silica was 500 μm/h, whereas for the non-exposed area, the etching rate was 20 μm/h. The resulting rectangular chamber's internal dimensions were 2 mm × 200 μm (length × width) with a height h that varied from 100 μm to 500 μm (Figure 1c).

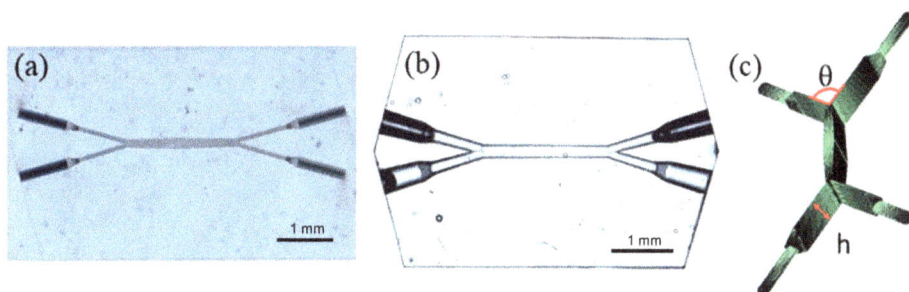

Figure 1. Microfluidic chip geometry; image from optical microscope of the chip after (**a**) femtosecond (fs) laser irradiation and subsequent; (**b**) chemical etching; (**c**) schematic of chip design, where h is the chamber height and θ is the separation angle.

The suitable structure for the diffusion study in a microfluidic chip is the double Y configuration as shown in Figure 1, showing optical microscopic images of the device after femtosecond laser irradiation (Figure 1a) and then after chemical etching (Figure 1b). The final microfluidic device consists of two inlets and two outlets at both ends of a long chamber in which the interface interaction occurs.

To complete the characterization of the chip, the behavior of diffusive mass transfer was studied by varying the separation angle between the inlet/outlet channels, θ, and the pumping pressure of the fluids, p. To avoid turbulence, the inlet tubes and the interaction chamber were designed considering the continuity of the fabrication process and the equality in resistance for both the inlets and outlets of the chip. A microfluidic pump (OB1, Elveflow, Paris, France) was connected to the reservoirs of the solutions. Polytetrafluoroethylene (PTFE) tubing was inserted into the reservoirs and drove the fluids into the chips by Polyether ether ketone (PEEK) tubing with an outer (inner) diameter of 360 μm (150 μm) by using an appropriate adapter. The latter tubing was connected to the glass chip using ultraviolet (UV) glue. The materials for both tubings as well as the reservoirs and the glue were selected due to their extraordinary chemical inertness to organic solvents.

2.2. Materials

To visualize the interaction zone between the two parallel flows, an optical technique was used. Using a coloured (1 mM Rhodamine 6G in ethanol) and a transparent solution (pure ethanol) it was possible to understand how the geometric and microfluidic parameters influence the diffusive mass transfer between two interfacing laminar flows.

The dye solution was prepared starting from R6G powder (Sigma-Aldrich) dissolved in filtered ethanol. The solution was filtered once more to prevent undissolved dye particles and impurities from entering the chip and creating turbulence. The Rhodamine solution was stored in glass vials in a dry, cool, and dark environment to prevent degradation.

3. Results

3.1. Data Analysis

Starting from the work of Werts et al. [7], we developed a simple and useful method to extract and analyze data using a conventional optical microscope. We obtained an empty channel image to use as a reference (Figure 2a), and subsequently we imaged the flowing colorants, varying the pumping pressure (Figure 2b). Using a custom image management algorithm (developed in MATLAB), the two aforementioned data images can be subtracted to indicate the difference i.e., the dye solution flow, and to eliminate electronic noise caused by the intrinsic roughness of the device. The visible roughness of the chip, as well as the discontinuation between the chamber and the outlets, in Figure 2a,b is caused by the laser writing pattern of the device, and even though it can be improved by post-fabrication annealing, there was no visible turbulence in the chip due to this effect. As it has been reported in the past, the roughness of microchannels formed by this technique have sub-micrometer roughness [19].

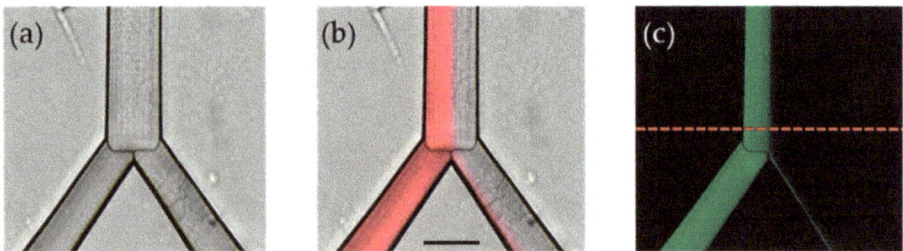

Figure 2. Visual analysis technique of laser-fabricated microfluidic chip: (**a**) reference image, (**b**) colored image with the flowing dye solution; (**c**) negative image obtained by subtracting (**a**) from (**b**). The red dashed line indicates the output of the chip position chosen to extract the intensity profile. Scale bar corresponds to 200 μm.

In this way, we obtained a clear negative image (Figure 2c) from which the cross-section intensity profile is extracted at the fixed position near the outlet of the channel (red dashed line in Figure 2c). The typical intensity profile of the bi-colour image (Figure 2c) is shown in Figure 3a. The signal is then processed with a low-pass filter (Figure 3b) and normalized (Figure 3c) to be comparable with other intensity profiles.

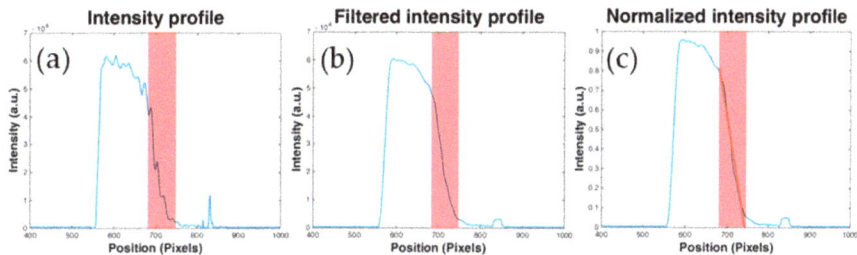

Figure 3. Intensity profile of the flow behavior processed with MATLAB algorithm. The x-axis represents the position in the image [pixels], while the y-axis is the fluorescence intensity in arbitrary units (a.u.). The red region indicates the diffusion zone between the two liquids. (**a**) original, (**b**) processed by a low-pass filter, and (**c**) normalized intensity profile. The slope of the red line represents the diffusion behavior.

At this point, in order to qualitatively define the conditions for the lowest diffusion of R6G inside the channel, we calculated the slope of the cross-section intensity profile with linear interpolation. The left region of high intensity in the normalized profile in Figure 3c indicates the presence of the dye, while the region of low intensity on the right indicates the absence of it. The transitional region highlighted in red between the high and low intensity regions represents the diffusion zone. It is important to note that the value of the slope in the diffusion region is inversely proportional to the amount of diffused dye between the liquids (ideally infinite slope indicates zero diffusion).

3.2. Preliminary Considerations

Reynolds number (Re) and Peclet number (Pe) are two dimensionless values that define fluidic and diffusive mixing characteristics of a microfluidic system, respectively. They are described as [20,21]

$$\text{Re} = \frac{\rho \bar{v} \ell}{\mu}, \quad \text{Pe} = \frac{\bar{v} \ell}{D} \tag{1}$$

where, ρ is the density, $\bar{v}$ is the mean flow velocity, μ is the dynamic viscosity of the fluid, ℓ is the characteristic length of the microfluidic channel, and D is the diffusivity. A flow with Re lower than 2300 is considered laminar, while a high Pe number defines the number of channel widths required to completely mix two fluids by diffusion.

For ethanol, $\rho = 789$ kg/m^3 and $\mu = 1.2$ mPa·s. For Rhodamine 6G, $D = 3 \times 10^{-10}$ m^2/s [22]. The characteristic length of the rectangular channel was calculated to be 1.33×10^{-4} m for $h = 100$ μm and 2.85×10^{-4} m for $h = 500$ μm.

In a microfluidic device, we can assume the Hagen–Poiseuille equation to describe the relationship between the pressure drop ($\Delta P = P_{\text{in}} - P_{\text{out}}$) and the flow rate ($Q$) of pressure-driven flow [23]:

$$\Delta P_{\text{total}} = R_{\text{total}} Q = R_{\text{total}} \bar{v} S \tag{2}$$

where R is the hydrodynamic resistance of the microfluidic system and S is the cross-sectional area of the microfluidic chamber. For a single microfluidic chip, we can consider R and S as constants, meaning that the fluid velocity increases with the pumping pressure.

Starting from Stokes equations, it is possible to calculate the fluidic resistance for a fluid with viscosity μ flowing inside a [23]:

- cylindrical channel (tubings) with length L and internal radius r,

$$R_{\text{tubing}} = \frac{8\mu L}{\pi r^4} \tag{3}$$

- rectangular channel (glass chip) with length L, height h, and width w,

$$R_{chip} = \frac{12\mu L}{1 - 0.63\left(\frac{h}{w}\right)} \cdot \frac{1}{h^3 w} \qquad (4)$$

In the present work, since all the elements are connected in series, we can consider $R_{total} = R_{tubing} + R_{chip}$. Considering that $R_{tubing} \gg R_{chip}$ due to the tubing's comparable radius to the chip but a comparatively much greater L, we can neglect the resistance of the glass chip and calculate from Equation (3) that $R_{total} = 1.11 \times 10^{11}$ mbar·s/m^3 for $r = 75$ µm and $L = 11.5$ cm. Assuming that the pressure given by the pump is equal to ΔP_{total}, we can calculate the flow rate and subsequently, the flow velocity of the chips with different h and varying pumping pressures, as reported in Table 1.

Table 1. Calculated flow rates, velocities, Reynolds number (Re), and Peclet number (Pe) for different pumping pressures. $\bar{v}_1$ and $\bar{v}_2$ correspond to the flow velocity of glass chips with heights of 100 µm and 500 µm, respectively.

ΔP (mbar)	Q (µL/min)	$\bar{v}_1$ (mm/s)	$\bar{v}_2$ (mm/s)	Re	Pe
25	13.5	22.5	4.5	2	7507.5
50	27.0	45.0	9.0	3	15,015.0
100	54.1	90.0	18.0	6	30,030.0
200	108.1	180.0	36.0	12	60,060.0

3.3. The Effect of Pumping Pressure

The flow velocity is directly dependent on the pumping pressure according to the Hagen–Poiseuille equation in Equation (2), and it is the only parameter other than the geometrical characteristics of the chip that can affect the diffusion of R6G in ethanol. Diffusion is a time-dependent process, and it is obvious that a slowly flowing solution (i.e., in the case of $\Delta P = 25$ mbar) exhibits greater diffusion, as seen in Figure 4.

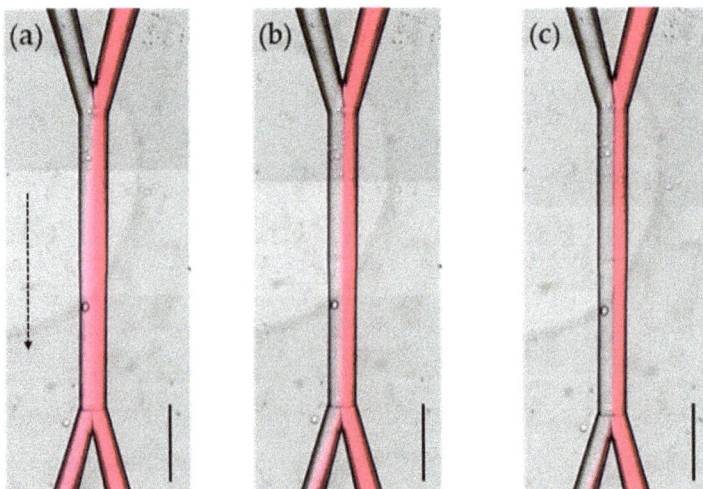

Figure 4. Diffusion behavior inside a microfluidic chip with 30° incident angle and $h = 500$ µm, at increasing pumping pressure (a) $\Delta P = 25$ mbar; (b) $\Delta P = 75$ mbar; (c) $\Delta P = 200$ mbar. Scale bars correspond to 500 µm. Arrow indicates the flow direction.

3.4. The Effect of Angle Between Inlets

We studied the effect of the angle between the two inlets. All the microfluidic chips were fabricated with a constant chamber height (h = 500 μm). Three different angles θ = 30°, 60°, and 80° were chosen and the results are reported in Figure 5. Angles greater than 80° exhibited significant diffusion due to the trajectory of the fluids, so we excluded them for the purposes of this work.

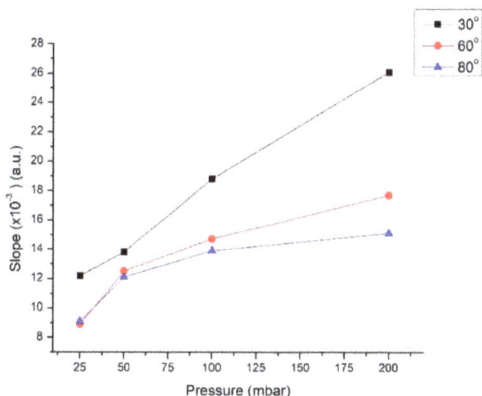

Figure 5. Slope of the linear interpolation of the intensity profile of the visual diffusion analysis (see Figure 3c) in the chip as a function of the pair inlet pressure at the different separation angles θ = 30°, 60°, and 80°.

For each angle value, the slope of the intensity profile in the diffusion region increases monotonously with the (identical) pumping pressure at the two inlets: the greater the pressure, the greater the velocity of the fluids, leading to an increased value of the Pe coefficient and reduced diffusive mixing. It is clear that for a given pumping pressure, a smaller separation angle results in reduced mixing between the two parallel flowing fluids.

Increasing the separation angle of the inlets, the fluids undergo active mixing in the first part of the chamber due to a greater change in trajectory upon entering the main chamber. In the case of 60° and 80°, a saturation of the slope value is observed. However, in the case of smaller angles, the parallel flows enter into the chamber with minimum turbulence. For this reason, we determined that a separation angle of θ = 30° is the most suitable for minimum diffusion flow.

3.5. The Effect of Chamber Height

In Figure 6, we report the slope of the diffusion zone for channel heights of h = 100 μm and 500 μm for chips with a length of 2 mm and an angle of 30° between inlets. It is evident that the channel height has a significant impact on the diffusion process in this microfluidic platform, with the slope in the diffusion region for the 100 μm tall chamber being twice that of the 500 μm tall channel in any pressure measured. In other words, the diffusion is less pronounced in the shorter chamber, irrespective of the pumping pressure. A reasonable explanation for this observation is that by decreasing the height of the chamber, the interaction area of the two fluids decreases as well. Considering the length of the chamber is 2 mm, the interaction area of the fluids is 0.2 mm^2 for the 100 μm high chamber and 1 mm^2 for the 500 μm.

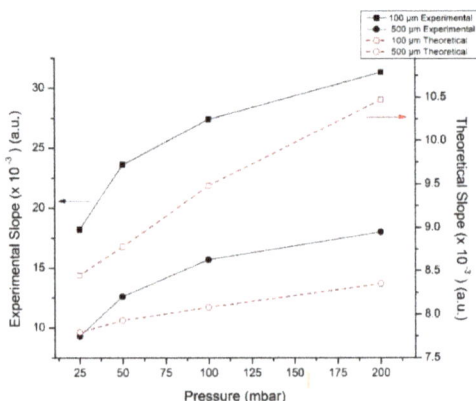

Figure 6. Theoretical and experimental slope of the intensity profile in the diffusion region versus the pumping pressure for two different chamber heights: 100 μm and 500 μm.

Another parameter that justifies this drastic change in the slope is the fluid velocity, which in the case of $h = 500$ μm is much lower than that of $h = 100$ μm (Table 1) at constant pressure. As mentioned before, the diffusion is proportional to the residence time of the interacting fluids in the chamber and consequently inversely proportional to the flow velocity.

The parametric study of the effect of the height the diffusion was simulated by COMSOL, as well as the theoretical data, are also presented in Figure 6. For the purposes of this study, we assumed the same mean velocities as in Table 1. Considering the diffusion coefficient of R6G in ethanol, we were able to extract the slope of the concentration gradient, following the same procedure as for the experimental data.

Although the same general trends are observed in the theoretical simulations, there is a discrepancy in the absolute values of slopes. The resistance of the chip and the roughness of the walls were neglected for the purposes of the theoretical study, which can explain the lower diffusive mixing that is predicted by the simulations compared to the experimental data in Figure 6.

4. Conclusions

In this work, we performed a parametric study of the geometry of a double Y-shaped microfluidic chip in order to minimize the diffusive mass transfer between two laminar flows. This first approach on the characterization of such chips examines the behavior of two ethanolic solutions, but the results can be translated in any kind of application of multifunctional complex microfluidic systems with a similar configuration. The optical technique that was developed for the purposes of this study is simple, straightforward, and can be replicated using widely used and easily accessible equipment such as an optical microscope with a CCD camera and image processing software.

The FLICE manufacturing technique has enabled the fabrication of complex 3D geometries of buried microfluidic chips in glass while avoiding the problematic process of sealing two substrates together by welding. Being able to use glass as a substrate for microfluidics removes the limitations that are created by the incompatibility of the majority of elastomers, such as PDMS, and allows for a wider variety of applications in flow chemistry.

For the visualization of the diffusion, a R6G solution and a transparent solution were used, enabling the detection of mixing by a simple optical microscopy image analysis. As a result, we determined that an angle of 30° or lower between the two inlet streams is optimal for non-mixing flow. Also, we found that the height of the interaction chamber has a major impact on the diffusion, with the smaller height of 100 μm being preferable. In future work, we will perform chemical reactions at the interface of the parallel laminar flows in the laser-inscribed buried branching network.

Author Contributions: G.B., S.M.E., L.C., G.L., and R.R. conceived the idea of laser forming the non-mixing microfluidic chip in glass. V.I., S.B., L.C., V.B., and S.M.E. designed the geometry of the microfluidic chip. V.I., S.B., L.C., A.N.G., and T.L.P. aided in the characterization and analysis of the microfluidic chip. All authors discussed the experimental implementation and results and contributed to writing the paper.

Funding: This work was funded by the H2020 Marie Skłodowska-Curie ITN PHOTOTRAIN project, FP7 DiamondFab CONCERT Japan project and DIAMANTE MIUR-SIR grant.

Acknowledgments: The authors thank Sara Lo Turco and Simone Varo for enlightening scientific discussions.

Conflicts of Interest: The authors declare no conflict of interest.

References

1. Weigl, B.H.; Yager, P. Microfluidic Diffusion-Based Separation and Detection. *Science* **1999**, *283*, 346–347. [CrossRef]
2. Kuo, J.S.; Chiu, D.T. Controlling Mass Transport in Microfluidic Devices. *Annu. Rev. Anal. Chem.* **2011**, *4*, 275–296. [CrossRef] [PubMed]
3. McMullen, J.P.; Jensen, K.F. Integrated Microreactors for Reaction Automation: New Approaches to Reaction Development. *Annu. Rev. Anal. Chem.* **2010**, *3*, 19–42. [CrossRef] [PubMed]
4. Kenis, P.J.A.; Ismagilov, R.F.; Whitesides, G.M. Microfabrication Inside Capillaries Using Multiphase Laminar Flow Patterning. *Science* **1999**, *285*, 83–85. [CrossRef] [PubMed]
5. Yager, P.; Edwards, T.; Fu, E.; Helton, K.; Nelson, K.; Tam, M.R.; Weigl, B.H. Microfluidic diagnostic technologies for global public health. *Nature* **2006**, *442*, 412–418. [CrossRef] [PubMed]
6. Suh, Y.K.; Kang, S. A Review on Mixing in Microfluidics. *Micromachines* **2010**, *1*, 82–111. [CrossRef]
7. Werts, M.H.V.; Raimbault, V.; Texier-Picard, R.; Poizat, R.; Franais, O.; Griscom, L.; Navarro, J.R.G. Quantitative full-colour transmitted light microscopy and dyes for concentration mapping and measurement of diffusion coefficients in microfluidic architectures. *Lab Chip* **2012**, *12*, 808–820. [CrossRef] [PubMed]
8. Ren, K.; Zhou, J.; Wu, H. Materials for Microfluidic Chip Fabrication. *Acc. Chem. Res.* **2013**, *46*, 2396–2406. [CrossRef] [PubMed]
9. Mcdonald, J.C.; Duffy, D.C.; Anderson, J.R.; Chiu, D.T. Review General Fabrication of microfluidic systems in poly (dimethylsiloxane). *Electrophoresis* **2000**, *21*, 27–40. [CrossRef]
10. Lee, J.N.; Park, C.; Whitesides, G.M. Solvent Compatibility of Poly(dimethylsiloxane)-Based Microfluidic Devices. *Anal. Chem.* **2003**, *75*, 6544–6554. [CrossRef] [PubMed]
11. Kirby, B.J. *Micro- and Nanoscale Fluid Mechanics*; Cornell University: Ithaca, NY, USA, 2010; ISBN 9780521119030.
12. Ramponi, R.; Osellame, R.; Cerullo, G. *Femtosecond Laser Micromachining*; Springer: Cham, Switzerland, 2012.
13. Osellame, R.; Hoekstra, H.J.W.M.; Cerullo, G.; Pollnau, M. Femtosecond laser microstructuring: An enabling tool for optofluidic lab-on-chips. *Laser Photonics Rev.* **2011**, *5*, 442–463. [CrossRef]
14. Lo Turco, S.; Di Donato, A.; Criante, L. Scattering effects of glass-embedded microstructures by roughness controlled fs-laser micromachining. *J. Micromech. Microeng.* **2017**, *27*, 065007. [CrossRef]
15. Taylor, R.; Hnatovsky, C.; Simova, E. Applications of femtosecond laser induced self-organized planar nanocracks inside fused silica glass. *Laser Photonics Rev.* **2008**, *2*, 26–46. [CrossRef]
16. Hnatovsky, C.; Taylor, R.S.; Simova, E.; Rajeev, P.P.; Rayner, D.M.; Bhardwaj, V.R.; Corkum, P.B. Fabrication of microchannels in glass using focused femtosecond laser radiation and selective chemical etching. *Appl. Phys. A* **2006**, *84*, 47–61. [CrossRef]
17. Sugioka, K.; Xu, J.; Wu, D.; Hanada, Y.; Wang, Z.; Cheng, Y.; Midorikawa, K. Femtosecond laser 3D micromachining: A powerful tool for the fabrication of microfluidic, optofluidic, and electrofluidic devices based on glass. *Lab Chip* **2014**, *14*, 3447–3458. [CrossRef] [PubMed]
18. Sugioka, K.; Cheng, Y. Femtosecond laser processing for optofluidic fabrication. *Lab Chip* **2012**, *12*, 3576–3589. [CrossRef] [PubMed]

19. Bellouard, Y.; Said, A.; Dugan, M.; Bado, P. Fabrication of high-aspect ratio, micro-fluidic channels and tunnels using femtosecond laser pulses and chemical etching. *Opt. Express* **2004**, *12*, 2120–2129. [CrossRef] [PubMed]
20. Okuducu, M.; Aral, M.; Okuducu, M.B.; Aral, M.M. Performance Analysis and Numerical Evaluation of Mixing in 3-D T-Shape Passive Micromixers. *Micromachines* **2018**, *9*, 210. [CrossRef] [PubMed]
21. Kamholz, A.E.; Weigl, B.H.; Finlayson, B.A.; Yager, P. Quantitative Analysis of Molecular Interaction in a Microfluidic Channel: The T-Sensor. *Anal. Chem.* **1999**, *71*, 5340–5347. [CrossRef] [PubMed]
22. Hansen, R.L.; Zhu, X.R.; Harris, J.M. Fluorescence correlation spectroscopy with patterned photoexcitation for measuring solution diffusion coefficients of robust fluorophores. *Anal. Chem.* **1998**, *70*, 1281–1287. [CrossRef] [PubMed]
23. Bruus, H. *Theoretical microfluidics*; Oxford University Press: New York, NY, USA, 2008.

MDPI

St. Alban-Anlage 66

4052 Basel

Switzerland

Tel. +41 61 683 77 34

Fax +41 61 302 89 18

www.mdpi.com

Micromachines Editorial Office

E-mail: micromachines@mdpi.com

www.mdpi.com/journal/micromachines